NANOTHERAPEUTICS

From Laboratory to Clinic

NANOTHERAPEUTICS

From Laboratory to Clinic

Ezharul H. Chowdhury
MONASH University (Sunway campus), Malaysia

CRC Press is an imprint of the
Taylor & Francis Group, an **informa** business

CRC Press
Taylor & Francis Group
6000 Broken Sound Parkway NW, Suite 300
Boca Raton, FL 33487-2742

© 2016 by Taylor & Francis Group, LLC
CRC Press is an imprint of Taylor & Francis Group, an Informa business

No claim to original U.S. Government works

Printed and bound in India by Replika Press Pvt. Ltd.

Printed on acid-free paper
Version Date: 20160121

International Standard Book Number-13: 978-1-4987-0751-0 (Hardback)

This book contains information obtained from authentic and highly regarded sources. Reasonable efforts have been made to publish reliable data and information, but the author and publisher cannot assume responsibility for the validity of all materials or the consequences of their use. The authors and publishers have attempted to trace the copyright holders of all material reproduced in this publication and apologize to copyright holders if permission to publish in this form has not been obtained. If any copyright material has not been acknowledged please write and let us know so we may rectify in any future reprint.

Except as permitted under U.S. Copyright Law, no part of this book may be reprinted, reproduced, transmitted, or utilized in any form by any electronic, mechanical, or other means, now known or hereafter invented, including photocopying, microfilming, and recording, or in any information storage or retrieval system, without written permission from the publishers.

For permission to photocopy or use material electronically from this work, please access www.copyright.com (http://www.copyright.com/) or contact the Copyright Clearance Center, Inc. (CCC), 222 Rosewood Drive, Danvers, MA 01923, 978-750-8400. CCC is a not-for-profit organization that provides licenses and registration for a variety of users. For organizations that have been granted a photocopy license by the CCC, a separate system of payment has been arranged.

Trademark Notice: Product or corporate names may be trademarks or registered trademarks, and are used only for identification and explanation without intent to infringe.

Visit the Taylor & Francis Web site at
http://www.taylorandfrancis.com

and the CRC Press Web site at
http://www.crcpress.com

Contents

Preface xiii
Acknowledgments xv
Author xvii

**Chapter 1 Emergence of nanotherapeutics: Challenges
in classical drug transport versus macromolecular
drug design** 1
1.1 Administration of small-molecule drugs: Traffic routes
toward the bloodstream 1
    1.1.1 Barriers to the oral route 1
    1.1.2 Barriers to sublingual and buccal routes 2
    1.1.3 Barriers to the rectal route 3
    1.1.4 Barriers to the intranasal route 3
    1.1.5 Barriers to the pulmonary route 3
    1.1.6 Barriers to intramuscular and subcutaneous routes 4
    1.1.7 Barriers to the transdermal route 4
1.2 Fates of the small-molecule drugs in blood 4
    1.2.1 Plasma drug concentration on intensity
of therapeutic action 4
    1.2.2 Drug transport from blood capillaries
to extracellular fluid 5
    1.2.3 Drug transport from extracellular fluid to cells 6
    1.2.4 Elimination of drugs from the body 7
1.3 Major problems associated with traditional formulations
of small-molecule drugs 8
1.4 Alteration of pharmacokinetics of small-molecule drugs
with macromolecules 9
    1.4.1 Enhancement of drug solubility and stability 9
    1.4.2 Prolongation of retention time in blood 10
    1.4.3 Facilitating fast drug release 12
    1.4.4 Enabling sustained release 12
    1.4.5 Passive targeting by exploiting leaky vasculature
architecture 13

v

	1.4.6	Active targeting through receptor-mediated delivery....... 13
	1.4.7	Magnetic targeting 13
1.5	Protein-based macromolecular drugs 15
	1.5.1	Proteins as independent therapeutic drugs......................... 15
	1.5.2	Chemically modified and carrier-bound prcteins.............. 15
	1.5.3	Proteins as drug carriers 17
1.6	DNA/RNA-based macromolecular drugs.. 17
1.7	Macromolecules for prodrug therapy.. 17
	1.7.1	ADEPT.. 18
	1.7.2	GDEPT.. 18
1.8	Macromolecules for vaccine delivery 19
1.9	Nanoparticles for photodynamic therapy...................................... 19
1.10	Macromolecules for image-guided drug delivery......................... 19

**Chapter 2 The ultimate destinations for delivery and release
of nanotherapeutics 21**
2.1	Sustained-release formulations 21
2.2	Intracellular delivery and release.. 21
	2.2.1	Endocytosis of nanoparticles 22
	2.2.2	Influences of physicochemical properties on cellular uptake of nanoparticles 22
2.3	Factors involved in drug release from nanoparticles..................... 24
	2.3.1	Biodegradability of pure drug particles.............................. 24
	2.3.2	Hydrophilicity and biodegradability of drug carriers....... 24
	2.3.3	Stimuli responsiveness of drug carriers or drug carrier complexes 25

**Chapter 3 Diversity of bioactive nanoparticles from biological,
chemical, and physical perspectives 29**
3.1	Viral vectors... 29
	3.1.1	Retroviral vectors 29
	3.1.2	DNA virus vectors.. 30
3.2	Nonviral vectors 31
	3.2.1	Lipid-based nonviral vectors 31
	3.2.2	Polymer-based nonviral vectors.. 33
	3.2.3	Inorganic carriers... 35
3.3	Hybrid particles 38
	3.3.1	Lipid–polymer hybrid nanoparticles.................................... 38
	3.3.2	Organic–inorganic hybrid nanoparticles............................. 39
	3.3.3	Inorganic hybrid nanoparticles 39
3.4	Genetically engineered drug carriers.. 39
3.5	Bioconjugation schemes for functionalization of and ligand attachment to nanoparticle surface.. 40

Contents

Chapter 4 Fabrication strategies for biofunctional nanoparticles 43
4.1 Chemical synthesis and engineering 43
    4.1.1 Production of drug nanoparticles: Top-down approaches ... 43
    4.1.2 Production of nonviral vectors: Bottom-up approaches 44
4.2 Recombinant DNA, hybridoma, and phage display techniques 56
    4.2.1 Synthesis of protein-based nanoparticles 56
    4.2.2 Generation of monoclonal antibodies 57
    4.2.3 Production of viral vectors 58

Chapter 5 Interactions and orientation of therapeutic drugs
            in the vicinity of nanoparticles 61
5.1 Dendrimer–drug interactions 61
5.2 Amphiphilic block copolymer–drug interactions 62
    5.2.1 Drug loading into polymeric micelles 63
    5.2.2 Polymeric micellar drug conjugate 63
    5.2.3 Electrostatic complexation with DNA/siRNA 63
5.3 Liposome–drug interactions 64
5.4 Inorganic nanoparticle–drug interactions 65

Chapter 6 Variable interactions of nanoparticles with blood,
            lymph, and extracellular and intracellular components ... 67
6.1 Serum proteins with affinity to nanoparticles 67
    6.1.1 Surface hydrophobicity 69
    6.1.2 Surface charge 69
    6.1.3 Size and curvature of nanoparticles 69
    6.1.4 Proteins with affinity for specific chemical groups
           of nanoparticles 70
6.2 Fates of the serum protein–coated nanoparticles 70
    6.2.1 Removal by macrophage, thrombosis,
           and hypersensitivity 70
    6.2.2 Aggregation 71
    6.2.3 Dissociation of complex and leakage of drugs 71
6.3 Interactions of nanoparticles with interstitial fluid and lymph 74
6.4 Extracellular matrix–nanoparticle interactions 75
6.5 Interactions between nanoparticles and cell components 76

Chapter 7 Pharmacokinetics and biodistribution
            of nanoparticles 77
7.1 Influence of particle size 77
7.2 Influence of plasticity of nanoparticles 79
7.3 Influence of protein corona formed around nanoparticles 79
    7.3.1 Opsonin-facilitated phagocytosis 79
    7.3.2 Dysopsonin-enhanced blood circulation time 79
    7.3.3 Uptake by nonphagocytic cells 81

viii *Contents*

7.4 Influence of charge and hydrophilicity 81
7.5 Influence of endogenous membrane coating.................................... 82
7.6 Influence of ligand coating... 82
7.7 Influence of coating of CD47 as a "self" marker 83
7.8 Extravasation from blood through vascular endothelium............. 84
    7.8.1 Permeability of vascular endothelia 84
    7.8.2 Different routes of traffic across continuous endothelium ... 85
    7.8.3 Deregulated vascular endothelium 86
    7.8.4 Vascular endothelium as a target for drug delivery 87
7.9 Transport across the interstitium 87
7.10 Cellular uptake, metabolism, and excretion.................................... 89

**Chapter 8 Specific roles of nanoparticles in various steps
of drug transport... 91**
8.1 Protection of nucleic acid– and protein-based drugs against
    degradation.. 91
    8.1.1 Determinants of polyplex stability 91
    8.1.2 Determinants of stability of lipoplex and other lipid-
        based complexes 92
8.2 Passive targeting to facilitate endothelial escape............................ 94
8.3 Drug delivery via the lymphatic system 95
8.4 Targeting cell surface receptors and facilitated uptake 97
    8.4.1 Monoclonal antibody–mediated targeting 97
    8.4.2 Carbohydrate-mediated targeting 99
    8.4.3 Peptide-mediated targeting 99
    8.4.4 Aptamer-mediated targeting 100
    8.4.5 Transferrin receptor–mediated targeting 100
    8.4.6 Folate-mediated targeting 102
8.5 Endosomal escape 104
    8.5.1 Fusogenic lipids or peptides 105
    8.5.2 Endosomal buffering 105
8.6 Nuclear targeting.. 106

**Chapter 9 Nanotechnology approaches to modulate transport,
release, and bioavailability of classical and emerging
therapeutics 109**
9.1 Controlled release and bioavailability of oral
    nanoformulations... 109
    9.1.1 Pharmaceutical techniques for controlling
        drug release 109
    9.1.2 Strategies for drug release from nanoparticles 112
9.2 Sustained release and bioavailability of ocular drugs.................. 114
    9.2.1 Different routes for controlled release
        of ophthalmic drugs... 114

Contents   *ix*

9.2.2 Strategies for controlled release and enhanced bioavailability of ophthalmic drugs.....115
9.3 Sustained release and bioavailability of dermal drugs 119
    9.3.1 Roles of nanoparticles in sustained release 120
9.4 Sustained release and bioavailability of pulmonary drugs 122
    9.4.1 Challenges of controlled release from nanoformulations 123
    9.4.2 Controlled release and bioavailability of pulmonary drugs 124
9.5 Intracellular and extracellular transport vehicles 126
    9.5.1 Gene, siRNA, and ODN......126
    9.5.2 Synthetic peptides and recombinant proteins 132
    9.5.3 Conventional small drugs......137
    9.5.4 Combinations of drugs and nucleic acids......138

Chapter 10 Nanotechnology in the development of innovative treatment strategies 143
10.1 Gene therapy 143
    10.1.1 Transgene expression......143
    10.1.2 Knockdown of endogenous gene expression 144
    10.1.3 Inhibition of endogenous miRNA functions..................147
    10.1.4 Gene therapy applications......147
10.2 Protein- and DNA-based prophylactic vaccines 151
    10.2.1 Subunit vaccines 151
    10.2.2 DNA vaccines......151
    10.2.3 Vaccine delivery......152
10.3 Immunotherapy 156
    10.3.1 Strategies of cancer immunotherapy 156
10.4 Photodynamic therapy......159
    10.4.1 Nanoscale drug delivery systems in PDT......160
10.5 Image-guided therapy 164
    10.5.1 Personalized medicine......164
    10.5.2 Nanotheranostics in cancer......165
    10.5.3 Nanotheranostics in cardiovascular diseases 170

Chapter 11 Nanoparticles for therapeutic delivery in animal models of different cancers 171
11.1 Brain cancer 171
    11.1.1 Taxane-loaded poly(ε-caprolactone) nanoparticles..........171
    11.1.2 Doxorubicin-loaded surfactant-coated nanoparticles......172
    11.1.3 hTRAIL gene-loaded cationic albumin–conjugated PEGylated PLA nanoparticles......172
    11.1.4 Nanoliposomal irinotecan......173
    11.1.5 Bcl2L12-targeting spherical nucleic acids......173

x   *Contents*

11.2   Breast cancer.. 173
      11.2.1   Anti-HER2 liposome formulations of DOX
               and topotecan..174
      11.2.2   Rapamycin-loaded nanoparticles of elastin-like
               polypeptide diblock copolymers...174
      11.2.3   Sorafenib-loaded nanoliposomal ceramide...................... 175
      11.2.4   Mdr-1 and surviving siRNA-loaded
               poly(β-amino esters).. 175
      11.2.5   DOX-loaded folate-targeted pH-sensitive
               polymer micelle 176
11.3   Colon cancer 176
      11.3.1   Taxol-containing liposomes 176
      11.3.2   Endostar-loaded PEG–PLGA nanoparticles...................... 177
      11.3.3   PEG-coated Bcl-2 siRNA lipoplex.. 177
      11.3.4   Thymidylate synthase shRNA-expressing Ad.................. 177
      11.3.5   Oxaliplatin-encapsulated chitosan micelle....................... 178
11.4   Lung cancer 178
      11.4.1   EGFR siRNA-containing anisamide–PEG–LPD
               nanoparticles... 178
      11.4.2   Bcl-x SSO-loaded anisamide–PEG–LPD
               nanoparticles... 179
      11.4.3   DOTAP/cholesterol–tumor suppressor
               plasmid complexes 179
      11.4.4   Poly-(γ-L-glutamylglutamine)–PTX nanoparticles 180
      11.4.5   DOX-encapsulated PEG–PE micelle 181
11.5   Ovarian cancer... 181
      11.5.1   EphA2 siRNA-encapsulated neutral liposomes............... 181
      11.5.2   Liposomal EphA2 siRNA-loaded silica particles............. 181
      11.5.3   Multifunctional DOX-carrying lipid-based
               nanoformulations 182
      11.5.4   Hybrid micelles carrying cisplatin and PTX 182
      11.5.5   LHRH peptide- and PTX-conjugated dendrimer/
               CD44 siRNA 183
11.6   Pancreatic cancer 183
      11.6.1   uPAR-targeted, gemcitabine-loaded iron oxide
               nanoparticles... 183
      11.6.2   DACHPt-loaded polymeric micelle..................................... 184
      11.6.3   PEGylated human recombinant hyaluronidase PH20 184
      11.6.4   HER-2 siRNA-loaded immunoliposome 185
11.7   Skin cancer... 185
      11.7.1   PTX/ETP-loaded lipid nanoemulsions............................... 185
      11.7.2   Nanoliposomal siRNA targeting B-Raf and Akt3 186
      11.7.3   Antigenic peptide–encapsulated PLGA nanoparticles.... 186

Contents xi

11.7.4 MART-1 mRNA-loaded mannosylated nanoparticles 187
11.7.5 Anti-CD47 siRNA-encapsulated liposome–
protamine–HA nanoparticles 187

**Chapter 12 Nanoparticles for therapeutic delivery in animal
models of other critical human diseases............................ 189**
12.1 Arthritis 189
12.1.1 Liposomally conjugated MTX (G-MLV) 189
12.1.2 SOD enzymosomes 190
12.1.3 VIP–SSM.. 190
12.1.4 CPT–SSM–VIP 190
12.1.5 Betamethasone-encapsulated PEGylated PLGA
nanoparticles 191
12.1.6 Gold nanoparticles 191
12.1.7 PEGylated cyclodextrin–methylprednisolone conjugate... 191
12.2 Cardiovascular diseases 192
12.2.1 Nanotherapeutics in atherosclerosis............................... 192
12.3 Diabetes... 193
12.3.1 Oral insulin delivery using nanotechnologies................ 194
12.3.2 Nasal insulin delivery using nanotechnologies 197
12.3.3 Pulmonary insulin delivery.. 198
12.3.4 Transdermal insulin delivery 199
12.3.5 Subcutaneous insulin delivery 199
12.4 Neurodegenerative diseases 199
12.4.1 Nanoparticles as neuroprotective and therapeutic
drugs for neurodegenerative diseases............................. 200
12.5 Degenerative retinal diseases 202
12.5.1 Neurotrophic factor therapy 202
12.5.2 Antioxidant therapy 203
12.5.3 Anti-inflammatory therapy... 203
12.5.4 Inhibiting choroidal neovascularization.......................... 203
12.5.5 Retinal gene therapy 204
12.6 Inflammatory bowel diseases 205
12.6.1 TNF-α siRNA-encapsulated polymeric nanoparticles..... 205
12.6.2 TNF-α–neutralizing nanobodies....................................... 205
12.6.3 PHB gene–carrying Ad and PHB-entrapped PLA
nanoparticles 206
12.6.4 NF-κB decoy ODN-loaded chitosan–PLGA
nanoparticles 206
12.6.5 Map4k4 siRNA-encapsulated glucan shells 206
12.6.6 β_7 integrin-targeted, CyD1 siRNA-loaded liposomes....... 206
12.6.7 IL-10 gene-encapsulated NiMOS...................................... 207
12.6.8 Mesalamine (5-ASA)-loaded nanoparticles 207

xii *Contents*

12.7 Obstructive respiratory diseases 208
    12.7.1 Nanomedicine for allergic asthma 208
    12.7.2 Nanomedicine for respiratory syncytial virus 209
12.8 Hepatic fibrosis and infections 209
    12.8.1 Hepatic fibrosis 210
    12.8.2 Hepatic infections 211
12.9 Malaria 212
    12.9.1 Curcumin nanoparticles 212
    12.9.2 Curcumin-entrapped chitosan nanoparticles 213
    12.9.3 Artemisinin- and curcumin-loaded liposomes 213
    12.9.4 β-Artemether–loaded liposomes 213
    12.9.5 Chloroquine-encapsulated immunoliposomes 213
    12.9.6 Liposome-coupled TNF-α 214
    12.9.7 ARM-loaded lipid nanoparticles 214
    12.9.8 Primaquine-loaded nanoemulsions 215
12.10 Regeneration of tissues 215
    12.10.1 Nanotechnology in wound healing 215
    12.10.2 Bone regeneration 217

Chapter 13 Nanomedicine in clinical trials 219
13.1 Different phases of clinical trials 219
13.2 Nanoparticulate drug delivery systems in clinical trials 220
13.3 Monoclonal antibodies as therapeutics in clinical trials
    (selected) 220

Chapter 14 Approved and commercialized nanomedicine 253

Chapter 15 Current safety issues: Biodegradability, reactivity,
    and clearance 287
15.1 Nanoparticle interaction with blood cells 287
    15.1.1 Hemolysis of red blood cells 287
    15.1.2 Platelet aggregation and activation 288
    15.1.3 Macrophage uptake and immune responses 288
15.2 Deformation of cellular membrane 288
    15.2.1 Disturbance of phospholipid bilayer 288
    15.2.2 Interactions with membrane proteins and blocking
    of ion channels 289
15.3 Lysosomal rupture and release of contents 289
15.4 Disruption of cytoskeleton 289
15.5 Damage to nuclear DNA and proteins 290

References 291
Index 315

Preface

Regardless of the administration routes, delivery of small-molecule drugs to their target sites of action historically poses one of the biggest challenges because of their homogeneous tissue distribution, renal clearance, and lack of target specificity. Nanotherapeutics have evolved as novel drug formulations at dimensions of roughly 1–100 nm by virtue of the integration of nanotechnology with medicine for treating and preventing critical human diseases effectively and precisely. The favorable pharmacokinetics with prolonged circulation time, selective endothelial permeability at several target tissues, and high specificity for biological targets are the attractive attributes of nanotherapeutics that drive the pharmaceutical industries to conduct a large number of preclinical and clinical trials, with enormous successes seen in the past in getting approval and commercialization of nanotechnology-based medical products.

To my knowledge, this is the first book that comprehensively discusses the current shortcomings for delivery of classical (small) drugs, macromolecular therapeutics, and recombinant vaccine via the common intravascular and extravascular routes; describes the synthetic/chemical engineering methods as well as recombinant, hybridoma, and phage display technologies to fabricate different types of nanoparticulate carriers and drugs; reveals the diversified approaches undertaken by harnessing nanotechnology to overcome the multistep extracellular and intracellular barriers and to facilitate the development of novel strategies for therapeutic delivery and imaging; and elaborates the preclinical and clinical trials of potential nanoparticle-based products in animal models and patients, respectively, and the approval/commercialization of nanotherapeutics, addressing all relevant human diseases.

Features

- Bridges the gap for the first time between the preclinical development in the laboratory and the clinical evaluation and regulatory approval of nanotherapeutics.

xiii

xiv

- Integrates all interdisciplinary components of nanotherapeutics from design/fabrication to final development.
- Explains the strategic contributions of nanocarriers to improving the local delivery, sustained release, and pharmacokinetics of small drugs and macromolecular drugs depending on the routes of delivery.
- Presents the emerging approaches of nanomedicine in relation to their preclinical evaluations in animal models of diverse human diseases.
- Provides an extensive database on the potential nanomedicine products under clinical trials and the approved ones in relation to their architectures, sponsors, and disease targets.
- Depicts the critical and important issues with 71 completely new, vivid, and beautiful images.

Acknowledgments

My deepest gratitude goes first and foremost to our Creator and Sustainer who is the source of light (or energy) of the invisible and visible universe, both grasping and managing it. As a human being, I always feel that I could have been born with a physical or intellectual disability or brought up in an underprivileged or noncompetitive environment, ending up without flourishing myself. In addition, moving academically in the right direction based on one's interests or gifted capability is not something always in our hands. In other words, I like to put myself among those many fortunate people who should be immensely thankful to Him. I would also like to thank CRC Press for the invitation, giving me a wonderful opportunity to fulfill one of my professional and personal aspirations in this small but beautiful life. Like any other son, I am always greatly indebted to my beloved parents. I have also been remarkably inspired by my wife and growing children during the entire writing period.

Author

Dr. Ezharul H. Chowdhury currently holds an associate professor position and is a cluster leader of biomedical engineering under the Advanced Engineering Platform at Monash University (Sunway Campus). Before this appointment, he was a senior lecturer at the International Medical University (IMU), since September 2008, an independent assistant professor at the Tokyo Institute of Technology (Tokyo Tech), and a visiting professor at the Shizuoka Cancer Center Institute since 2006. He earned his doctor of engineering degree in 2003 at Tokyo Tech, where he also carried out a postdoctoral study and subsequently served as a team leader.

Dr. Chowdhury has pioneered the development of a range of pH-sensitive inorganic nanoparticles as smart tools for efficient and targeted intracellular delivery of genetic materials, gene-silencing elements, proteins, and classical anticancer drugs. He is currently applying this smart nanotechnology for the treatment of cancer, particularly breast carcinoma, and cardiovascular diseases, such as diabetes. His team is now conducting preclinical trials with some interesting nanoformulations of classical anticancer drugs, therapeutic gene(s), and small interfering RNAs (siRNAs). Dr. Chowdhury's research team is one of the leading groups in the world, having original and unique contributions in this multidisciplinary field. Currently, he is supervising eight PhD students and one master's student. His research projects have so far been funded internally through institutional grants as well as externally by the Japanese Government's Ministry of Education, Culture, Sports, Science & Technology; the Japan Society for the Promotion of Science; the Malaysian Ministry of Science, Technology, and Innovation; and the Ministry of Higher Education. He works as an editorial board member for five international journals. His outstanding contributions have so far produced more than 65 publications in international journals of high repute, with more than 15 average citations per article and 5 Japanese and United States patents.

chapter one

Emergence of nanotherapeutics
Challenges in classical drug transport versus macromolecular drug design

1.1 Administration of small-molecule drugs: Traffic routes toward the bloodstream

Despite the low bioavailability of drugs compared to the intravascular injection through veins (intravenously) or arteries (intra-arterially), the extravascular administration through oral, nasal, intramuscular, subcutaneous, dermal, pulmonary, and rectal routes offers clear advantages in terms of patient compliance and safety. The unique anatomical, biochemical, and cellular features of each of the extravascular routes account for its distinctive barriers to drug delivery en route to the bloodstream.

1.1.1 Barriers to the oral route

The oral route confers the most common, least invasive, and easiest way of administration of a drug, which subsequently passes through the stomach to the small and large intestines of the gastrointestinal (GI) tract. The small intestine is more valuable than the other two segments (stomach and large intestine) to facilitate drug absorption through its large surface area accommodating a dynamic population of epithelial cells including goblet cells to secrete mucus, endocrine cells to produce hormones, and absorptive cells (enterocytes) to absorb drugs or nutrients from the GI lumen to the blood. The first rate-limiting step is the dissolution from the solid dosage forms, such as tablets in the aqueous media of variable pHs of the GI tract, with incomplete dissolution leading to poor absorption and bioavailability. The dissolved drug molecules might additionally undergo degradation by the extreme pH or hydrolytic enzymes of the GI lumen, as illustrated in Figure 1.1. Moreover, the free drugs could be subjected to metabolism by the enzymes present in the gut wall or even in the liver after being absorbed and transported through the portal vein. The loss as the drug passes through the GI tract and the liver is known as the "first-pass effect." The epithelial lining usually offers impedance to the absorption of polar ionized drugs while the unabsorbed drugs travel further for fecal excretion.

1

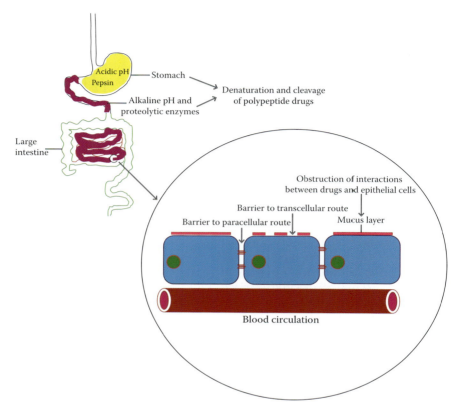

Figure 1.1 Existing barriers to oral delivery of small-molecule and macromolecular drugs.

In the small intestine, a number of transporters critical for absorption of dietary constituents and drugs are expressed on the brush border membrane of enterocytes. Among the transporters expressed in enterocytes, P-glycoprotein (P-gp, also known as multidrug resistance 1) and multidrug resistance-associated protein 2 (MRP2) actively extrude drugs back into the intestinal lumen, thereby limiting the rate of drug absorption, while peptide transporters 1 and organic anion-transporting polypeptide 1A2 can enhance the extent of drug bioavailability (Benedetti et al. 2009).

1.1.2 *Barriers to sublingual and buccal routes*

Drugs can be delivered to the systemic circulation through the oral cavity via either the sublingual route by placing a solid dosage form under the tongue for subsequent absorption via the sublingual mucosa or the buccal route by

Chapter one: Emergence of nanotherapeutics   3

putting the dosage form between the gums and the cheek to be absorbed via the buccal mucosa (the lining of the cheek). While the sublingual mucosa is relatively permeable, giving rapid absorption, the buccal mucosa is considerably less permeable with consequential limited bioavailability of a drug. The oral mucosa consists of an outermost layer of stratified squamous epithelium, the most important barrier to drug absorption, below which a basement membrane, the lamina propria, and the submucosa sequentially exist. The buccal epithelium is composed of 40–50 cell layers, whereas the sublingual epithelium possesses somewhat fewer layers. Although both of the routes can bypass the first-pass effect, they possess disadvantages, such as difficulty in holding the dosage form within the mouth and limited capacity for drug absorption because of the lack of adequate surface area.

1.1.3   Barriers to the rectal route

The rectum is an alternative to the oral route for drug administration in the cases when drugs could be destroyed by extreme pH or enzymes of the GI tract, or under the circumstance of nausea, vomiting, or convulsion. The rate and extent of drug absorption are often lower than the oral absorption, probably owing to the relatively small surface area of the rectum for drug absorption. The composition of the rectal formulation could be a determinant for drug release from it during the absorption process. The partial avoidance of hepatic first-pass metabolism after rectal delivery of some drugs appears to be responsible for their higher bioavailability than for oral delivery. Indeed, part of the rectal blood supply bypasses the hepatic portal circulation and dumps directly into the inferior vena cava, thus avoiding loss of some of the drugs as happens in oral delivery. The low water content of the rectum makes lipophilic drugs particularly suitable for delivery through this route.

1.1.4   Barriers to the intranasal route

The large surface area of the nasal cavity with a high degree of vascularization allows rapid absorption of lipophilic drugs through the nasal mucosa but permits very limited bioavailability of polar drugs. The drugs that are not easily absorbed across the nasal membrane can be subjected to rapid clearance from the nasal cavity owing to the mucociliary clearance mechanism or enzymatic metabolism in the nasal epithelium. However, like oral mucosa, nose mucosa bypasses the GI or hepatic first-pass effect.

1.1.5   Barriers to the pulmonary route

The respiratory system consists of the conducting airways, which are further divided into nasal cavity, nasopharynx, oropharynx, larynx, trachea, bronchi, bronchioles, and the respiratory region that comprises respiratory bronchioles, alveolar ducts, and alveolar sacs. The drug transport in the upper airways is inefficient because of a smaller surface area, a lower regional blood flow, and a high filtering capacity removing up to 90% of the drug particles. Additionally, ciliated cells present in this region cause propulsion of the secreted mucus upward, thus clearing the foreign drug molecules. In contrast, the respiratory region accounting for more than 95% of the lung's surface area is directly connected to the systemic circulation. However, the pore size and tight junction depth of both alveolar and endothelial cells can influence the transepithelial drug transport. The success of the pulmonary administration system mainly depends on the delivery devices, such as nebulizers, metered-dose inhalers, and dry powder inhalers.

1.1.6   Barriers to intramuscular and subcutaneous routes

Unlike the GI tract, the absorption rate of most of the injected drugs from muscle and subcutaneous tissue to blood is proportional to blood flow, since the capillary wall with a much more loosely knit structure than the epithelial lining of the GI tract offers little impedance to the transport of even polar ionized drugs with a molecular weight of roughly 5000 to the bloodstream.

1.1.7   Barriers to the transdermal route

Transdermal drug delivery offers the advantage of avoiding first-pass metabolism and providing sustained drug release for a prolonged period. The skin that guards the underlying muscles, bones, ligaments, and internal organs comprises three primary layers: epidermis, dermis, and hypodermis. Although the epidermis contains no blood vessels, blood capillaries can extend to the upper layers of the dermis to provide nutrients through diffusion for the nourishment of the cells present in the deepest layers of epidermis. Stratum corneum, the outer layer of the epidermis with a unique structure (i.e., a lipid-rich matrix and embedded keratinocytes), acts as the most important barrier to drug absorption. Chemical enhancers have led to limited success compared to electronically controlled devices involving ultrasound and electric fields in increasing transdermal transport of small molecules.

1.2   Fates of the small-molecule drugs in blood

1.2.1   Plasma drug concentration on intensity of therapeutic action

In general, the initial dose of a drug in plasma is directly proportional to the resulting concentration of a drug at its site of action as well as

Chapter one: Emergence of nanotherapeutics   5

the intensity of its actions. Factors that affect the concentration of the drug in plasma will also affect its concentration at sites of action. Thus, removal of the drug from plasma by the tissues that store, metabolize, or excrete it ultimately leads to a lower concentration of drug at the site of action.

1.2.2   Drug transport from blood capillaries to extracellular fluid

1.2.2.1   Influences of plasma level and molecular size of a drug

The primary factors in determining the concentration of a drug in plasma are as follows: (1) mode of drug administration, (2) uptake of drug by body tissues after escape from blood vessels, and (3) elimination of the drug from the body. Upon arrival either intravenously or through extravascular routes, the molecular size of a drug determines its ability to move from the blood vessel into the extracellular fluid compartment. Most of the conventional drugs are, however, sufficiently small to pass through the pores of vascular capillary membranes. Unlike the extravascular routes, the intravascular administration results in instantaneous availability of the drug in plasma with the concentration determined by the dose size and the rate of injection. The highest plasma drug concentration immediately after the intravascular injection is accompanied by the most rapid rate of drug entry into tissues. Unless additional doses are administered, the drug concentration in plasma declines progressively as the drug enters body tissues. Since the drug is continuously taken up by tissues (and also eliminated from the body), equilibrium can be reached at which concentrations of the drug in plasma and tissue compartments are equal with no net exchange of the drug. With continuing elimination of the drug from plasma, the gradient can be reversed with the net drug exchange now from the tissues to the plasma and an eventual decline of drug concentrations in both compartments.

1.2.2.2   Influence of blood flow

Blood flow can be the rate-limiting factor in the exchange between plasma and tissues for a drug with the ability to rapidly penetrate blood capillaries. Organs with a high blood perfusion rate, such as brain, kidney, liver, and heart, exhibit a faster drug uptake than the other tissues with a low perfusion rate. The accumulation of a drug in a tissue even with high affinity for the drug can be limited by a low blood flow to the tissue, especially when the plasma concentration of the drug is not maintained by repeated doses or continuous infusion. If administration of the drug is discontinued, blood flow can restrict the rate of drug removal from the tissue, allowing it to serve as a reservoir and maintaining low but persistent plasma levels of the drug (Benedetti et al. 2009).

6   *Nanotherapeutics*

1.2.2.3   *Influence of protein binding*

Almost all drugs are bound to some extent by plasma proteins, particularly albumin, by the weak forces through ionic, hydrogen, or van der Waals bonds. On the other hand, large protein molecules are markedly restricted to penetrate the blood vessels. As a result, protein-bound drugs do not directly participate directly in the equilibration of the drug concentrations across the capillary membranes. For a particular drug having some degree of affinity for plasma proteins, the total amount of the drug inside the blood capillary will be higher than that outside the capillary at equilibrium.

1.2.3   *Drug transport from extracellular fluid to cells*

A drug penetrates cellular membranes to gain access to the sites of action, storage, biotransformation, and excretion, which, in turn, serves as a major determinant of the drug exit rate from the plasma. Two different routes of penetration, namely, passive diffusion and active transport, are predominantly available in various types of cells for the drug molecules with variable physicochemical properties.

1.2.3.1   *Cellular uptake by passive diffusion*

Passive diffusion means transport of a substance down its concentration gradient across a membrane without investment of any energy. It is the lipophilic character of a drug that determines the rate of its passive diffusion across the lipid bilayer of a cell. Weak acids and bases must be in their nonionized forms in order to move via the process.

1.2.3.2   *Cellular uptake by active transport*

Active transport utilizes energy and a transport protein to carry a drug across the cell membrane. The main transporters involved in drug uptake across the sinusoidal membrane of liver hepatocytes include the organic anion transporter (OAT), organic cation transporter (OCT), and organic anion-transporting polypeptide families. The primary transporters for uptake of drugs into the renal proximal tubular cells include OAT1/3 and OCT2. OCT3 is involved with uptake of drugs into the heart and transport of drugs across the placenta. ABC efflux transporters, such as P-gp, multidrug resistance-associated proteins (MRPs), and breast cancer resistance protein (BCRP), which are present in the blood–brain barrier and the blood–cerebrospinal fluid barrier, effectively remove many polar drugs from the central nervous system, thus limiting brain uptake (Benedetti et al. 2009).

1.2.3.3   *Uptake by tumor cells*

Both passive diffusion and active transport are involved in the uptake of a drug by tumor cells. In contrast, efflux transporters expressed in tumor

Chapter one: Emergence of nanotherapeutics 7

cells can extrude the drug, preventing its intracellular accumulation and leading to the development of drug-resistant cancer cells with the final outcome of treatment failure in cancer patients. Generally, cancer cells become simultaneously resistant to multiple drugs of different chemical structures owing to overexpression of P-gp, MRP, and BCRP, which is known as multidrug resistance.

1.2.3.4   *Uptake by nonresponsive cells*

The uptake of a drug by nonresponsive tissues reduces the pharmacological effect of the drug by lowering its concentration in plasma as well as at sites of action. If the drug is simply stored in nonresponsive tissues without being metabolized or excreted, it preserves its pharmacological potential, which can be manifested once it reenters the circulation and is transported to the sites of action.

1.2.4   *Elimination of drugs from the body*

Kidneys and liver are the major excretory organs, while the pulmonary excretion via exhaled breath is mainly for gaseous and volatile substances.

1.2.4.1   *Chemical modification with functional groups*

Most of the drugs administered via any of the aforementioned routes undergo chemical modification before excretion in the urine or bile. In case of oral administration, the small intestine is responsible for the initial metabolic processing of many drugs as mentioned earlier before they transport through the portal circulation to the liver, which metabolizes not only the drugs absorbed from the GI tract but also the drugs that escape this first-pass metabolism and recirculate through it, and those that come after distribution to other organs or tissues via different routes. Other organs that produce drug-metabolizing enzymes include the nasal mucosa, the lungs, and the kidneys. The common chemical reactions catalyzed by the enzymes are oxidation, reduction, hydrolysis, and, subsequently, sulfation, glucuronidation, or glutathione conjugation, thereby introducing hydrophilic chemical groups to the drug molecules, providing a greater water solubility and increased molecular weight to the metabolites, reducing the membrane permeability and generally terminating the biological activities.

1.2.4.2   *Renal excretion*

Nephron, the functional unit of kidneys, controls the renal excretion of drugs through (1) glomerular filtration, (2) tubular secretion, and (3) tubular reabsorption. In the glomerulus, drugs of low molecular weight are filtered in Bowman's capsule through the pores present in the endothelium of glomerular capillaries, unless the drugs are tightly bound to plasma proteins. After filtration, the remaining plasma moves in the arterioles adjacent to the proximal tubule where drugs can be actively secreted from the arterial into the proximal tubule or reabsorbed in the opposite direction. Active uptake of anionic and cationic drugs and their extrusion into the urine mainly take place in the renal proximal tubules. Since the kidney tubular membrane is lipoidal, nonionized and lipid-soluble drugs are significantly reabsorbed into the plasma whereas ionized and polar compounds are predominantly excreted in urine.

1.2.4.3 Biliary excretion

Drug metabolites that are excreted in the bile can be either irreversibly excreted into feces through the GI tract or reabsorbed into the hepatic portal vein. Anions, cations, and nonionized molecules with molecular weight greater than 500–600 are excreted into the bile and lower-molecular-weight compounds are reabsorbed before being excreted in the GI tract from the bile duct. Conjugation with glucuronic acid, which leads to high-molecular-weight metabolites, facilitates biliary excretion. While the majority of small, lipophilic compounds enter the hepatocyte via the sinusoidal membrane by simple passive diffusion, more polar and bulky molecules require transport systems to cross the membrane. Once inside the hepatocyte, compounds can be either transported into bile after biotransformation or excreted into blood by basolateral transport proteins. The principal transporters involved in drug transport through the biliary canalicular membrane of hepatocytes into the bile duct include P-gp and MRP2 (Benedetti et al. 2009).

The fates of small-molecule drugs in the bloodstream after administration through various routes are schematically shown in Figure 1.2.

1.3 Major problems associated with traditional formulations of small-molecule drugs

The main problems encountered with administration of a small-molecule drug involve its even biodistribution throughout the body, absence of specific affinity toward a pathological site (organ, tissue, or cell), and lack of sufficient specificity for a cellular target, thus necessitating administration of a large dose of the drug to achieve high local concentration and the required therapeutic activity with eventual nonspecific toxicity and other adverse side effects particularly associated with anticancer drugs, which are cytotoxic and lack specific targeting, traveling across blood vessels freely. Thus, in order to ensure efficacy and safety, drugs are ideally required to be developed with precise cellular target specificity and delivered to their target sites selectively at an optimal rate.

Chapter one: Emergence of nanotherapeutics   9

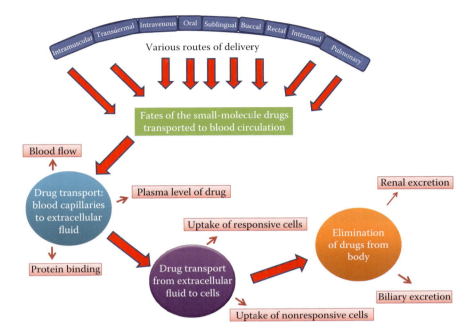

Figure 1.2 Fates of small-molecule drugs after reaching blood circulation through various routes.

1.4 Alteration of pharmacokinetics of small-molecule drugs with macromolecules

Macromolecular drug carriers could be designed to maximize the therapeutic efficacy of a conventional drug by enhancing solubility, stability, and blood retention time and by facilitating controlled release and diseased tissue–targeted delivery of the drug. A number of techniques have been proposed and developed so far to achieve these goals (Figure 1.3).

1.4.1 Enhancement of drug solubility and stability

The majority of drug candidates emerging from drug discovery process are relatively hydrophobic, since they exhibit superior binding affinity to receptor targets. However, poor water solubility hampers the dissolution of hydrophobic drug particles in GI fluids, a prerequisite for intestinal absorption, although some degree of apolar character is necessary for drug absorption. Moreover, for parenteral administration, these molecules are generally mixed with cosolvents, which have many undesirable side effects. Among the novel strategies currently being undertaken for

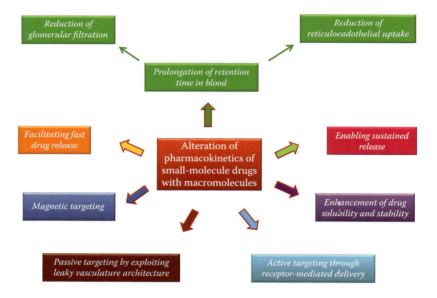

Figure 1.3 The roles of macromolecules in modifying pharmacokinetics of small-molecule drugs.

efficient delivery of such poorly water-soluble drugs, the most favorable approach is the exploitation of lipid-based nanocarriers to individually encapsulate the drug molecules, protecting them from the surrounding aqueous environment and preventing drug precipitation during formulation and the absorption process (Figure 1.4). The polymeric micelle is another potential tool to deliver hydrophobic drugs by embedding them within its hydrophobic core while its hydrophilic polymeric segment at the outer surface confers the colloidal stability in plasma (Figure 1.4). A relatively new but powerful approach is the reduction of drug particle size to form surfactant-stabilized drug nanoparticles or nanosuspensions (Figure 1.5), which have dramatic effects on enhancing absorption, dose escalation, and eliminating food effects and thus improving efficacy and safety (Merisko-Liversidge and Liversidge 2011).

1.4.2 Prolongation of retention time in blood

Prolonged retention time in blood provides drugs the chance to distribute to their target tissue/organ and allows increased duration for their pharmacological activities. This can be achieved by avoiding glomerular filtration and hepatic uptake of drug carriers unless the kidney or the liver is a drug target.

Chapter one: Emergence of nanotherapeutics 11

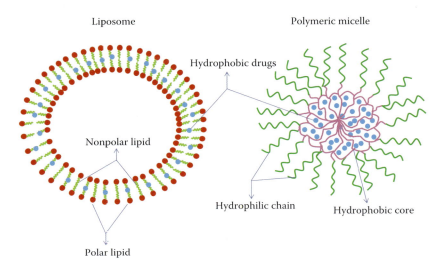

Figure 1.4 Liposome and polymeric micelle for encapsulating hydrophobic drugs.

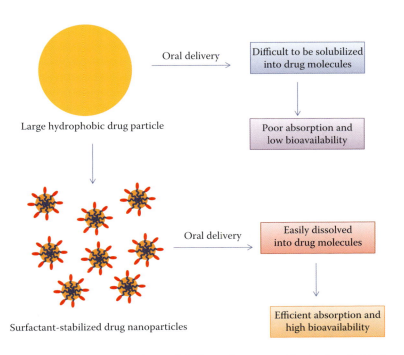

Figure 1.5 Drug nanosuspension: solubility enhancement for hydrophobic drugs.

1.4.2.1 Reduction of glomerular filtration

Since the diameter of the fenestrae (pores) in glomerular endothelium is as large as 70–90 nm, macromolecules below the size limit freely pass through the endothelium. However, the basement membranes that support glomerular endothelial cells pose an additional barrier restricting the molecules of the size of albumin to pass through these pores and allowing only the macromolecules with a molecular weight below 50,000 (approximately 6 nm in diameter). The charge of macromolecules is another factor influencing the glomerular filtration, with positively charged macromolecules filtered more effectively than anionic ones, since the basement membrane of the glomerulus is mainly composed of negatively charged polysaccharides (glycosaminoglycans) (Bohrer et al. 1978).

1.4.2.2 Reduction of reticuloendothelial uptake

The reticuloendothelial system is composed of a set of mononuclear phagocytic cells that engulf foreign particles, macromolecules, and microorganisms. Phagocytosis of drug carriers is accelerated after they are associated with complements and immunoglobulins (Opanasopit et al. 2002). Thus, to enable long-term retention in the systemic circulation, these nanoparticle carriers can be coated with hydrophilic polymers or proteins, such as polyethylene glycol (PEG) or albumin, respectively, to escape from the reticuloendothelial uptake by hindering the interactions of the particle surface with opsonins and cell membranes (Pasut and Veronese 2012).

1.4.3 Facilitating fast drug release

Nonsteroidal anti-inflammatory drugs (NSAIDs) represent a very useful class of medications for treating musculoskeletal and inflammatory diseases with associated GI and potential cardiovascular toxicity. By reducing drug particles approximately 10 times smaller than conventional formulations, it is possible to achieve comparable peak plasma concentrations (C_{max}) with lower overall extent of systemic exposure (area under the curve). Examples include submicron NSAIDs of diclofenac, indomethacin, naproxen, and meloxicam, with promising results obtained at various clinical stages of drug development (Atkinson et al. 2013).

1.4.4 Enabling sustained release

Controlled drug delivery from various nanoparticle formulations has evolved to tailor drug release profiles in response to the physiological need for a particular drug either at predetermined time intervals, such as in diabetes, or for an extended period, such as in chemotherapy, while avoiding the adverse effects attributed to the high systemic drug exposure. Sustained drug release could result from modulation of the

Chapter one: Emergence of nanotherapeutics 13

dissolution kinetics of nanoparticle core/shell structures and gradual diffusion of the drug from the core. The drug release kinetics is usually characterized by an initial burst release owing to the dissolution and diffusion of the weekly entrapped drug, followed by a slower and continuous release (Danhier et al. 2009). A sophisticated control in drug release could be achieved by using stimuli-sensitive polymers that respond to certain environmental signals, such as temperature, pH, light, electricity, or ionic strength (Wang et al. 2010).

1.4.5 *Passive targeting by exploiting leaky vasculature architecture*

An imbalance between pro- and antiangiogenic factors promotes pathological angiogenesis with immature and fragile blood vessels in a wide range of disorders such as solid tumors and inflammation tissues. An extensive vascular permeability to macromolecules and lack of lymphatic drainage result in the selective accumulation of the extravasated macromolecules with a certain range of radius in tumor tissues for long periods (Figure 1.6). This is known as the "enhanced permeability and retention" (EPR) effect (Maeda 2001; Maeda et al. 2000).

1.4.6 *Active targeting through receptor-mediated delivery*

Receptor-mediated endocytosis allows cell-specific delivery of a drug using the nanoparticles that have the potential of being modified either covalently or noncovalently with a variety of targeting ligands ranging from small organic molecules to proteins that include carbohydrates, nucleic acids (i.e., aptamers), synthetic peptides, transferrin, asialoglycoproteins, apolipoproteins, and antibodies (Figure 1.6). Because of the unique vascular structures of the liver with elevated blood flow and vascular permeability compared to other tissues/organs, the overall probability for receptor-mediated uptake of a ligand-anchored drug carrier is much higher in the liver.

1.4.7 *Magnetic targeting*

Superparamagnetic iron oxide nanoparticles with surface-bound or encapsulated drugs can facilitate the extracellular, intracellular, and site-specific delivery of drugs under the influence of a strong permanent magnet usually placed outside the body over the target site. The external magnetic field can retain the magnetic particles at the target site after intravenous or intra-arterial administration if the magnetic forces exceed the linear blood flow rates in arteries or capillaries. If intracellular localization of drugs, such as cytotoxic agents, is necessary, the magnetic carriers can

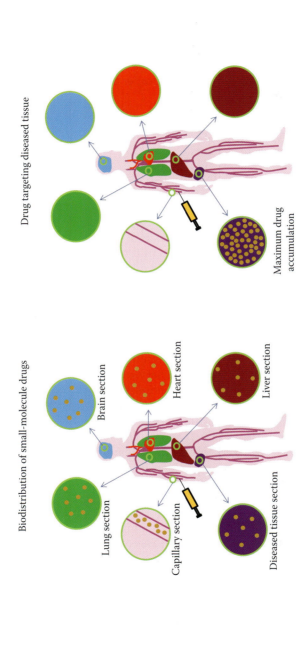

Figure 1.6 Role of active or passive targeting for diseased tissue–selective delivery of small-molecule drugs, preventing homogeneous distribution.

Chapter one: Emergence of nanotherapeutics 15

be coated with receptor-specific ligands for delivery into the target cells (Figure 1.7). Moreover, once the drug/carrier complexes are concentrated at the desired location, the drugs can be released either enzymatically or through changes in physiological pH, osmolality, or temperature (Wang et al. 2013).

1.5 Protein-based macromolecular drugs

1.5.1 Proteins as independent therapeutic drugs

Along with the advances in recombinant DNA technology, hybridoma technology, and large-scale production of proteins and peptides, there has been a significant understanding of the roles of peptides and proteins in basic research and clinical therapy. Therefore, the last two decades have seen a remarkable growth of recombinant therapeutic proteins and monoclonal antibody-based drugs for various clinical uses. However, although many protein therapeutics have been approved or are currently in advanced clinical testing, the development of more sophisticated delivery systems has not kept pace with this rapidly expanding class of therapeutics. Therefore, considering the intrinsic physicochemical and biological properties of these biopharmaceuticals, particularly including poor permeation through biological membranes, large molecular size, short plasma half-life, aggregation, and immunogenicity, development of approaches for efficient and convenient delivery of these therapeutic agents remains to be a major challenge (Shantha Kumar et al. 2006).

1.5.2 Chemically modified and carrier-bound proteins

Therapeutic proteins face several major challenges owing to several inherent shortcomings. Proteins often have short half-lives (owing to degradation by proteolytic enzymes, neutralization by antibodies, and faster renal clearance), a wide tissue distribution, and the potential for immunogenicity, and most of the time, they need to be dosed frequently. When frequent dosing is required, it can result in increased cost, complicated dosing regimens, and patient noncompliance, leading to reduced efficacy. In an effort to address the aforementioned challenges for delivery of protein therapeutics and improve patient compliance and drug efficacy, protein drugs are usually subjected to chemical modification through PEGylation, amino acid substitution, and reversible lipidization to enhance their plasma stability by decreasing the system clearance. In addition, implantable or injectable protein delivery systems could confer sustained release of the protein drugs over a period of several days, weeks, or even months.

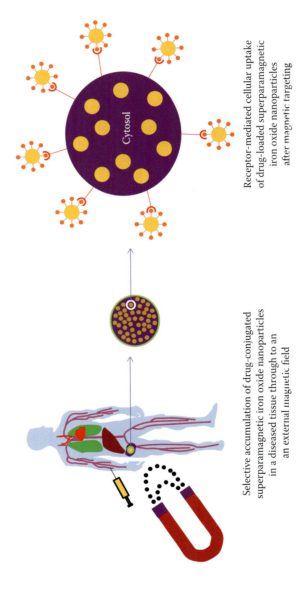

Figure 1.7 The principle of magnetic drug targeting.

1.5.3 Proteins as drug carriers

Serum proteins with desirable and established pharmacokinetic profiles along with potential affinity toward drugs could be an excellent candidate as drug carriers. Albumin has emerged as a potent tumor-targetable drug carrier considering its accumulation in the tumor interstitium because of the EPR effect and its binding affinity to albondin present on the endothelium and secreted protein acidic and rich in cysteine (SPARC) in the tumor interstitium (Merlot et al. 2014). The development of nab-paclitaxel or Abraxane, which is a paclitaxel-loaded albumin nanoparticle formulation approved by the Food and Drug Administration, is a major breakthrough in the field of macromolecule-based drug development. Abraxane shows better efficacy but less toxicity than traditional paclitaxel therapy.

1.6 DNA/RNA-based macromolecular drugs

Gene therapy can be defined as the transfer of functional gene(s) in order to be expressed or chemically synthesized anti-sense DNA/RNA sequences, such as oligodeoxyribonucleotides (ODNs) or small interfering RNA (siRNA), so as to silence gene expression into the cells for the purpose of treating hereditary disorders, such as hemophilia and cystic fibrosis, as well as acquired diseases, such as cancer and AIDS. A functional gene could be a better alternative to a therapeutic protein in the context of the duration of therapeutic effects and the associated cost. Nucleic acid can be introduced into humans by an ex vivo or in vivo approach. The former involves the removal of target cells from the patient, followed by their culture, expansion, and genetic manipulation by gene expression or knockdown before being reinfused back into the patient. On the other hand, the in vivo approach simply involves the injection of DNA/RNA into the patient. However, delivery of DNA and RNA, which are sensitive to extracellular (plasma) and intracellular nucleases, demands an efficient carrier to facilitate their safe transport into the cellular target site. Although recombinant viral vectors, particularly adenovirus, are quite efficient, they pose severe health risks in terms of potential immune responses and carcinogenicity. In contract, synthetic nonviral vectors are in general inefficient because of their limitations in overcoming certain extracellular and intracellular barriers. Enhanced serum stability of ODNs or siRNAs through chemical modifications of nucleic acid backbones often compromises their specificity for selective gene knockdown (Scholz and Wagner 2012).

1.7 Macromolecules for prodrug therapy

To enhance the therapeutic efficacy and reduce the off-target effects of anticancer drugs, two highly promising approaches, namely, antibody-directed

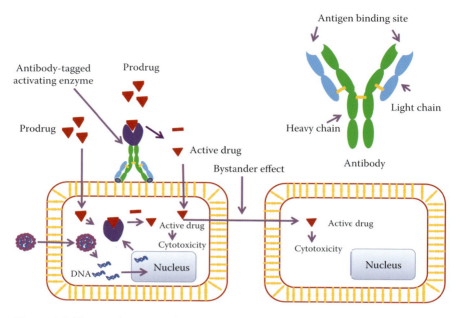

Figure 1.8 The mechanisms of ADEPT and GDEPT.

enzyme prodrug therapy (ADEPT) and gene-directed enzyme prodrug therapy (GDEPT), have emerged, allowing a selective release of cytotoxic agents from nontoxic prodrugs at the tumor site (Tietze and Schmuck 2011) (Figure 1.8).

1.7.1 ADEPT

This approach exploits antibody–enzyme conjugates (AECs) in order to bind to specific antigen on tumor cells before addition of a prodrug so that it can be subsequently activated by the enzyme of the targeted AEC, thereby generating a highly cytotoxic drug selectively at the tumor site. Since a large amount of the active drug can be liberated through the catalytic action of a single enzyme molecule, a high local concentration of the drug becomes available in the tumor tissue for efficient destruction of cancer cells. Furthermore, adjacent tumor cells that do not express the particular antigen could also be killed by the diffusible active drug through the bystander killing effect.

1.7.2 GDEPT

Unlike the systemic delivery of AECs, GDEPT (also known as suicide gene therapy) employs a suitable viral or nonviral vector via systemic administration for transducing or transfecting tumor cells, respectively, in order to express an exogenous enzyme that is capable of converting relatively nontoxic prodrugs into active cytotoxic drugs, thus leading to a specific and efficient killing of cancerous cells.

1.8 Macromolecules for vaccine delivery

Despite successful vaccine development against many infectious diseases including diphtheria, pertussis, tetanus, polio, measles, mumps, rubella, and smallpox, there is currently no effective vaccine available against the human immunodeficiency virus and influenza virus, which have the capacity to quickly change their surface antigens as a result of sequence variability and glycosylation (van Riet et al. 2014). Nanoparticles could be utilized for combined delivery of the conserved antigen(s) (or DNA that encodes the antigen) crucial for the fitness of the pathogen and the adjuvant(s) to help boost the immune responses to the antigen(s). The vaccine could be further improved by coating specific ligand(s) on the surface of the nanoparticles for specific delivery to the antigen-presenting cells that express the corresponding receptor(s) on their cell membranes.

1.9 Nanoparticles for photodynamic therapy

Photodynamic therapy (PDT) is a potent therapeutic approach to remove unwanted cells utilizing light-sensitive molecules, called photosensitizers (PS), which, after photon absorption, transform from a ground state to an excited state, transferring energy to tissue oxygen to activate it into singlet oxygen, a type of reactive oxygen species and inducing cell death through modification of intracellular components. For effective and safe PDT, nanoparticles should be exploited for delivery of PS in therapeutic concentrations to the target cells only, such as tumor cells or infected ones, to minimize the undesirable side effects in healthy tissues. Moreover, nanotechnology could be harnessed to prevent the aggregation of hydrophobic PS in an aqueous environment and improve their delivery to the target cells. There are cases where nanoparticles could themselves be the PS (Gupta et al. 2013).

1.10 Macromolecules for image-guided drug delivery

Nanotheranostics, which integrates imaging and therapeutic functions in a single platform utilizing "multifunctional" nanoparticles, holds great promises, since it could enable simultaneous noninvasive diagnosis and treatment of diseases with the possibility of monitoring drug release and distribution in real time, thus predicting and validating the effectiveness

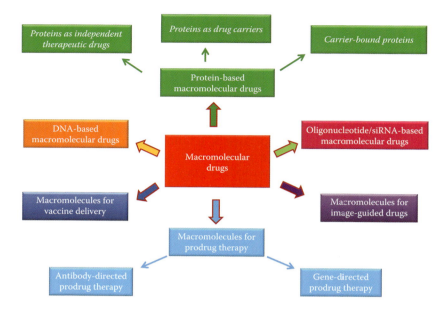

Figure 1.9 Different categories of macromolecular drugs.

of the therapy in individualized treatment protocols. Multifunctional nanocarriers loaded with contrast agents as well as therapeutic molecules would enable earlier detection with imaging technologies and treatment of diseases, which, in turn, would facilitate earlier assessment of the response, allowing screening of the patients who are potentially responding with a favorable outcome (Mura and Couvreur 2012).

The various forms of macromolecular drugs are summarized in Figure 1.9.

chapter two

The ultimate destinations for delivery and release of nanotherapeutics

2.1 Sustained-release formulations

Biodegradable and hydrophilic polymers have been extensively investigated for sustained release of small and macromolecular drugs. The entrapped drugs could be released slowly over a long period by controlling biodegradability and hydrophilicity of polymeric implants, microparticles, and nanoparticles. The first generation of such devices was mainly based on polymeric implants and used in the field of orthopedics (Nitsch and Banakar 1994). However, toxicity, inflammation, infection, and the need for additional surgical intervention for implant removal are the major shortcomings of these devices. Micro- and nanoparticles have recently been explored to fabricate controlled-release formulations for therapeutic molecules, especially costly recombinant proteins including growth factors that have limited half-lives requiring frequent administration with eventual patient noncompliance. These new approaches would enable local administration of the therapeutic molecules to provide the optimum drug release, achieve the desirable outcome locally, and prevent their transport to nontarget organs responsible for systemic toxicity. However, despite high potential in tailoring the release kinetics of associated drugs, polymeric particles face some important drawbacks, particularly in the context of their dependence on aggressive processing techniques using organic solvents, high temperatures, and freeze–thaw cycles for encapsulating therapeutic proteins, resulting in loss of protein activity. Moreover, each of the delivery routes, such as oral, ocular, dermal, and pulmonary routes, which have been extensively investigated for sustained drug release, poses its own unique barriers that are required to be overcome by the formulations to ensure expected therapeutic activity.

2.2 Intracellular delivery and release

Nanoparticles could be utilized to deliver small drugs, genes, oligodeoxyribonucleotides (ODNs), or small interfering RNAs (siRNAs) into cells.

Although small therapeutic agents can usually be transported passively or actively into the cells depending on their polarity and transporter availability, genes, ODNs, or siRNAs have limited permeability owing to their relatively larger size and possession of negative charges that can electrostatically repel the same charges present on cell membrane. Since small drugs generally lack sufficient specificity for their targets, distribution of such molecules to other organs apart from the target one might cause various side effects. In the case of cytotoxic drugs, nontargeted delivery as seen in traditional chemotherapy results in adverse effects on the body. Selective delivery of the small drugs to their target organs using ligand-anchored nanoparticles would enable recognition of specific receptors on the target cells before the endocytosis of the complexes and cytosolic release of the drugs, thus presenting an attractive approach to eliminate (or reduce) the off-target effects and enhance the therapeutic efficacy.

2.2.1 Endocytosis of nanoparticles

Endocytosis, a complex energy-dependent process for cellular uptake of diversified particles, is generally classified into phagocytosis, which is an actin-dependent process for uptake of large particles mainly by specialized mammalian cells such as macrophages, monocytes, and dendritic cells, and pinocytosis, which is used by almost all cells for uptake of solid and small particles and usually mediated by either clathrin-coated pits or caveolae. Clathrin-mediated endocytosis is mediated by specific receptors in a process whereby clathrin-coated pits invaginate into the cell forming vesicles with the particles ranging from 60 to 200 nm in diameter. Caveolae-mediated endocytosis is used to take up particles above 200 nm and involves specialized lipid rafts known as caveolae that are capable of producing invagination after interacting with a large number of signaling-associated proteins, such as receptor tyrosine kinases, G-protein–coupled receptors, and steroid hormone receptors.

2.2.2 Influences of physicochemical properties on cellular uptake of nanoparticles

Endocytosis of nanoparticles depends on the particle size and the charge, hydrophilicity, and availability of a ligand on the particle surface (Marin et al. 2013).

2.2.2.1 Size

Particle size determines the efficiency and mechanism of cellular uptake depending on cell type. Epithelial cells can internalize nanoparticles of a diameter smaller than 100 nm, while nonphagocytic cells internalize

Chapter two: Destinations for delivery and release of nanotherapeutics 23

nanoparticles with diameters from 200 nm to 1 µm preferably via a caveolae-mediated pathway. Endocytosis of the particles with a diameter less than 100 nm takes place rapidly via a receptor-mediated, vesicle-coated mechanism.

2.2.2.2 Charge, hydrophilicity, and presence of ligand

Depending on its constituents, a particle can carry a net positive or negative charge, or they can be hydrophilic or hydrophobic. Positively charged nanoparticles electrostatically interact with cell membrane that carries negatively charged (sulfate-containing) proteoglycans. However, the surface properties can be modified by adsorption of surfactants, hydrophilic polymers (such as polyethylene glycol [PEG]), protein (such as albumin), or mucoadhesive polymers (such as chitosan) on the nanoparticle surface. Covalent or noncovalent attachment of a ligand and coating of a hydrophilic polymer or protein on the nanoparticle surface is a common and effective approach for efficient drug delivery into the target cells through ligand–receptor interactions, while preventing nonspecific interactions with serum proteins and engulfment by macrophages with the help of surface hydrophilic groups (Figure 2.1).

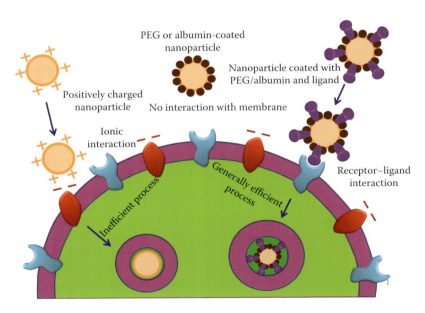

Figure 2.1 Effects of surface charge, hydrophilicity, and ligand attachment on cellular uptake.

2.3 Factors involved in drug release from nanoparticles

2.3.1 Biodegradability of pure drug particles

Nanosuspension is a colloidal suspension of surfactant-stabilized pure, solid-state hydrophobic drug particles that are formed by building particles from the molecular state through precipitation or by breaking down larger particles. Because of their small size and increased surface area, the dissolution of such drug nanoparticles into molecular drugs is substantially high, accounting for their efficient absorption rate across the intestinal epithelium and high bioavailability after oral administration. In case of intramuscular or ophthalmic applications, an additional but relevant benefit of using pure drug particles is high drug loading in a reduced administration volume (Balmayor et al. 2011).

2.3.2 Hydrophilicity and biodegradability of drug carriers

Drug molecules can be either entrapped within or immobilized onto the hydrophilic or biodegradable polymeric carrier. When the drug is entrapped within the polymeric particles, its release can be regulated by either hydrophilicity or biodegradability of the carrier. When hydrophilicity is the prime determinant of drug release, the hydrophilic segments will start swelling immediately after the particles come into contact with the aqueous environment, resulting in diffusion of the drug into the external environment. On the other hand, if degradation of the nanocarrier is the rate-limiting factor, the nanoparticle will be subjected to degradation as a result of its contact with the external environment, allowing a gradual release of the entrapped drug (Figure 2.2). US Food and Drug Administration–approved biodegradable poly(lactic-co-glycolic acid) nanoparticles have been widely used for controlled drug release. Among the inorganic carriers, pH-sensitive carbonate apatite, which was originally developed in our laboratory, releases the electrostatically bound drugs by undergoing self-dissolution at endosomal acidic pH (Balmayor et al. 2011).

Biodegradable polymers are degraded mainly by hydrolysis, oxidation, or enzymatic degradation.

2.3.2.1 Hydrolytic degradation

This process involves reaction of vulnerable bonds in the polymer with water molecules, resulting in shortening of its main or side chains. Polymers with the backbones susceptible to hydrolytic biodegradation under particular conditions include polyesters, polyamides, polyurethanes, and polyanhydrides.

Chapter two: Destinations for delivery and release of nanotherapeutics 25

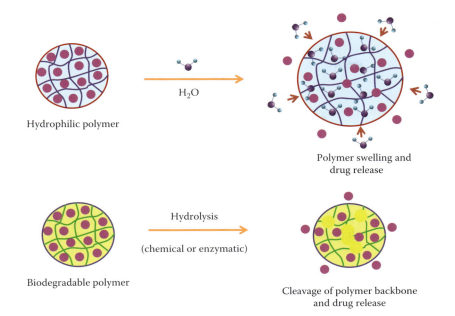

Figure 2.2 Effects of hydrophilicity and biodegradability of polymer on drug release.

2.3.2.2 Oxidation
The inflammatory response of leukocytes and macrophages produces reactive oxygen species such as nitric oxide, hydrogen peroxide, or superoxide, which can cause degradation of polymers. In addition, some polymers, such as polyethylene, polyether, and polyurethanes, can easily generate free radicals and are therefore more suitable to degradation by an oxidation reaction.

2.3.2.3 Enzymatic degradation
Hydrolytic enzymes (hydrolases), such as proteases, glycosidases, and phosphatases, are responsible for enzymatic degradation of certain polymers. Modification of a polymer might reduce its susceptibility to the enzymatic degradation. For example, acetylation of starch reduces its fast enzymatic degradation and thus improves its potential as a drug carrier.

2.3.3 Stimuli responsiveness of drug carriers or drug carrier complexes
Nanoparticles can be subjected to changes in either structural composition or conformation through decomposition, isomerization, polymerization,

or activation of supramolecular aggregation as a result of a chemical, biochemical, or physical stimulus, thus triggering the release of their cargo (drug) in specific intracellular or extracellular environment. If the bioactive agent is covalently linked to the nanocarrier, its release is subject to splitting of the linker with an environmental stimulus. A particular pathological trigger present in a diseased tissue would enable stimuli-responsive nanocarriers to release their cargo precisely while substantially reducing the drug-associated side effects. Both internal (physiological, pathological, and pathochemical conditions) or external (physical stimuli including heat, light, and magnetic and electrical fields) stimuli could contribute to the temporal or spatial pattern of drug release (Fleige et al. 2012).

2.3.3.1 pH-dependent drug release

The variable pH status in the GI tract, which varies from a very acidic to a basic (~2.0–8.0) state, has been harnessed for the design of orally active pH-responsive prodrugs and sustained-release systems. In cellular and subcellular levels, the pH even plays a critical role in normal or abnormal tissues. The inflamed, infected, or malignant tissues usually demonstrate lower pH values than their respective normal tissues. The extracellular pH in cancer tissues can fall to 6–7 or even lower owing to their underdeveloped vasculature with consequently prevailing anaerobic metabolism leading to lactic acid formation in the malignant cells (Figure 2.3). On the

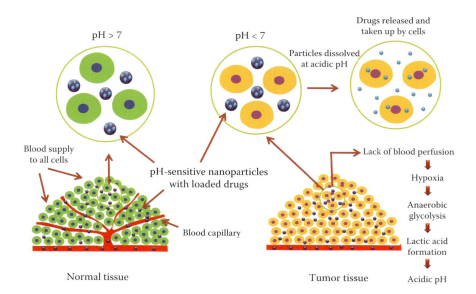

Figure 2.3 pH-responsive nanoparticles can release drugs at extracellular space of tumor.

Chapter two: Destinations for delivery and release of nanotherapeutics 27

other hand, after endocytosis of nanocarriers in both normal and abnormal cells, the early endosome gradually develops an acidic pH of approximately 5–6 while the lysosome, which can fuse with the late endosomes (matured from the early endosomes), could have a pH of approximately 4–5. Thus, on the basis of these extra- and intracellular pH gradients, smart delivery devices could be formulated in order to selectively release their drug cargo at the desirable location. The drug that is released from its carrier in extracellular space will passively diffuse through the plasma membrane (Figure 2.3). To render the bond between a drug and a nanocarrier or the polymeric backbone sensitive to a pH gradient, a number of covalent linkages including hydrazone, hydrazide, and acetal are often used. For example, poly(ethylene glycol)-block-poly(aspartate) was conjugated with doxorubicin via a pH-labile hydrazone bond for efficient release of the drug at acidic pH. For intracellular release of a gene-carrying plasmid or ODN/siRNA, nanocarriers, such as poly(ethyleneimine) (PEI), serve as proton sponges for consuming protons in endosomes, thus finally inducing endosomal membrane rupture with the result of cytosolic release of the carrier complex. As a fusogenic lipid, DOPE (1,2-dioleoyl-*sn*-glycero-3-phosphoethanolamine) can be carried with liposomal formulations to facilitate endosome escape of nucleic acids by changing at the acidic endosomal pH from its anionic state to a helical conformation that induces the fusion of the vector with endosomal membrane (Fleige et al. 2012; Hatakeyama et al. 2009).

2.3.3.2 *Enzyme-triggered drug release*

A number of enzymes, such as proteases, glucuronidase, or carboxylesterases, which are differentially expressed in either the extra- or intracellular compartment of normal and malignant cells, can be used as a biochemical stimulus to cleave an enzyme substrate segment present in the linker between a drug and a nanocarrier or within the scaffold of the carrier, facilitating controlled release of the drug. Cathepsin B, a lysosomal protease, has been intensively investigated for the development of enzyme-labile nanocarriers. One of the most frequently used linkers is the tetrapeptide Gly–Phe–Leu–Gly, a substrate of cathepsin B that is usually overexpressed in the lysosomes of tumor cells. Extracellular proteases, such as the matrix metalloproteases responsible for the proteolysis of the extracellular matrix and basement membranes, could similarly be harnessed as a biochemical trigger (Fleige et al. 2012).

2.3.3.3 *Redox potential–controlled drug release*

The higher concentration of glutathione in the intracellular compartment than the extracellular space causes a significant difference in redox potential (~100- to 1000-fold) across the plasma membrane of a normal cell and thus makes the extracellular and intracellular areas, respectively, oxidative and reductive. In addition, the poor vasculature in tumor tissues results in the formation of hypoxic areas that are environmentally reductive owing to the presence and action of reductases. On the other hand, redox-responsive nanocarriers can be equipped with a disulfide cross-linked moiety that is sensitive in reductive environment, breaking down into sulfhydryl moieties and releasing the drug either within normal cells or in cancerous tissues (Fleige et al. 2012).

2.3.3.4 Thermoresponsive drug release

A change in temperature can bring about a change in the hydrophilic–hydrophobic balance, that is, in the hydration state of a thermoresponsive nanocarrier resulting in collapse or disassembly of the system and releasing of the therapeutic payload from it. Heating a polymer with a lower critical solution temperature within the temperature range of 37°C–42°C (beyond which body anatomical structures could be disrupted and therapeutic proteins could be denatured) will cause the polymer to dehydrate, become more hydrophobic, and finally collapse. Thermoresponsive nanocarriers can be used to deliver and release cytotoxic agents in some tumor tissues that show slightly increased temperatures by externally applying heat (Fleige et al. 2012).

2.3.3.5 Photoresponsive drug release

Tissue-compatible radiation of ultraviolet, near-infrared, and infrared frequency can be used to induce conformational changes within the nanocarriers' chemical structures to spatially and temporally release encapsulated or conjugated bioactive molecules (Fleige et al. 2012).

2.3.3.6 Drug release from dual-responsive nanocarriers

Dual-responsive nanocarriers can release drugs responding to two different stimuli often involving the combination of pH and ionic strength or pH and thermoresponsiveness. An additional use of the dual-responsive systems is that one stimulus can be used to load the drug with the carrier while the second can trigger the drug release (Fleige et al. 2012).

chapter three

Diversity of bioactive nanoparticles from biological, chemical, and physical perspectives

3.1 Viral vectors

A virus is a biological entity that can penetrate into the cell membrane and subsequently transfer its genetic material into the nucleus of a host cell to express its own genetic material and replicate it by using the cellular machinery before spreading to the other cells. Therefore, in order to exploit the virus to carry a gene of interest, it must be modified by genetic engineering by replacing its pathogenic genes by the therapeutic gene while retaining its nonpathogenic structures (envelope proteins, fusogenic proteins, etc.) that will allow it to infect the cell. Currently, viral vectors are most often used in gene therapy trials because of their high transfection efficiency in vivo despite possessing some drawbacks, such as acute immune response, carcinogenicity, high production cost, and limited capacity of carrying genetic materials.

3.1.1 Retroviral vectors

A retrovirus vector is a spherical particle of approximately 80–100 nm in diameter, enclosed by a lipid bilayer (derived from the host cell plasma membrane) into which the viral envelope protein is inserted. Its internal structure is mainly composed of the products of the viral *gag* gene. Simple retroviruses have only three genes, *gag, pol,* and *env,* while the complex ones, such as HIV-1, encode a number of additional proteins that are involved in expression of the viral genome and enhancement of the viral replication. On infection, its single-stranded RNA genome enters the host cell along with the enzyme, reverse transcriptase, which catalyzes the synthesis of a DNA strand complementary to the viral RNA and degrades the RNA strand of the viral RNA–DNA hybrid replacing it with DNA. The resulting duplex DNA is translocated to the nucleus and incorporated into the host's genome by an integrase enzyme. The retroviral vector usually infects dividing cells since the duplex DNA is transported to the nucleus once the nuclear membrane is temporally disassembled during mitosis. This is of great advantage in tumor-specific expression of therapeutic genes. However, lentiviral vectors, most of which have been derived from HIV, can generally transduce nonmitotic cells by harnessing the cellular nuclear transport mechanism. These vectors can be pseudotyped through inclusion of other viral glycoproteins, such as the vesicular stomatitis virus G protein. As a result, they are not restricted to only CD4+ cells but rather can be used to transduce other cell types (Wu and Ataai 2000). The ability of retroviruses to integrate into the host cell chromosome raises the possibility of insertional mutagenesis within the result of oncogene activation and cancer development.

3.1.2 DNA virus vectors

Unlike retroviruses, DNA viruses contain single- or double-stranded DNA as the viral genome. The most prominent DNA viruses that are used extensively as nucleic acid carriers are the adenoviruses, adeno-associated viruses (AAVs), and herpes simplex viruses (HSVs).

3.1.2.1 Adenoviral vectors

The human adenovirus is a large virus (~150 nm in diameter) consisting of a capsid surrounding its double-stranded (ds) DNA genome and core proteins. The capsid shell comprises multiple copies of three major capsid proteins (hexon, penton base, and fiber) and four minor/cement proteins that are organized with icosahedral symmetry (Wu and Ataai 2000). The linear dsDNA genome of approximately 36 kilobases (kb) provides ample space for inserting large sequences compared to the retroviruses. They can infect both dividing and nondividing cells. Recombinant adenovirus vectors are usually constructed by deleting E1 and E3 genes from the genome of wild-type adenovirus, making them replication defective with the maximum capacity of carrying the desirable gene of approximately 8.0 kb. Unlike the retroviral genome, the adenovirus HSV genome remains episomal after translocating to the nucleus, hence eliminating the possibility of opportunistic insertional mutagenesis of the host genome.

3.1.2.2 Adeno-associated viral vectors

AAVs are small, single-stranded, nonpathogenic DNA viruses that require a helper virus for replication and completion of life cycle. AAVs are composed of the rep gene, which encodes four nonstructural proteins with roles in viral genome replication, transcription, and packaging, and the cap gene, which encodes the three structural proteins of the AAV capsid, which is ~25 nm in diameter and composed of 60 subunits arranged in icosahedral symmetry. The therapeutic expression cassette carrying the

Chapter three: Diversity of bioactive nanoparticles 31

desirable therapeutic gene replaces rep and cap, leaving the viral inverted terminal repeats as the only viral sequences. Like adenoviruses, they do not lead to insertional mutagenesis. The vector can be pseudotyped and can transduce both dividing and nondividing cells. One major limitation is that the insert size of the vector is restricted to just over 4 kb (Kotterman and Schaffer 2014).

3.1.2.3 Herpes simplex virus
HSV is an enveloped dsDNA virus of a 150-kb genome with the largest capacity of loading ~30 kb of foreign genes. HSV has four different components: a core that consists of a single linear molecule of dsDNA, an icosahedral capsid of 100 nm in diameter surrounding the core and consisting of 162 capsomeres (12 pentons and 150 hexons), an envelope with the outer layer composed of a lipid bilayer carrying viral glycoproteins and tegument, and a protein-filled region between the capsid and the envelope. The virus offers additional advantages, such as the ability to infect a wide range of cells because of the wide expression pattern of the cellular receptors recognized by the virus and the absence of insertional mutagenesis (Wu and Ataai 2000).

3.2 Nonviral vectors
Considering the potential adverse effects of viral particles along with their limitations in the delivery of small interfering RNAs (siRNAs) and small-molecule drugs, the development of nonviral vectors has received tremendous interest. Moreover, thanks to the advancement of synthetic and genetic engineering approaches, there is currently remarkable flexibility in modifying the nonviral particles, such as surface functionalization with hydrophilic proteins such as polyethylene glycol (PEG) to modulate their pharmacokinetics and receptor targeting ligands to promote diseased tissue–specific delivery.

3.2.1 Lipid-based nonviral vectors

3.2.1.1 Liposomes
A liposome is a spherical vesicle composed of amphiphilic phospholipids that self-assemble into the closed bilayer sphere encapsulating an aqueous interior while shielding the hydrophobic groups from both the interior and exterior aqueous environment (Martins et al. 2007) (Figure 3.1). Depending on their size and number of bilayers, they can be classified into multilamellar, large unilamellar, and small unilamellar vesicles. Small unilamellar vesicles are 20–200 nm, large unilamellar vesicles are 200 nm to 1 μm, and giant unilamellar vesicles are >1 μm (Balazs and Godbey 2011). The unilamellar liposome has a single phospholipid bilayer sphere and the

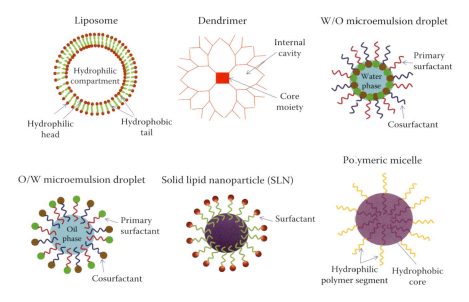

Figure 3.1 Different types of nonviral organic particles.

multilamellar vesicle resumes an onion-like structure wherein several unilamellar vesicles form on the inside of the other with a smaller size, making the overall structure of concentric phospholipid spheres separated by layers of water. While the vesicle size regulates the circulation half-life of liposomes, both size and number of bilayers influence the efficiency of drug encapsulation in the liposomes (Akbarzadeh et al. 2013). The unique advantages imparted by these lipid vesicles include their diverse range of morphologies, compositions, and abilities to envelope and protect many different types of therapeutics; lack of immunogenic response; low cost; and variable release kinetics. Some major problems limiting the manufacture and development of liposomes include instability, poor batch-to-batch reproducibility, difficulties in sterilization, and low drug loading. The head groups and hydrophobic hydrocarbon tails in the amphilic lipid molecules are connected via a backbone linker such as glycerol. The cationic lipids typically used in the formation of cationic liposomes confer a positive charge through one or more amines to the polar head group. Because of the polyanionic backbone of nucleic acids, such as DNA, oligodeoxyribonucleotide, or siRNA, cationic lipids are commonly utilized for gene delivery purposes, while the use of anionic or neutral liposomes has been restricted to the delivery of other therapeutics. The cationic lipids that are included in the commercially available transfection reagents include *N*-[1-(2,3-dioleyloxy)propyl]-*N,N,N*-trimethylammonium chloride (DOTMA), 1,2-bis(oleoyloxy)-3-(trimethylammonio)propane (DOTAP),

Chapter three: Diversity of bioactive nanoparticles 33

3β[N-(N',N'-dimethylaminoethane)-carbamoyl] cholesterol (DC-Chol), and dioctadecylamidoglycylspermine (DOGS) (Balazs and Godbey 2011). Liposome properties can be manipulated considerably based on lipid composition, surface charge, size, and the method of preparation. Thus, the choice of bilayer components determines the "rigidity" or "fluidity" as well as the charge of the bilayer. For example, unsaturated phosphatidyl-choline species from natural sources give much more permeable and less stable bilayers, whereas the saturated phospholipids with long acyl chains form a rigid and impermeable structure (Akbarzadeh et al. 2013).

3.2.1.2 Solid lipid nanoparticles

Solid lipid nanoparticles (SLNs) are submicron colloidal carriers of 50 to 1000 nm and consist of a solid hydrophobic core with a surrounding monolayer of phospholipids (Figure 3.1). Drugs are dissolved or dispersed in the solid high melting fat matrix of the core while the hydrophobic chains of phospholipids are associated with the fat matrix. The physical stability of SLNs can be increased by the addition of an emulsifier, such as poloxamer, polysorbate, or lecithin up to 0.5% to 5% depending on the type and concentration of the lipid (Martins et al. 2007).

3.2.1.3 Lipid nanoemulsions

Lipid nanoemulsions (LNEs) are fine oil/water (o/w) dispersions in the size range of 50 to 200 nm. They are basically composed of a liquid oily core and a surfactant layer, thus differentiating them from the SLNs that contain a solid lipid in the core. Self-emulsifying drug delivery systems (SEDDS), which represent a particular type of LNEs, are isotropic mixtures of oils, surfactants, solvents, cosolvents/cosurfactants, and solubilized drug substance that rapidly and spontaneously form fine oil-in-water emulsions (Figure 3.1). SEDDS formulations are physically stable with high drug entrapment capacity (Martins et al. 2007).

3.2.1.4 Nanostructured lipid carriers

Nanostructured lipid carriers are the second-generation lipid nanoparticles that are prepared from a blend of a solid lipid with a liquid lipid (oil) in such a proportion that the mixture should be solid at least at 40°C. The lipid matrix confers more flexibility for modulating drug release, increasing the drug loading, and preventing its leakage (Martins et al. 2007).

3.2.2 Polymer-based nonviral vectors

3.2.2.1 Polymeric micelles

Micelles are spherical colloidal particles having a hydrophobic or cationic polymer segment to encapsulate poorly water-soluble drugs or neutral-ize anionic nucleic acids, respectively, forming the core and a hydrophilic shell to stabilize the particles in the aqueous environment (Figure 3.1) and evade nonspecific capture by the mononuclear phagocyte system (MPS). Contrast agents can be entrapped within the hydrophobic core or linked covalently to the surface of micelles for imaging, or pH-sensitive drug-binding linkers can be introduced in the interior for controlled drug release. The small particle size (<50 nm in diameter) and adequate stability of micelles provide obvious benefits over liposomes, allowing their prolonged circulation in blood with enhanced therapeutic or imaging efficacy particularly in relation to cancer treatment and diagnosis. Many existing solvents, such as Cremophor EL (BASF) or ethanol, for solubilization of water-insoluble pharmaceuticals are toxic, thus limiting the therapeutic doses and restricting the treatment options. Polymeric micelles thus offer a safer alternative for parenteral administration of poorly water-soluble drugs. Multifunctional polymeric micelles can be developed to facilitate simultaneous drug delivery and imaging. Covalent cross-linking of the micellar core can be designed to effectively enhance the stability of the micelles, avoiding their dissociation in the blood and tuning the rate of drug release (Rosler et al. 2001).

3.2.2.2 Dendrimers

Dendrimers are a unique class of polymeric macromolecules synthesized in a nearly perfect three-dimensional geometrical pattern with repeated branching around the central core via divergent or convergent synthesis by a series of controlled polymerization reactions. The high-generation dendrimers can have numerous cavities within the branches to hold therapeutic and diagnostic agents (Figure 3.1). The contemporary dendrimers can encapsulate the drugs or contrasting molecules inside the core. Dendrimers used in drug delivery and imaging are usually 10 to 100 nm in diameter. The multiple functional groups present on their surface enable modification with ligands for targeted delivery. Polyamidoamine dendrimers, for example, that are synthesized by the repetitive addition of branching units to an amine core, which can be either ammonia or ethylene diamine, can function as vehicles for delivery of drugs, genetic material, and imaging probes (Samad et al. 2009).

3.2.2.3 Hydrogels

Hydrogels are three-dimensional and cross-linked networks of hydrophilic homopolymers or copolymers with their highly porous structure easily tunable by controlling the density of cross-links in the gel matrix and the affinity for the surrounding water molecules. The porosity of hydrogels permits loading of drugs into the gel matrix and subsequent drug release depending on the diffusion coefficient of the drug molecules through the gel network (Hoare et al. 2008). The chemically cross-linked networks of hydrogels have permanent junctions, while the physical

Chapter three: Diversity of bioactive nanoparticles   35

networks have transient junctions arising from ionic interactions, hydrogen bonds, or hydrophobic interactions (Ahmed 2015).

3.2.2.4   Other polymeric nanoparticles

Apart from the micelles, dendrimers, and hydrogels, there are other polymeric nanoparticles usually prepared from dispersion of preformed polymers, such as solvent evaporation, nanoprecipitation, emulsification/solvent diffusion, salting out, dialysis, and supercritical fluid technology. The most common natural polymers used in preparation of polymeric nanoparticles include chitosan, gelatin, alginate, and albumin, while among the synthetic polymers, polylactides, polyglycolides, poly(lactide co-glycolides) (PLGA), polyanhydrides, polycyanoacrylates, polycaprolactone, polyglutamic acid, polymalic acid, poly(N-vinyl pyrrolidone), poly(methyl methacrylate), poly(acrylic acid), polyacrylamide, and poly(methacrylic acid) are frequently employed. Biodegradable PLGA polymers that were approved by the Food and Drug Administration and the European Medicines Agency for uses in drug delivery systems via parenteral routes have received considerable attention. Depending on the methods of formulation, both small-molecule drugs and macromolecular therapeutics can be encapsulated into the PLGA nanoparticles. The subsequent drug release from the nanoparticles is greatly influenced by the preparation method, the particle size, and the ratio of lactide to glycolide moieties (Danhier et al. 2012).

3.2.3   Inorganic carriers

3.2.3.1   Apatite-based nanoparticles

Calcium phosphate precipitation is a common method for preparation of hydroxyapatite particles to carry genetic materials into the mammalian cells. However, uncontrolled growth of the particles usually leads to large-size aggregates that are inefficient for subsequent cellular uptake. Extensive efforts were made to reduce the particle size by dropwise mixing of Ca solution to inorganic phosphate-containing buffer, rapid mixing of the two reactants followed by short incubation to minimize the growth, or particle preparation in the presence of block copolymers capable of simultaneously interacting with nascent nanoparticles and preventing interactions with neighboring nanoparticles. Recently, substitution to an extent of either phosphate with carbonate (carbonate apatite) (Chowdhury et al. 2006) or calcium with magnesium (Ca–Mg phosphate) (Chowdhury et al. 2004) was shown to dramatically decrease the particle diameter with a concomitant increase in the dissolution rate at endosomal acidic pH, resulting in both enhanced intracellular delivery and release of genes, siRNAs, proteins, and anticancer drugs. By virtue of having distinct cation and anion binding domains, the surface of apatite particles can interact with diverse macromolecules having net negative and positive charges, respectively, as required for modulating their pharmacokinetic properties and achieving tissue targetability (Chowdhury et al. 2005).

3.2.3.2 Gold nanoparticles

Gold (Au) nanoparticles have natural affinities toward a wide range of organic molecules. The gold–sulfur interaction, for example, is a strong quasi-covalent bond that can be dissociated under reductive conditions in the presence of cytoplasmic high glutathione concentrations (as a result of oxidation of thiol groups), thus enabling stimuli-responsive drug delivery. Furthermore, surface modification of the Au nanoparticles allows them to prevent self-aggregation or bind to the specific targeting drugs or biomolecules. Since Au has special affinity to sulfur, thiol-containing hydrophilic polymer or protein with or without ligand can be immobilized on the Au nanoparticles (Figure 3.2). Their unique chemical, physical, and photophysical properties can be exploited to control the transport and controlled release of therapeutic agents (Pissuwan et al. 2011). Au nanoparticles are most often produced by chemically reducing Au salt in the presence of surface stabilizers that prevent the aggregation of the preformed nanoparticles. Monodisperse Au nanoparticles can be formed with core

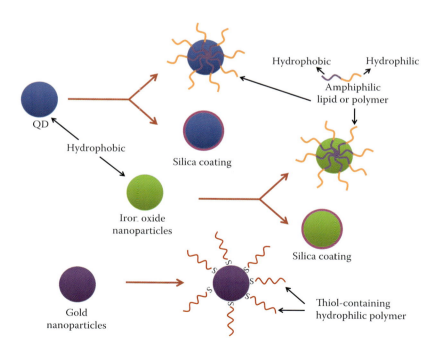

Figure 3.2 Strategies to prevent aggregation of inorganic nanoparticles.

Chapter three: Diversity of bioactive nanoparticles   37

sizes ranging from 1 to 150 nm and different shapes, such as nanospheres, nanoshells, or nanorods. Since the nanoparticles are commonly synthesized via citrate reduction, which makes them negatively charged because of citrate adsorption, one of the approaches for complexation between them and the anionic DNA or siRNA is to incorporate a positively charged polymer, such as poly(ethylene)amine between the Au-citrate core and the siRNA in a layer-by-layer format (Guo et al. 2010). Because of their unique optical and electronic properties, these nanoparticles present attractive opportunities for both therapeutic and bioimaging purposes.

3.2.3.3   Silica nanoparticles

Silica nanoparticles that can be formulated in variable particle size, porosity, crystallinity, and shape to tune their nanostructures with many possible surface modifications offer a desirable platform for biomedical imaging, therapeutic delivery, or photodynamic therapy. Most of the synthetic techniques rely on the sol–gel processing at 25°C with the optimized reactant-to-solvent ratios or the use of templates so as to control particle size distribution. Mesoporous silica nanoparticles having a very large surface area with controllable pore size and volume allow a large amount of drugs to be loaded into the particles. The nanoparticles are frequently functionalized with 3-aminopropyl triethoxysilane, 3-mercaptopropyl tri-methoxysilane, and various PEG silanes for improved bioavailability and targetability (Liberman et al. 2014).

3.2.3.4   Quantum dots

Quantum dots (QDs) that can be synthesized from various types of semi-conductor materials via colloidal synthesis or electrochemistry are semi-conductor nanocrystals of 2 to 10 nm in diameter. The most common QDs that are used as contrast agents in bioimaging with much greater resolution than existing fluorescent dyes include cadmium selenide, cadmium telluride, indium phosphide, and indium arsenide. Depending on the sizes, these particles absorb light of particular wavelengths and emit at wavelengths of different energy. The band gap energy that determines the energy and the color of the emitted (fluorescent) light is inversely proportional to the size of the QD. Thus, different QDs emit different fluorescent light of the wavelength ranging from 400 to 1350 nm. Assembly of QD particles of various size and composition can be used to support derivation of multicolored images for in vivo imaging and related applications. Being hydrophobic in nature, QDs have the natural tendency of forming self-aggregates. This aggregation phenomenon can be prevented by coating the QD surface with inert silica or an amphiphilic lipid or polymer whose hydrophobic segment will interact with the particles with the hydrophilic block facing the aqueous phase (Figure 3.2). The surface of QDs can also be modified via conjugation of various cell-recognizable molecules for targeted delivery. QDs are therefore suitable for simultaneous drug delivery and in vivo imaging (Ho and Leong 2010).

3.2.3.5 Magnetic nanoparticles

By virtue of intrinsic magnetic properties, magnetic nanoparticles, particularly superparamagnetic iron oxide nanoparticles (SPIONs), have been explored for diverse biomedical applications, such as magnetic resonance imaging (MRI) and targeted drug delivery under the influence of an external magnetic field. Organic ligands, such as PEG, dextran, amphiphilic molecules, and aminosilanes (Figure 3.2) are commonly used as surface protectants to stabilize the magnetic nanoparticles, although such modification usually decreases the surface magnetic moment of the metal atoms present at the surface of the particles (Mody et al. 2014). Addition of bioactive molecules to the SPION surface can increase the targeting specificity of the nanoparticles for both imaging and targeted drug delivery. Currently, a number of SPIONs have been approved for medical imaging and therapeutic applications, for example, Lumiren for bowel imaging, Feridex IV for liver and spleen imaging, Combidex for lymph node metastases imaging, and ferumoxytol for iron replacement therapy (Veiseh et al. 2010).

3.2.3.6 Carbon nanotubes

Carbon nanotubes (CNTs) that belong to the fullerene family of carbon allotropes are graphene sheets rolled into an open-ended or capped cylinder-shaped structure. A single graphene sheet results in a single-walled nanotube (SWNT) with a diameter varying between 0.4 and 2 nm while several graphene sheets make up multiwalled carbon nanotubes with diameters in the range of 1–3 nm for the inner tubes and 2–100 nm for the outer tubes. CNTs can be surface functionalized to enhance their solubility in the aqueous phase or enable them to bind to a desired therapeutic material or a target tissue. In addition to their potential ability of acting as carriers for a wide range of drugs, the large surface area with feasibility of manipulating their surfaces and physical dimensions has enabled their exploitation in the photothermal destruction (thermal ablation) of cancer cells. Therapeutic molecules can noncovalently adsorb into or onto CNTs or to the charged groups chemically attached earlier to CNT surface, through ionic interactions. With the use of a linker, both the drug and CNT can react to form covalent bonds (Madani et al. 2011).

3.3 Hybrid particles

3.3.1 Lipid–polymer hybrid nanoparticles

To combine the beneficial features of both polymeric nanoparticles and liposomes, the concept of lipid–polymer hybrid nanoparticles (LPNs) has

Chapter three: Diversity of bioactive nanoparticles 39

emerged. LPNs consist of (1) a polymer core in which the therapeutic molecules are encapsulated; (2) an inner lipid layer enveloping the polymer core to confer biocompatibility to the polymer core, to function as a molecular fence minimizing drug leakage during the LPNs preparation, and to slow down the polymer degradation rate by limiting inward water diffusion so as to ensure sustained release kinetics of the payloads; and (3) an outer lipid–PEG layer that serves as a stealth coating prolonging blood circulation time of the LPNs as well as providing steric stabilization (Zhang et al. 2008).

3.3.2 Organic–inorganic hybrid nanoparticles

For improving image contrast in either MRI or other types of medical imaging systems, hydrophilic iron oxide–based magnetic nanoparticles (for MRI), gold nanoparticles, and luminescent QDs (for optical imaging) can be encapsulated in the interior or on the outer membrane of liposomes and their hydrophobic nanoparticles can be inserted into the hydrophobic interior of the liposomal membrane. Coating of liposomes on the rigid mesoporous silica nanoparticles was shown to allow more effective loading and sustained release of a drug compared with conventional liposomes. Polymeric nanohybrids prepared by encapsulating magnetic nanocrystals and anticancer drugs in an amphiphilic block copolymer (PEG-PLGA) using a nanoemulsion method were found advantageous for simultaneous drug delivery and imaging (Sailor and Park 2012).

3.3.3 Inorganic hybrid nanoparticles

Simultaneous imaging by MRI and photothermal therapy could be achieved using gold-based hybrid nanoparticles that consist of a silica nanosphere core surrounded by a gold nanoshell, with embedded magnetite nanoparticles. In a similar fashion, gold-coated single-walled CNTs facilitated high-contrast photoacoustic and photothermal imaging (Sailor and Park 2012).

3.4 Genetically engineered drug carriers

On the basis of the primary amino acid sequences, genetically engineered drug nanocarriers can be either polymeric carriers, which include elastin-like polypeptides, silk-like polypeptides, extended recombinant polypeptide polymers, and silk–elastin-like polypeptides, or nonpolymeric carriers with defined tertiary and quaternary structure that have been developed from viral proteins and vault proteins. Protein polymers consisting of natural or unnatural repetitive amino acid sequences are generally biosynthesized in either prokaryotes or eukaryotes. By precisely changing their amino acid sequences in the repetitive sequences utilizing recombinant DNA technology, libraries of polymers with different charges, hydrophobicity, or secondary structures can be created in order to establish the best-performing nanocarriers. Among the nonpolymeric carriers, the vault nanoparticle, the most bulky ribonucleoprotein complex abundantly present in most eukaryotes, is highly promising because it has a spacious internal volume for the encapsulation of drug molecules and forms a "dynamic" nanostructure capable of dissociating into halves at a low pH environment, a property that strongly influences intracellular drug release (Shi et al. 2014).

3.5 Bioconjugation schemes for functionalization of and ligand attachment to nanoparticle surface

Electrostatic interactions confer the simplest method for complexation between nanoparticles and hydrophilic macromolecules or targeting moieties (ligands). However, such noncovalent bonding is not strong enough to prevent their complete or partial dissociation particularly in the presence of serum proteins. Thanks to the available cross-linking

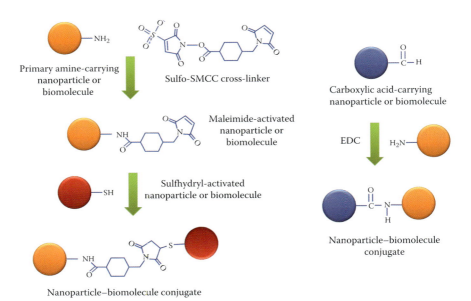

Figure 3.3 Cross-linking reactions for surface functionalization and ligand attachment.

Chapter three: Diversity of bioactive nanoparticles 41

agents (cross-linkers) that possess two or more reactive ends, nanomaterials or biomolecules having primary amines, sulfhydryls, or carboxylic groups can be coupled to each other through covalent bonding, thus facilitating immobilization of ligands or hydrophilic proteins/polymers on the nanoparticles. The two most frequently used schemes, which are based on sulfo-SMCC [sulfosuccinimidyl-*trans*-4-(*N*-maleimidomethyl) cyclohexane-1-carboxylate] reactive toward amino and sulfhydryl groups and EDC [1-ethyl-3-(3-dimethylaminopropyl)carbodiimide] reactive toward carboxyl and amine groups, are depicted in Figure 3.3.

chapter four

Fabrication strategies for biofunctional nanoparticles

4.1 Chemical synthesis and engineering

Depending on the starting materials, the particle synthesis processes can be categorized into "top-down" and "bottom-up." In the top-down approach that includes milling methods, such as wet and jet milling, and homogenization methods, such as rotor stator or high-pressure homogenization (HPH), solid particles that are substantially larger than the targeted nanoparticles are mechanically broken down into smaller sizes. The wet milling and HPH methods usually employed to produce drug nanoparticles are scalable and capable of producing nanoparticles with a narrow size distribution. However, these processes require huge energy and significant time, which could be up to a few days and potentially introduce contamination from the milling media or the homogenization chamber. In the bottom-up approach that includes the majority of the preparation methods, nanoparticles are formed from the molecular level enabling better control on particle properties, such as size, morphology, and crystallinity than the top-down processes (Zhang et al. 2011).

4.1.1 Production of drug nanoparticles: Top-down approaches

4.1.1.1 Wet milling

Wet milling, which is more efficient than dry milling (e.g., jet milling) to obtain a size in the nanometer range, has been successfully used to produce a few Food and Drug Administration–approved drugs, such as Rapamune, Emend, Tricor, and Megace ES. The milling media usually consisting of glass, zirconium oxide, or highly cross-linked polystyrene resin and a slurry containing raw drug particles and stabilizer are circulated through a milling chamber to generate the shear force of impact leading to the disintegration of the raw material into nanosized particles in the size range of 100–400 nm range.

4.1.1.2 High-pressure homogenization

HPH, the second most important technique to produce drug nanocrystals, is scalable like the wet milling process, although unlike the latter, it usually requires shorter processing time (from less than 30 minutes to a few hours) and is less prone in generating process impurities because of abrasion and wearing of the equipment. A slurry feeding stream, usually composed of drug coarse particle and stabilizer, is pressurized and allowed to pass through a relief valve where cavitation, high shear force, and collision between the particles are induced by a sudden release of pressure. The particle size reduction is thus caused by cavitation forces, shear forces, and collision (Möschwitzer 2013). In general, several homogenization cycles are required to achieve the minimal particle size. The homogenization pressure determines the flow rate, number of cycles, solid loading, and amount of materials to be processed. This technique has been used to produce inhaled budesonide and salbutamol sulfate nanoparticles (Zhang et al. 2011).

4.1.2 Production of nonviral vectors: Bottom-up approaches

4.1.2.1 Liposome

4.1.2.1.1 Thin-film hydration This is the most common method for preparing multilamellar vesicles (MLVs), that is, by dissolving the phospholipids in the organic solvents, such as dichloromethane, chloroform, ethanol, and chloroform–methanol mixture; evaporating the solvent under vacuum at 45°C–60°C to eventually form a thin and homogeneous lipid film at the inner surface of the flask walls; and finally mixing with the film a solution of distilled water, phosphate buffer, and normal saline buffer containing selected drug molecules. A vigorous shaking above the phase transition temperature (T_m) of the lipids leads to a dispersion of the lipid multilayers in the aqueous solution, resulting in the formation of liposomes with heterogeneous size distribution and lamellarity (Figure 4.1). The time for hydration of the dried film and agitation conditions are critical in determining the amount of the aqueous drug solution that could be entrapped within the internal compartments of the MLVs (Popovska et al. 2013).

4.1.2.1.2 Sonication The principle of sonication involves the use of pulsed, high-frequency sound waves (sonic energy) either with a bath-type or a probe-type sonicator under an inert atmosphere including nitrogen or argon in order to agitate and disrupt a suspension of the MLVs prepared using the thin-film hydration method, thus producing small unilamellar vesicles (SUVs) with diameter in the range of 15–50 nm. The probe tip sonicator that delivers high energy to the lipid suspension could degrade the lipid by overheating and release titanium, which must subsequently be removed by centrifugation. The bath sonicators that are most widely used for the preparation of SUVs in a temperature-controlled water bath is advantageous in minimizing thermodegradation of the lipid as well

Chapter four: Fabrication strategies for biofunctional nanoparticles 45

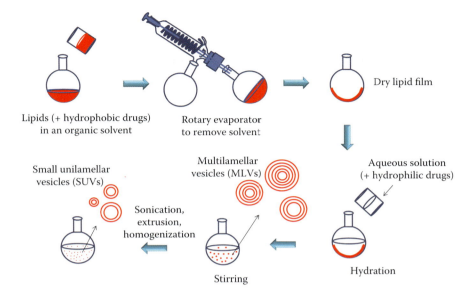

Figure 4.1 Thin-film hydration for liposome fabrication.

as the entrapped substance (Uhumwangho and Okor 2005). The size of the SUVs is mainly regulated by the sonication conditions and the vesicle membrane composition. Among the major limitations of sonication are oxidation of unsaturated bonds in the fatty acid chains of phospholipids, hydrolysis to lysophospholipids and free fatty acids, and denaturation or inactivation of liposome-encapsulated thermolabile substances, such as DNA, small interfering RNAs (siRNAs), or proteins (Popovska et al. 2013).

4.1.2.1.3 Extrusion This is another method of converting MLVs prepared using the convection method into SUVs by repeatedly passing the former under high pressure through the polycarbonate membrane filters with very small pore size (0.8–1.0 µm). Usually, the extrusion is started with filters having relatively large pores, followed by a filtration through smaller pores. By choosing the filters with appropriate pore sizes, liposomes of desirable and homogeneous size distributions could be fabricated (Popovska et al. 2013; Uhumwangho and Okor 2005).

4.1.2.1.4 Detergent removal This method involves the solubilization of lipids in an aqueous solution of a detergent (surfactant) with a high critical micelle concentration (CMC) and the drugs to be encapsulated and the subsequent removal of the detergent via either dialysis or column chromatography, leading to the formation of large unilamellar vesicles (ULVs) of diameter 8–200 nm (Uhumwangho and Okor 2005).

4.1.2.1.5 Reverse-phase evaporation This technique involves the rapid injection of an aqueous solution of a drug into an organic solvent that contains the dissolved lipid, resulting in the formation of water droplets in the organic solvent (i.e., a "water-in-oil" emulsion). After the emulsion is dried down to a semisolid gel in a rotary evaporator, the gel is subjected to vigorous mechanical agitation to induce a phase reversal from water-in-oil to oil-in-water dispersion with the consequence of generation of ULVs (diameter, 0.1–1 μm) in aqueous suspension (Uhumwangho and Okor 2005).

4.1.2.2 Solid lipid nanoparticles

4.1.2.2.1 HPH This can be performed at an elevated temperature (hot HPH) or at or below room temperature (cold HPH) with the cavitation and turbulences harnessed to reduce the particle size. In the case of the hot HPH method, the lipid is heated at around 5°C–10°C above its melting point and added together with a drug to an aqueous surfactant solution of the same temperature. A hot pre-emulsion thus formed by high-speed stirring is converted into a nanoemulsion when processed in three to five cycles with a temperature-controlled high-pressure homogenizer and finally recrystallizes into solid lipid nanoparticles (SLNs) upon cooling down at room temperature. On the other hand, in the cold HPH technique, which is suitable for encapsulating temperature-labile or hydrophilic drugs, the lipid and drug are melted together and then rapidly cooled down under liquid nitrogen, forming solid lipid micropartices. The presuspension formed by high-speed stirring of the particles in a cold aqueous surfactant solution is homogenized in five cycles at or below room temperature, thereby producing SLNs. In general, cold HPH produces a larger particle size with broader size distribution compared to hot HPH (Martins et al. 2007).

4.1.2.2.2 Microemulsion Microemulsion is usually composed of a lipophilic phase, surfactant, cosurfactant, and water. For preparation of SLNs, solid lipids and a mixture of surfactant, cosurfactant, and water are separately heated at a temperature above the melting point of the lipids (65°C–70°C) before their mixing under mild stirring conditions. The resultant hot microemulsion is subsequently dispersed under stirring in excess cold water (2°C–3°C) in the typical ratio of microemulsion to cold water (1:10 to 1:50) or transferred dropwise into cold water with gentle stirring, producing SLNs (Parhi and Suresh 2012) (Figure 4.2).

4.1.2.2.3 Solvent emulsification–evaporation In this technique, both the lipid and hydrophobic drug are first dissolved in a water-immiscible organic solvent and the mixture is then emulsified in an aqueous phase using a high-speed homogenizer. The evaporation of the organic solvent

Chapter four: Fabrication strategies for biofunctional nanoparticles   47

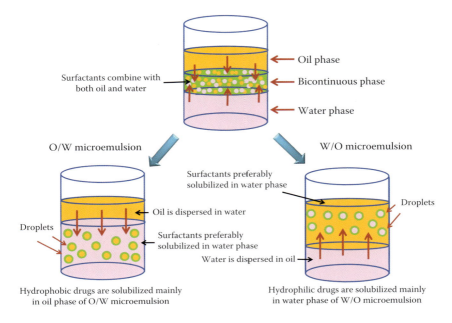

Figure 4.2 Preparation process of microemulsion.

by mechanical stirring at room temperature and reduced pressure (e.g., rotary evaporator) causes precipitation of the lipid as SLNs (Figure 4.3). Since the method avoids the thermal stress, highly thermolabile drugs can be encapsulated into the particles, although the use of organic solvent poses a limitation because of its potential cytotoxicity effects (Parhi and Suresh 2012).

4.1.2.2.4   Solvent emulsification–diffusion   The solvent used in this technique must be partially miscible with water. Initially, both the solvent and the aqueous solution (water) containing the stabilizer are mutually saturated at higher temperature in order to solubilize the lipid in the subsequent step. The lipid and drug are dissolved in the water-saturated solvent and later subjected to emulsification with the aqueous phase using a mechanical stirrer. After formation of the oil-in-water emulsion, an excess amount of water (dilution medium) is added to the system allowing diffusion of the solvent into the continuous phase, thus precipitating the lipid as SLNs. The diffused solvent can be eliminated by either vacuum distillation or lyophilization (Parhi and Suresh 2012).

4.1.2.2.5   Melting dispersion   Instead of dissolving a solid lipid in an organic solvent, a solid lipid is melted first in this technique. Afterward,

Figure 4.3 Solvent emulsification–evaporation method.

a drug is incorporated into the lipid melt as a solid or solution by vigorous vortex mixing and the mixture is emulsified into a small volume of aqueous phase heated above the melting point of the lipid. The dispersion is later cooled down to room temperature producing SLNs (Parhi and Suresh 2012).

4.1.2.2.6 *Double emulsion* In the first step of the method, hydrophilic drugs are generally dissolved in an aqueous solvent and the mixture is dispersed in a lipid-containing emulsifier/stabilizer to produce the primary emulsion (water-in-oil). The addition of an aqueous solution of a hydrophilic emulsifier to the primary emulsion under the stirring condition leads to the formation of a double emulsion (water-in-oil-in-water) (Figure 4.4) and lipid precipitation in the form of SLNs. The technique does not necessitate the melting of the lipid, providing the scope of nanoparticle surface modification for sterically stabilizing SLNs. However, two emulsification steps result in a highly polydisperse droplet distribution with poor control in the final size and structure of SLNs (Parhi and Suresh 2012).

4.1.2.2.7 *Solvent injection* The solid lipid is dissolved in a water-miscible solvent or a water-miscible solvent mixture. Subsequently, the lipid solvent mixture was injected through an injection needle and stirred into an aqueous solution with or without surfactant, causing the lipid precipitation as SLNs. The presence of emulsifier within the aqueous phase helps stabilize the SLN until the solvent diffusion is completed (Parhi and Suresh 2012).

Chapter four: Fabrication strategies for biofunctional nanoparticles 49

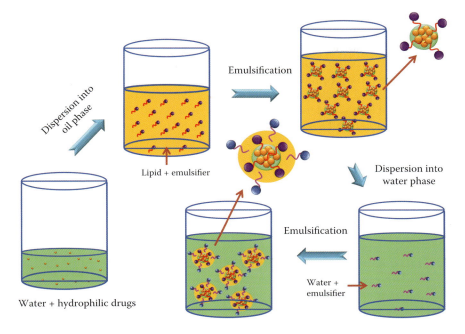

Figure 4.4 Double emulsion method.

4.1.2.2.8 Supercritical fluid technology A fluid is considered to be supercritical when its pressure and temperature exceed their respective critical values. As the pressure is raised, the density of the gas as well as its ability to dissolve compounds increases. Thus, the solvation power of a gas can be altered by controlling the changes in temperature and pressure. Out of the many gases, carbon dioxide (CO_2) is known as the best supercritical fluid since it is safe, inert, and miscible in organic solvents, for example, dimethyl sulfoxide or dimethyl formamide. Supercritical fluid technology can be categorized into several processes for synthesis of nanoparticles, such as rapid expansion of supercritical solution (RESS), gas antisolvent/supercritical antisolvent (GAS/SAS), particles from gas saturated solution (PGSS), aerosol solvent extraction solvent (ASES), solution-enhanced dispersion by supercritical fluid (SEDS), and supercritical fluid extraction of emulsions (SFEE). Out of them, SAS and PGSS are frequently used for preparation of SLNs (Parhi and Suresh 2012). SAS exploits the ability of CO_2 to dissolve in organic solvent and reduce the solvation power of a solid lipid in solution (supersaturation), thus causing the solid to precipitate as SLNs. PGSS, on the other hand, involves solubilizing of CO_2 in a melted or liquid-suspended lipid, leading to a so-called gas-saturated solution/suspension that is further expanded through a nozzle where the CO_2, which is more volatile, escapes, leaving dry lipid fine particles as SLNs (Byrappa et al. 2008).

4.1.2.3 Nanostructured lipid carriers

HPH is the most popular method for synthesis of both SLNs and nanostructured lipid carriers (NLCs). However, while a solid lipid is used for synthesis of SLNs, a blend of solid lipid with a liquid lipid (oil) is required for production of NLCs according to the protocol described above.

4.1.2.4 Dendrimers

Dendrimers can be prepared in either the divergent or the convergent method. In the former (divergent) approach, the molecule assembles from the core to the periphery through the stepwise attachment of repeating layers around a central while the latter (convergent) method involves synthesis of the individual segments of a dendrimer, termed *dendrons*, and coupling them to a polyfunctional core molecule in the final step. The divergent approach is successful for the production of large quantities of dendrimers, since each generation-adding step results in doubling of the molar mass of the dendrimer. Although very large size dendrimers can be prepared through this process, incomplete growth steps and side reactions ultimately lead to a mixture of pure and side products that are virtually impossible to be purified from each other. The convergent approach that was originally developed to overcome the weaknesses of the divergent method is composed of two stages, with the first one being a repeated coupling of protected/deprotected branch to produce a focal point functionalized dendron and the second one involving a divergent core anchoring step to produce various multidendron dendrimers. The dendrons of desired generation are typically constructed by the same activation/deactivation strategies as for the divergent growth, before the dendron core is deactivated and the individual dendrons are finally coupled to the core molecule (Figure 4.5). The major advantage of the technique is that the growth of each dendron can be more carefully monitored and controlled, thus severely lowering the risk of generating side products. Although the product mixture may contain dendrimers with a lower number of dendrons attached to them, the purification procedure is more straightforward as the partly derivatized dendrimers are much smaller than the desired ones. In addition, the method enables preparation of well-defined, multifunctional dendrimers through inclusion of dendrons having multiple groups of different functionality (Carlmark et al. 2009; Samad et al. 2009).

4.1.2.5 Nanocapsules and nanospheres

Nanocapsules consisting of a solid material shell surrounding a core, which is liquid or semisolid at room temperature (15°C–25°C), allow

Chapter four: Fabrication strategies for biofunctional nanoparticles   51

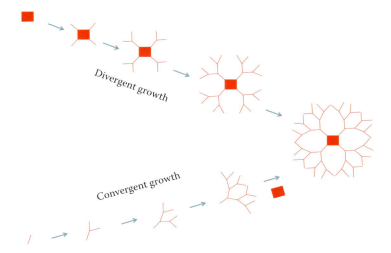

Figure 4.5 Dendrimer synthesis: divergent and convergent methods.

incorporation of either lipophilic drugs in the core composed of oil or hydrophilic drugs in the aqueous core. Generally, the polymer shell surrounding the liquid core is formed as a result of polymerization at the interface between the dispersed and continuous phase of the emulsion, or through precipitation of a preformed polymer at the surface of emulsion droplets. Nanospheres that are entirely solid nanoparticles of spherical shape allow the drugs to be entrapped inside the nanoparticle or adsorbed on their surface (Vauthier and Bouchemal 2009).

4.1.2.5.1 Two-step procedure   Usually, biodegradable, nontoxic, and nonimmunogenic polymers are employed for fabrication of nanocapsules and nanospheres in two major steps. The first step is the preparation of an emulsified system, such as emulsions, miniemulsions, nanoemulsions, and microemulsions, while the second step includes the formation of nanoparticles via the precipitation or the gelation of a polymer or through the polymerization of monomers.

There are a few other methods that enable nanoparticle formation without the emulsification step by inducing the precipitation of a polymer or the self-assembly of a polymer and macromolecules to form nanogels or polyelectrolyte complexes (Vauthier and Bouchemal 2009).

Conventional methods for preparing emulsified systems require two immiscible phases and a surface active agent in order to achieve the dispersion of one phase in the other (continuous phase) (Vauthier and Bouchemal 2009). Most of the new emulsification methods are based on the high-energy mechanical processes as in colloid mills in which a rotor turns at high speeds or extrusion machines where the dispersed phase is forced to permeate through a microfiltration device, allowing the preparation of emulsions with droplets of well-defined size and a narrow size distribution. Spontaneous emulsification method is based on a solvent displacement phenomenon in which the organic phase prepared with a water-miscible solvent, that is, acetone or ethanol, rapidly diffuses in the aqueous phase without stirring. This process, which avoids the use of a strong energetic method, is widely used for the preparation of nanocapsules by nanoprecipitation and interfacial polymerization. The various emulsions formed thereby can be utilized for the preparation of different polymer nanoparticles similarly as SLNs and NLCs.

On the other hand, different methods are employed for the formation of polymeric nanoparticles from the emulsions.

Polymer precipitation by solvent removal: Polymers in the emulsion droplets can be precipitated as nanoparticles by removing the polymer solvent via different methods, such as solvent evaporation, fast diffusion after dilution, or reverse salting out. Thus, simple oil-in-water emulsions lead to the production of nanospheres upon removal of the polymer solvent. Oil-containing nanocapsules can be obtained by simply adding oil to the emulsion droplets while water-containing nanocapsules can be produced by applying the solvent removal methods to a multiple emulsion, such as water1-in-oil-in-water2. Lipophilic drugs are generally encapsulated in nanospheres and oil-containing nanocapsules, whereas hydrophilic peptides, proteins, and nucleic acids can be encapsulated in water-containing nanocapsules. Solvent removal methods by either evaporation or diffusion have been discussed earlier for synthesis of SLNs. In the case of the reverse salting out technique, the emulsion is formulated with a polymer solvent, such as acetone, which is normally totally miscible with water. High concentrations of salts (magnesium chloride, calcium chloride, or magnesium acetate) or sucrose that can retain water molecules and thus modify the miscibility properties of the water with acetone are adjusted in the aqueous phase for the salting out effects. Finally, the precipitation of the polymer in the emulsion droplets can be induced by dilution of the emulsion with a large excess of water (reverse salting out) so as to produce a sudden drop of the concentration of salt or sucrose in the continuous phase and allow the solvent (acetone) to migrate out of the droplets (Vauthier and Bouchemal 2009).

Gelation of the emulsion droplets: Gelification of the water-soluble polymers dissolved in the emulsion droplets leads to the formation of nanoparticles via different processes depending on the gelling properties of the polymers. In the case of a polymer such as agarose

Chapter four: Fabrication strategies for biofunctional nanoparticles   53

that can be formed into the gel by cooling down its solution temperature, emulsions are first prepared at high temperature and the nanoparticles are subsequently formed as a result of the gelation of the emulsion droplets after cooling down of the emulsion. Polymers such as alginate and pectin can be induced to form gels (e.g., hydrogel) by either adding a second component or modifying the pH of the polymer solution. In this procedure, two emulsions, one containing the gelling polymer and the other containing the gelling agent or the pH controlling agent in the dispersed phase, are mixed together under strong agitation to induce the gelation of the polymer and thus the formation of nanoparticles (Vauthier and Bouchemal 2009).

In situ polymerization: Addition of a monomer to the emulsion instead of a polymer solution results in the polymerization leading to generation of nanoparticles. For instance, the in situ polymerization of alkylcyanoacrylates, which is spontaneously initiated by hydroxyl groups of water or nucleophilic groups of the molecules dissolved in the reaction mixture system, can be used to produce nanospheres or nanocapsules. Interfacial polycondensation, which involves polymerization between two monomers (each dissolved in one of a pair of immiscible phases) at the interface of the oil droplets formed during the mixing of the two phases, can produce nanocapsules with envelopes of variable thickness and porosity depending on the concentrations and molecular weights of the monomers (Vauthier and Bouchemal 2009).

4.1.2.5.2   One-step procedure

Nanoprecipitation or solvent displacement: Nanoprecipitation, which is one of the simplest methods, is performed using three basic ingredients, namely, the polymer, the polymer solvent miscible in water and easy to remove by evaporation, and the nonsolvent of the polymer. The polymer solvent used can be acetone (most common) or blends of either acetone and a small amount of water or ethanol and acetone. The rapid diffusion of the polymer solution in the nonsolvent results in the instantaneous formation of nanoparticles (Vauthier and Bouchemal 2009).

Formation of polyelectrolyte complexes: Nanospheres or nanoplexes can be formed by electrostatically neutralizing or significantly reducing the positive charges of cationic polymers (polyamines) and, consequently, their solubility, with the help of negatively charged nucleic acids (DNA, siRNA, or oligodeoxyribonucleotide). The crucial factor determining the nanoplex formation with desirable size and surface charge includes the ratio of the number of amine groups of the poly-cation to that of phosphate groups of the nucleic acid.

Ionic gelation: Ionic nanogels can be produced by inducing the gela-tion of a charged polysaccharide in a very dilute aqueous solution using the concentrations of a gelling agent (small ions of opposite charges) below the gel point and stabilizing the resultant small clus-ters through complexation with oppositely charged polyelectrolytes. Thus, the gelation of anionic alginate is induced with calcium and later stabilized with polycations, such as polylysine or chitosan. The nanogel formed thereby is of a more compact structure compared to the complex of alginate and a cationic polymer formed in absence of the gelling agent (Vauthier and Bouchemal 2009).

4.1.2.6 Assembly of polymeric micelles

Polymeric micelles can be formed in an aqueous solution by self-assembly of amphiphilic block copolymers having a hydrophilic block, such as PEG or poly(ethylene oxide), that can form hydrogen bonds with the aque-ous surroundings forming a tight shell around the micellar core usually consisting of either a hydrophobic block to interact with and thus retain hydrophobic drugs or a cationic polymer segment to condense anionic nucleic acids through electrostatic interactions. The size and morphol-ogy of the micelles can be tuned by changing their chemical composi-tion, total molecular weight, and block length ratios. The performance of such self-assembled drug delivery systems can be improved by incorpo-rating one or more functional groups, such as a peptide comprising cys-teine residues either on the hydrophobic (or cationic hydrophilic) block allowing cross-linking of the micelle core or on the hydrophilic part of the molecule enabling cross-linking of the micellar corona (Rösler et al. 2001).

4.1.2.7 Lipid–polymer hybrid nanoparticles

4.1.2.7.1 Two-step method In the two-step method, the preformed polymeric nanoparticles are added either to a dried thin lipid film that is earlier prepared by dissolving the lipid in an organic solvent (e.g., chloro-form) and evaporating the lipid solution in a rotary evaporator or to lipid vesicles that are prepared before by hydration of the thin lipid film. In either of the cases, after production of lipid–polymer hybrid nanoparticles (LPNs) by either vortexing or ultrasonication of the mixed polymer/lipid suspension at a temperature higher than the gel-to-liquid transition tem-perature of the lipid, the LPNs are separated from the nonadsorbed lipid by centrifugation and usually subjected to homogenization or extrusion steps to obtain monodisperse nanoparticles. The physical characteristics of LPNs are usually dictated by the characteristics of the preformed lipid vesicles (Hadinoto et al. 2013).

Chapter four: Fabrication strategies for biofunctional nanoparticles 55

4.1.2.7.2 One-step method The one-step method involves simply the mixing of polymer and lipid solutions (instead of polymeric nanoparticles or lipid vesicles) leading to their self-assembly into LPNs in a process involving either nanoprecipitation or emulsification–solvent–evaporation where the lipid functions as stabilizers in place of the ionic/nonionic surfactants. In addition to being simpler and faster than the two-step method, the one-step technique confers significant provision for bringing in a lot of variations in the formulated LPNs. The molar ratio of lipid to polymer plays the most important role in the size distribution and colloidal stability of LPNs along with their drug encapsulation efficiency and drug release kinetics (Hadinoto et al. 2013).

4.1.2.8 Synthesis of inorganic nanoparticles

The most popular and common procedures used to produce inorganic nanoparticles are based on wet chemical approaches.

4.1.2.8.1 Sol–gel The sol–gel process is a wet-chemical technique that can be used for the synthesis of inorganic nanoparticles by first forming a colloidal suspension (sol) and then gelation of the sol to an integrated network in a continuous liquid phase (gel). It thus consists of two sequential reactions: hydrolysis of metallic alkoxides (metal-organic compounds having organic ligands attached to metal) that results in a sol and subsequent condensation and polymerization to form a viscous gel of metaloxide particles (Faramarzi and Sadighi 2013). Silica particles, for example, can be prepared in solution by the hydrolysis and polycondensation of silicon alkoxide. The use of mineralizers (acids or bases) allows for control of the rates of hydrolysis and condensation independently, thus allowing for growth control of the final particles. Addition of dispersants may further control the growth rates and size of the particles (Fadeel et al. 2010).

4.1.2.8.2 Chemical reduction Reduction of an ionic salt in an appropriate medium with the help of a reducing agent, such as sodium borohydride, hydrazine hydrate, or sodium citrate in the presence of a surfactant as a protective agent or a phase transfer agent, gives rise to nanoparticles. The size, shape, and dispersity of such nanoparticles can be manipulated by the variable concentrations of the reductants. For instance, gold nanoparticles can be generated by the reduction of $HAuCl_4$ in surfactant solutions (Faramarzi and Sadighi 2013).

4.1.2.8.3 Spray-drying This method basically consists of spraying a homogenized solution of inorganic precursor molecules and relevant additives within a chamber at a temperature at or above the boiling point of the solvent. At first, the precursor solution is atomized through a nozzle into droplets using flowing gas and, subsequently, the droplets are sprayed into a chamber. Finally, a flow of hot air or nitrogen is introduced for the quick evaporation of the droplets and the eventual generation of the nanoparticles (Fadeel et al. 2010).

4.1.2.8.4 Microemulsions There are two methods for generation of inorganic nanoparticles in microemulsions. In the "one-microemulsion" method, the reaction is initiated by introducing a triggering agent or a reactant into a single microemulsion that contains another reactant precursor, eventually leading to particle formation. This method is driven by the diffusion-based process, since the second trigger/reactant diffuses into the emulsion droplets containing the reactant. In the "two-microemulsion" method, two reactants that are dissolved in the aqueous nanodroplets of two separate microemulsions are allowed to mix though the fusion–fission events between the nanodroplets as a result of the Brownian motion of the micelles and the resultant intermicellar collisions (Malik et al. 2012).

4.1.2.8.5 Nanoprecipitation When two or more salts react with each other in an aqueous solution, one of the products could undergo precipitation owing to supersaturation in a process involving nucleation, formation of primary particles, growth, and aggregation of the particles. Calcium phosphate precipitation has traditionally been used to generate hydroxyapatite particles as carriers of genetic materials, although controlling the particle growth kinetics remained an unresolved issue. Substitution of calcium or phosphate ions of the apatite molecules with magnesium or carbonate, respectively, has successfully been used to retard the particle growth with consequential generation of nanosize apatite particles (Chowdhury et al. 2004, 2006).

4.2 Recombinant DNA, hybridoma, and phage display techniques

4.2.1 Synthesis of protein-based nanoparticles

Proteins that are frequently produced by recombinant DNA technology can serve as both therapeutic agents and drug carriers. The amino acid sequences of proteins offer unique opportunities to form nanostructures by virtue of their secondary, tertiary, and quaternary structures. In addition, site-directed mutagenesis at the primary amino acid sequence confers an additional advantage in tailoring the protein structure in order to improve the drug binding affinity and biodistribution profile of a protein (Shi et al. 2014). Recombinant DNA technology, which typically involves polymerase chain reaction (PCR) (or RT-PCR) for amplification of a desired DNA (or mRNA) sequence, cloning of the amplified sequence into an expression vector (plasmid), transformation or transfection of the plasmid, and finally biosynthesis and purification of the desired protein, also allows for a precise control over the hydrophilicity/hydrophobicity and biorecognizable motifs of the protein, thus enabling the delicate regulation of its cellular and tissue interactions (Frandsen and Ghandehari 2012). Moreover, since the biosynthesis ends up with the polypeptide chains being the exact copies of the recombinant gene, the protein polymers are largely monodisperse, producing distinct pharmacokinetic profiles.

4.2.2 Generation of monoclonal antibodies

Monoclonal antibodies are produced from a single B-lymphocyte and bind to the same epitope on an antigen. Generation of monoclonal antibodies is generally based on hybridoma and phase display techniques (Ohlin and Borrebaeck 1996).

4.2.2.1 Hybridoma technique

After an animal is immunized against a specific epitope on an antigen, B-lymphocytes are isolated from its spleen and fused with an immortal myeloma cell line. The resultant hybridoma cells are cultured in selective medium where only the hybridomas survive and initially produce polyclonal antibodies derived from many different primary B-lymphocyte clones. The cell culture medium is subsequently screened from many hundreds of different wells for the specific antibody activity as expected and the desired hybridoma from the positive well is either grown further for production of the target monoclonal antibody or stored for future use (Figure 4.6).

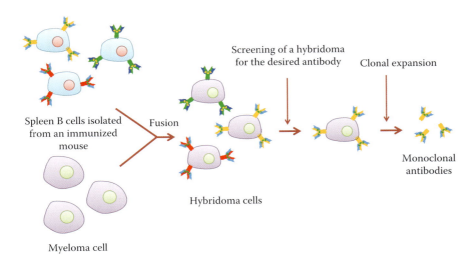

Figure 4.6 Hybridoma technology for production of monoclonal antibodies.

4.2.2.2 Phage display

After B-lymphocytes are isolated from the human blood, the mRNAs are extracted from the lymphocytes and converted with RT-PCR into cDNAs, amplifying all the VH and VL segments. These segments are then cloned into a vector next to the PIII protein of a bacteriophage and used to infect *Escherichia coli*, resulting in secretion and generation of a library of the bacteriophage particles containing the VH and VL segments as part of the bacteriophage coats. Specific VH and VL segments against the antigen are then selected and used to reinoculate *E. coli* with the bacteriophage. Since the antibody genes are cloned simultaneously with selection, they can be further engineered for increasing their affinity and modulating their specificity or effector function (Figure 4.7).

4.2.3 Production of viral vectors

Recombinant viral vectors are usually generated by deleting essential viral genes in order to make them unable to replicate into the hosts after administration as well as to provide the space in the genomes required for the insertion of desirable therapeutic genes. The packaging cell lines that provide *in trans* the missing viral functions are used for production of such vectors. For instance, the packaging cell line used for the generation of retroviral vectors carries three viral genes, namely, gag, pol, and

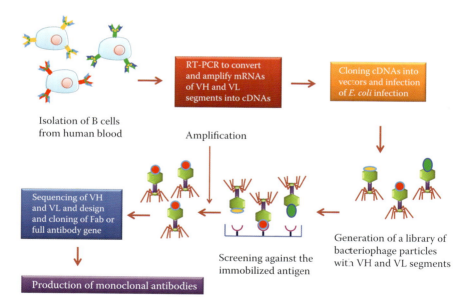

Figure 4.7 Phage display technology for production of monoclonal antibodies.

Chapter four: Fabrication strategies for biofunctional nanoparticles

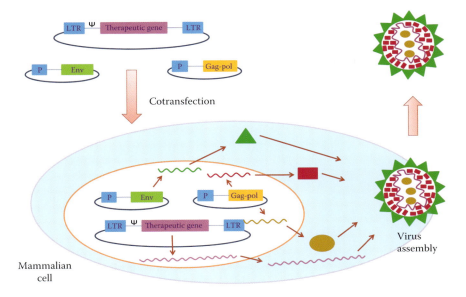

Figure 4.8 Production of lentiviral vectors.

env, that supply the functions required for the replication of viral genome, assembly, and packaging of the viral particles. Thus, the final vector contains the gene of interest (transgene) and the *cis*-acting elements, such as retroviral long terminal repeat or psi (ψ) packaging signal (Wu and Ataai 2000) (Figure 4.8). However, the absence of viral encoding genes (required for self-multiplication) is often accompanied by a dramatic reduction in the production yield. Generally, the factors affecting the efficiency of the vector production include the nature of the vector, packaging cell line used, culture conditions, multiplicity of infection, and level of inhibitory metabolic by-products. Metabolic engineering of cell-cycle regulation in the packaging cell line could extend the cell-cycle phases during which vector production is significant. Additionally, control-nutrient feeding strategies, genetic modifications, or apoptosis-inhibiting chemicals could be used to prevent or limit the apoptosis of the cell lines, induced by the viral infection or some environmental stimuli (Günzburg and Salmons 1995).

chapter five

Interactions and orientation of therapeutic drugs in the vicinity of nanoparticles

5.1 Dendrimer–drug interactions

The existence of large numbers of charged primary amine groups on the surface of dendrimers assists in electrostatic binding with ionizable drugs. In addition, several classes of dendrimers have also ionizable tertiary amine groups available for ionic complexation with the drugs at the branching points in the core. Since the pK_a values of the primary and tertiary amines are, respectively, 10.7 and 6.5, the solvent pH has an important role in drug binding in addition to the influences of dendrimer size, surface structure, and functionality of the drug molecules (D'Emanuele and Attwood 2005).

Although the interior structures of polyamidoamine and polypropyleneimine, the most widely used dendrimers, can promote both hydrophobic encapsulation and hydrogen bonding of a drug with the tertiary amines, the primary mechanism of drug complexation within the dendrimer structure is via electrostatic interactions with the surface primary and inner tertiary amines. The larger void volume available for drug accommodation and the stronger electrostatic affinity of higher-generation dendrimers toward the drug contribute generally to the higher drug loading into their structures (Kaminskas et al. 2012) (Figure 5.1).

The attachment of polyethylene glycol (PEG) chains to the dendrimer surface could increase the encapsulation efficiency of hydrophobic drugs by increasing the overall volume of the drugs for complexation through hydrogen bonding and electrostatic interactions. While the longer PEG chains in general accelerate more drug loading, very long chains (larger than approximately 5000 Da) can also reduce the encapsulation efficacy because of the formation of large PEG structures, thus filling the interior space of the dendrimer and reducing the volume for drug encapsulation (Kaminskas et al. 2012). A greater control over drug release can be achieved by covalently linking a drug through an amide or ester linkage with the surface of a dendrimer (Figure 5.1). The insolubility of

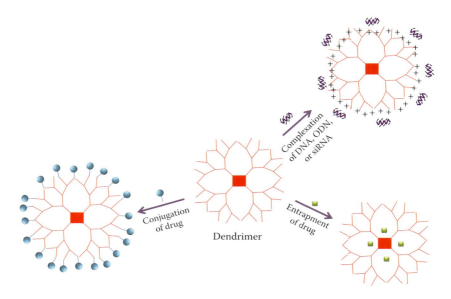

Figure 5.1 Dendrimer–drug interactions.

a drug–dendrimer complex that happens because of coupling or complexation of a large number of drugs to the dendrimer surface could be addressed through the attachment of PEG chains on the dendrimer surface.

The primary amines of a dendrimer can also be used to electrostatically complex on its surface with the negatively charged phosphate backbone of DNA, oligodeoxyribonucleotide (ODN), or small interfering RNA (siRNA) for cellular delivery (Figure 5.1).

5.2 Amphiphilic block copolymer–drug interactions

Amphiphilic block copolymers (ABCs) have been extensively used in solubilization of hydrophobic drugs, sustained-release formulation, and nucleic acid delivery. Among the different ABC-based nanodelivery systems, the polymeric micelle has attracted much attention because of its unique spherical core/shell structure characterized by a hydrophilic block (shell) interfacing the surrounding aqueous environment and a hydrophobic block (core) that is chemically tethered to the former and serves as a nano-depot for accommodation of hydrophobic drugs or nucleic acids (DNA, ODN, or siRNA) through hydrophobic and electrostatic forces (Xiong et al. 2011).

Chapter five: Interactions and orientation of therapeutic drugs

5.2.1 Drug loading into polymeric micelles

Drug entrapment into polymeric micelles depends on the miscibility between polymers and drugs with an increase in the miscibility accelerating drug accommodation into the micelles and the extent of hydrophobic interaction between drug and the micellar core. Introduction of hydrotropes, which are small molecules with both hydrophobic and hydrophilic moieties into the micelle core, could increase the miscibility between the core and the hydrophobic drugs (Figure 5.1). The length of the hydrophobic block and the type and degree of substitution on it differentially affect the loading efficiency in the polymeric micelles depending on the (hydrophobic) drug molecules. However, more drug loading as a result of stronger hydrophobic interactions is often accompanied by slower drug release.

5.2.2 Polymeric micellar drug conjugate

Early efforts were directed to reduce premature drug release by preparing drug–polymer conjugates through relatively stable linkers (e.g., amide and ester bond). In this case, the excessive stability of the polymeric prodrug may lead to the inactivity of the final product. The design of pH-responsive polymeric drug conjugate micelles has provided an exciting opportunity to achieve the site-specific release of incorporated drug from its carrier. This method involves formation of an acid-labile linkage between the therapeutic agent and the micelle-forming copolymer, a linkage that is stable at physiological pH, but which will be cleaved at the acidic pH of a tumor extracellular space or its endosomes, leading to the site-specific release of the parent chemotherapeutic agents from their micellar nanoconjugates (Figure 5.2).

5.2.3 Electrostatic complexation with DNA/siRNA

A polymeric micelle can be formed to act as a nonviral vector by chemically linking a hydrophilic block to serve as the shell, with a polycationic segment to form the core by neutralizing the negative charges present on DNA, ODN, or siRNA (Figure 5.2). An extensively used model of this type is poly(ethylene oxide)–poly(L-lysine) [PEO–P(Lys)], the stability of which can be improved by replacing some of the lysine residues in the core with thiol groups that can readily form disulfide linkages through cross-links with other neighboring thiol groups, thereby developing a network in the core after DNA complexation. Since the cross-linked core is cleavable inside the cell that possesses an increased level of glutathione, DNA can be selectively released only within a cell, but not in blood or any other extracellular space. Copolymers of PEO–poly(ε-caprolactone) (PEO–PCL)

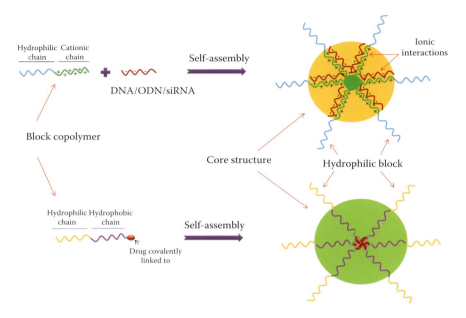

Figure 5.2 Polymeric micelle–drug interactions.

with the grafted polyamine on the PCL block can similarly be used to form a stable micelle core encapsulating siRNA with a high affinity. It is noteworthy that siRNA or ODN, being much shorter and thus fewer negative charges than plasmid DNA, has less affinity for ionic interactions with a vector.

5.3 Liposome–drug interactions

Liposome with a lipid bilayer can accommodate hydrophilic drugs into its aqueous interior and hydrophobic drugs inside the lipid bilayer. Electrostatic interaction could enhance hydrophilic drug encapsulation through the use of anionic or cationic lipids, enabling the charged liposome to interact with drugs having the opposite charge. The incorporation of cholesterol could contribute to the encapsulation efficiency of hydrophilic drugs by reducing the rotational freedom of the phospholipid hydrocarbon chains and thus decreasing the loss of the entrapped hydrophilic compounds. Liposomal membrane composition affects both partitioning and encapsulation efficiency of the lipophilic drugs, since these drugs are encapsulated within the lipid membrane depending on their solubility in the phospholipid bilayer. Being hydrophilic in nature, a protein usually resides in the aqueous compartment. However, if a protein can transform into a "molten globule" state with an unfolded intermediate conformation

Chapter five: Interactions and orientation of therapeutic drugs   65

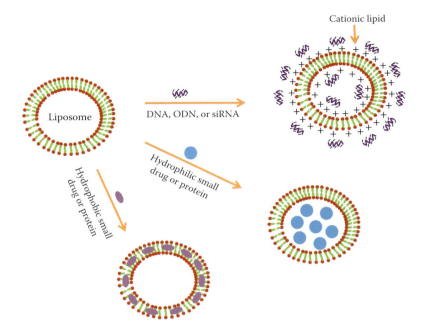

Figure 5.3 Drug–liposome interactions.

exposing the hydrophobic part, it may exist either in the aqueous interior or in the lipophilic bilayer. Additionally, protein incorporation in a liposome might depend on the liposomal charge states and the pH and ionic strength of the hydration medium. Cationic liposomes are widely used as a carrier of anionic nucleic acids, which can associate by ionic interactions with the cationic outer or inner surface of the liposomes (Rawat et al. 2008) (Figure 5.3).

5.4   *Inorganic nanoparticle–drug interactions*

Three main strategies undertaken for drug loading into inorganic nanoparticles include direct attachment of a drug to the nanoparticles via covalent linkage, adsorption of the drug onto the nanoparticle surface by electrostatic interactions, and drug entrapment or encapsulation. Existence of a suitable functional group in a drug molecule could offer an opportunity for covalently linking the drug to the nanoparticle through a biodegradable linker. On the other hand, ionic interactions of a drug with the nanoparticle are commonly based on the particle surface charge that, in turn, can be tuned by either surface enrichment with metal ions or surface stabilization (or modification) with functional groups, such as amine or thiol ligands (Figure 5.4). Thus, Au and SiO_2 nanoparticles

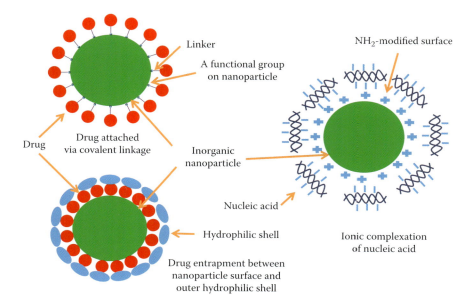

Figure 5.4 Inorganic nanoparticle–drug interactions.

modified with NH$_2$-terminal ligands or CdS nanoparticles enriched with metal ions could electrostatically bind with the negatively charged DNA molecules. The carbonate apatite nanoparticle that has been developed in our laboratory is another interesting candidate as a potential drug carrier, since its surface, which carries both positive and negative charges, allows adsorbing a drug (small drug, DNA, siRNA, or protein) regardless of whether it possesses positive or negative charge(s). Finally, for effective drug entrapment, a hydrophobic drug could be loaded onto the nanoparticle while an associated outer hydrophilic shell would avoid drug leaking and stabilize the particles in the surrounding aqueous environment. Inorganic nanoparticles with hollow interiors and porous shells are promising for encapsulation and eventually sustained release of small drugs (Ojea-Jiménez et al. 2013) (Figure 5.4).

chapter six

Variable interactions of nanoparticles with blood, lymph, and extracellular and intracellular components

Because of their high surface area/volume ratio and surface reactivity, nanoparticles can interact efficiently with the macromolecules and the cells present in blood, lymph, extracellular matrix, cell membrane, and intracellular compartment. Understanding the mechanisms of nanoparticle interactions with the molecular and cellular components of the body would be instrumental in developing strategies to prevent the unwanted interactions, thereby enhancing efficacy and eliminating toxicity of nanotherapeutics.

6.1 Serum proteins with affinity to nanoparticles

Immediately after coming into contact with nanoparticles, plasma proteome, which consists of tens of thousands of proteins, leads to the formation of "protein corona" on the surface of the nanoparticles. Corona formation is a dynamic process constantly changing according to changes in the surroundings with time. Affinity and abundance of the proteins determine the overall process wherein the abundant proteins first bind to the nanoparticle surface and is subsequently replaced by less abundant proteins with higher affinities for the nanoparticles. The proteins that bind nanoparticles abundantly are albumin, transferrin, fibrinogen, alpha-2-macroglobulin, C5 complement, apolipoprotein A-I, and haptoglobin, while those that interact moderately are ceruloplasmin, plasminogen, IgD, and retinol-binding protein (Karmali and Simberg 2011). Among the most abundant proteins, albumin and fibrinogen can bind to many different types of nanoparticles. The factors that regulate these interactions are discussed in Sections 6.1.1 through 6.1.4 and also illustrated in Figures 6.1 and 6.2.

67

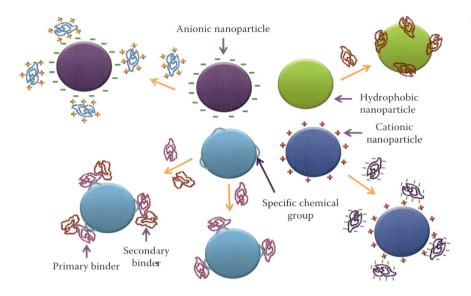

Figure 6.1 Modes of protein interactions with nanoparticles.

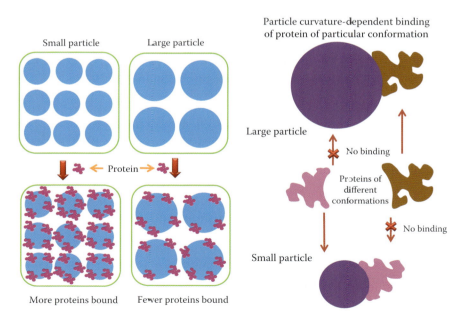

Figure 6.2 Effect of size and curvature on protein binding.

Chapter six: Variable interactions of nanoparticles

6.1.1 Surface hydrophobicity

Surface hydrophobicity can dictate the type and the amount of a protein that could bind with the nanoparticle surface (Figure 6.1). In general, hydrophobic nanoparticles bind more plasma proteins than the hydrophilic counterparts. Liposomes rich in cholesterol were found to bind lesser amount of proteins than the cholesterol-free ones. In addition, liposomes containing neutral saturated lipids with carbon number greater than 16 could bind more blood proteins than those having the C14 counterparts, particularly because of the stronger affinities of IgG and albumin for the hydrophobic domains. Apolipoprotein A-I has special affinity for the hydrophobic surfaces of liposomes and polymeric nanoparticles (Karmali and Simberg 2011).

6.1.2 Surface charge

The surface charge of an individual particle type plays another important role in protein interaction affecting the composition of the protein corona (Figure 6.1). Anionic liposomes including those composed of phosphatidyl serine could bind more serum proteins, such as apolipoprotein E or complement (C1q), than their neutral counterparts. In the case of negatively charged polymeric nanoparticles, plasma protein adsorption increased with increasing surface charge density of the nanoparticles. On the other hand, significant plasma protein binding was observed in cationic lipid-carrying liposomes apparently as a result of their ionic interactions with most of the negatively charged plasma proteins. Fibrinogen, an opsonin, showed a higher affinity to the cationic liposomes with a high surface charge density, whereas apolipoproteins and the C4b-binding protein demonstrated a stronger binding with those having a low charge density. Using polystyrene nanoparticles, it was revealed that the cationic particles predominantly adsorbed the proteins with a isoelectric point (p*I*) value of below 5.5, such as albumin, while the anionic particles electrostatically associated with the proteins having a p*I* of more than 5.5, such as IgG. Cationic charge neutralization by anionic albumin was proposed to protect the positively charged nanoparticles from causing adverse effects (Karmali and Simberg 2011; Moros et al. 2013).

6.1.3 Size and curvature of nanoparticles

Particle size determines the curvature of nanoparticles, which, in turn, could influence the interactions of the nanoparticles with proteins of variable conformations at the nano–bio interface (Figure 6.2). Thus, dextran-coated iron oxide or polystyrene nanoparticles of ~250 nm could attract more dextran-specific IgM than the larger particles (600 nm) (when normalized to the total surface area of the particles) to efficiently induce the classical IgM-dependent complement activation efficiently. Since a decrease in particle size can increase the total surface area, small particles should bind more proteins than the larger ones at a fixed particle weight. Consequently, gold particles of 30 nm were found to bind more proteins (almost twofold) than those of 50 nm. Nanoparticles of the NIPAM/BAM (*N*-isopropylacrylamide-co-*N-tert*-butylacrylamide) (50:50) copolymer with size between 70 and 700 nm also showed the same effect; that is, the amount of bound plasma proteins increased with increasing the surface area at a constant particle weight. On the contrary, the thickness of the protein corona formed in the presence of common human blood proteins, such as albumin, fibrinogen, g-globulin, histone, and insulin, was found to progressively increase with an increase in the size of gold nanoparticles, which range from 5 to 100 nm, indicating that the higher surface curvature of the smaller particles reduces the amount of protein that adsorbs onto their surface (Karmali and Simberg 2011; Moros et al. 2013).

6.1.4 Proteins with affinity for specific chemical groups of nanoparticles

Some of the serum proteins can specifically bind to the particular chemical groups present on the nanoparticle surface. Therefore, the hydroxyl groups of immobilized dextran could mediate the binding of the C3 complement through its thioester group with the nanoparticles. Similarly, mannose-binding lectins could bind to sugar moieties of dextran-embedded nanoparticles and mannose-binding protein could associate with phosphatidylinositol liposomes. There are proteins (primary binders) that directly bind to the nanoparticles and simultaneously invite other proteins (secondary binders) to bind through them. For example, kininogen could promote the indirect binding of plasma prekallikrein and FXII apparently to a foreign surface, because they circulate in a complex. Similarly, complement factors C5–C9 and C1q are recruited toward the nanoparticles after binding of IgG in a classical pathway or C3 in an alternative pathway (Karmali and Simberg 2011).

6.2 Fates of the serum protein–coated nanoparticles

6.2.1 Removal by macrophage, thrombosis, and hypersensitivity

Since the dose of a drug that can reach the site of action or the diseased tissue is directly proportional to its concentration in blood, engulfment and subsequent removal of the nanocarriers by the macrophages of the mononuclear phagocytic system (MPS) as a result of nanoparticle surface

Chapter six: Variable interactions of nanoparticles   71

coating by plasma proteins can lead to improper biodistribution, therapeutic inefficacy, and toxicity of the drug-associated nanoparticles (Figure 6.3). In addition, interactions of antibodies and complement or clotting factors with the nanoparticles can promote activation of defense responses including clotting and complement cascade, thus causing toxicity in the form of thrombosis and hypersensitivity. Furthermore, upon binding to the nanoparticles, proteins could change their conformation exposing new epitopes and ultimately inducing new injury. Decorating the surface of nanoparticles with hydrophilic polymers, such as polyethylene glycol (PEG), is an attractive approach currently employed to improve the pharmacokinetics and decrease the macrophage recognition of many types of nanoparticles by preventing plasma protein interactions (Karmali and Simberg 2011).

6.2.2   Aggregation

Nanoparticles can be subjected to aggregation upon exposure to serum proteins owing to surface charge neutralization, exposure of hydrophobic domains, or bridging of individual particles by the protein molecules. Strongly cationic or anionic particles are usually prone to a higher level of aggregation. Thus, sulfated polystyrene particles were found to be much more vulnerable to aggregation than their carboxylated counterparts. Poly-L-lysine nanoparticles were also demonstrated to undergo aggregation in serum, which was prevented by surface PEGylation. The major consequence of such aggregation is the accelerated clearance of the resultant large particles from the circulation by MPS (Karmali and Simberg 2011).

6.2.3   Dissociation of complex and leakage of drugs

Unlike the polymeric and inorganic nanoparticles, liposomal and micellar systems are generally unstable in the presence of serum, releasing much of the loaded drugs in blood before reaching their target sites. This could be explained for cationic liposomes by their strong interactions with anionic plasma proteins with resultant aggregation and clearance of the carriers from the circulation by MPS, or disintegration of the cargo-loaded carriers and potential destruction of the nucleic acid cargo by nuclease, and for micelles by the affinity of plasma proteins, such as lipoproteins for the hydrophobic domains of the amphiphilic carriers with the similar consequences of rapid clearance and drug leakage (Figure 6.4). PEG coating of lipoplexes (Nicolazzi et al. 2003) and inclusion of strong polycations, such as protamine in the lipoplex formulation (Mizuarai et al. 2001), could overcome to some extent the negative effects of plasma proteins.

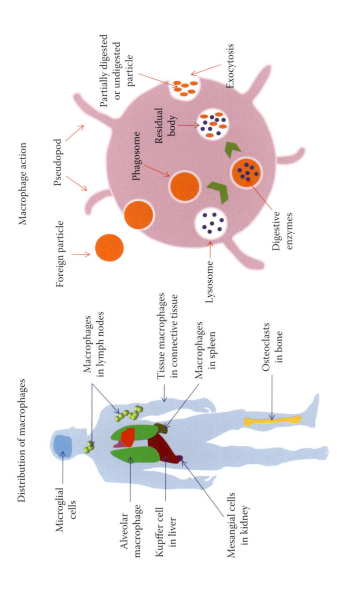

Figure 6.3 Removal of nanoparticles from circulation by MPS.

Chapter six: Variable interactions of nanoparticles 73

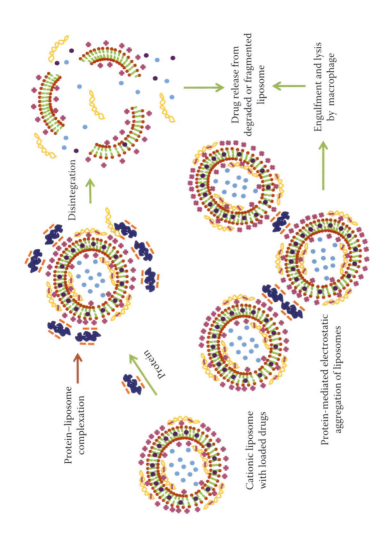

Figure 6.4 Mechanisms of drug leakage from liposomes.

6.3 Interactions of nanoparticles with interstitial fluid and lymph

The surface charge and hydrophobicity of a nanocarrier can play important roles for its lymphatic transport through the interstitial space after systemic (subcutaneous or intravenous) injection (Figure 6.5). Negatively charged carriers, such as dendrimers, poly(lactic-co-glycolic acid) (PLGA) nanospheres, and liposomes, showed higher lymphatic uptake than their neutral or positively charged counterparts probably because of the net negative charge of the interstitial matrix that causes electrostatic repulsion of the anionic carriers and lets them move faster. Therefore, highly negatively charged nanoparticles were found to be retained for a prolonged period in the lymph nodes. In contrast, positively charged particles face more resistance to move through the negatively charged interstitium as a result of the electrostatic attraction force. On the other hand, hydrophobicity of lipid-based nanocarriers promotes their binding with opsonins and subsequent phagocytosis after transport through the lymphatic vessels, thus facilitating their uptake by lymph nodes. In addition, large molecules and colloids that are too large to enter blood capillaries can

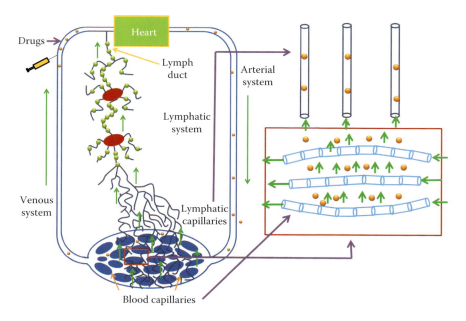

Figure 6.5 Transport of nanoparticles through blood vessels and the lymphatic system.

Chapter six: Variable interactions of nanoparticles 75

also be absorbed by the lymphatic system after the interstitial administration. Coating the particle surface with a hydrophilic layer may reduce their nonspecific interactions with the interstitial surroundings and allow improved migration through the interstitium, leading to increased lymphatic uptake and transport toward one or more lymph nodes where a fraction will be retained by either phagocytosis or by simple mechanical filtration. Conversely, since the hydrophilic coating of the nanoparticle surface reduces adsorption of proteins (such as opsonins), their uptake by macrophages and eventual retention by lymph nodes might also be reduced. PEGylated interferon (IFN) was more efficiently absorbed from the subcutaneous injection site than non-PEGylated IFN, with approximately 20% and 8% of the injected dose recovered in thoracic lymph over 30 hours after subcutaneous and intravenous administration, respectively. However, PEGylation of IFN with a longer chain of PEG showed incomplete absorption from the subcutaneous injection site with approximately 29% of the injected dose recovered after intravenous administration (Kaminskas et al. 2013).

6.4   Extracellular matrix–nanoparticle interactions

Extracellular matrix (ECM), which is mainly composed of a network of positively charged collagen fibers and negatively charged hyaluronic acid, is a crucial barrier to the diffusion of macromolecular drugs and nanoparticle-based carriers before interacting with target cells. Both the size and surface charge of nanoparticles as well the configuration and the physicochemical properties of ECM regulate the movement of the particles within the tissues. The diffusion of nanoparticles through the spaces between ECM network structures is regulated through the steric hindrance in the dense ECM network and the electrostatic interactions of matrix components with the nanoparticles (Figure 6.6).

The tumor ECM similarly consists of a highly interconnected network of collagen fibers, proteoglycans, and glycosaminoglycans (GAGs). Small chemotherapeutic drugs whose size is usually below 1–2 nm diffuse rapidly in the tumor matrix. However, the diffusion of nanoparticles is considerably hindered by interactions with the ECM components. In many tumors, particles larger than 60 nm in diameter are unable to effectively diffuse through the collagen matrix after extravasation from blood vessels and therefore remain concentrated around the vessels heterogeneously causing only local effects. Tumors rich in collagen hinder diffusion more than those with a low collagen content. Moreover, collagen fibers, being slightly positive, can interact with anionic nanoparticles, whereas sulfated GAGs, being highly negative, can bind cationic particles, thus inhibiting their transport by forming aggregates (Stylianopoulos et al. 2010) (Figure 6.6).

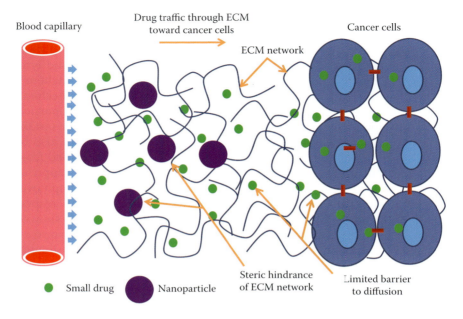

Figure 6.6 Diffusion of therapeutics: small drug versus nanotherapeutics.

6.5 Interactions between nanoparticles and cell components

The protein corona formed on the nanoparticle surface significantly alters its chemical properties and thus influences nanoparticle interactions with the cell membrane. In contrast to neutral or anionic nanoparticles, cationic nanoparticles are known to interact more readily with the highly anionic proteoglycans composed of a membrane anchor, a core protein, and GAGs at the extracellular side. GAGs that consist of the repeating disaccharide units, such as heparan sulfate, chondroitin sulfate, or keratan, each with high densities of carboxyl and sulfate groups, make the cell membrane negatively charged at physiological pH (Vercauteren et al. 2012). Consequently, the delivery vehicles developed so far for drug and gene delivery applications are mostly based on cationic nanoparticles and are frequently used to promote cellular entry for drug and gene delivery applications. Thus, coating of mesoporous silica nanoparticles (MSNPs) with cationic poly(ethyleneimine) (PEI) was reported to considerably enhance cellular uptake of PEI-MSNP compared to the unmodified MSNP or the PEG-coated MSNP. Hydrophilic coating of the nanoparticle surface to inhibit nonspecific cellular uptake while decorating the surface with moieties to bind specific receptors or antigens on the cell membrane is therefore critical for the design of targetable nanoparticles.

chapter seven

Pharmacokinetics and biodistribution of nanoparticles

Once the drug-loaded nanoparticles or macromolecular drugs reach the bloodstream through a delivery route, the diversified interactions of the nanotherapeutics of variable surface properties, with the molecular and cellular components of blood, lymph, and the extracellular matrix (ECM), notably influence their biodistribution profile.

7.1 Influence of particle size

Particle size plays an important role in biodistribution of nanotherapeutics because of the differential permeability and blood transfusion rates of different tissues or organs (Figure 7.1). Thus, 24 hours after intravenous administration of gold nanoparticles, it was revealed that a higher amount of 15-nm gold nanoparticles accumulated in all the tissues including blood, liver, lung, spleen, kidney, brain, heart, and stomach, and the 15- and 50-nm nanoparticles could cross the blood–brain barrier and the 200-nm gold particles were detected slightly in blood, brain, stomach, and pancreas only. However, particles of all sizes were mainly found in liver, lung, and spleen (Sonavane et al. 2008). Using uncoated silicon-based particles, it was reported that the number of particles that accumulate in non–mononuclear phagocyte system (MPS) organs reduced monotonically with the increase in the diameter of the spherical beads, providing additional support that smaller particles provide more uniform tissue distribution. Indeed, biodistribution of inorganic nanoparticles (such as, gold, silica, silver, and titanium dioxide nanoparticles along with quantum dots and carbon nanotubes) is usually size dependent after systemic administration with the small nanoparticles (10–20 nm) having wide distribution because of their ability to cross the tight junctions between the endothelial cells of blood vessels and rapid renal clearance owing to their filtration capability through the glomeruli of kidney. On the other hand, large particles (e.g., >1 μm) are also subjected to rapid clearance by the macrophages of MPS and thus accumulated mainly in liver, spleen, and bone marrow. The particles whose size falls in between (>20 nm and <1 μm) have more prolonged circulation time since they are less sensitive to the clearance by kidney and MPS (Figure 7.1). Conjugation of polyethylene

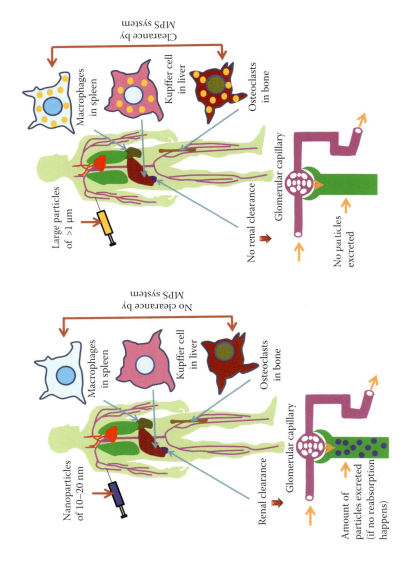

Figure 7.1 Fates of too small and too large particles after systemic administration.

Chapter seven: Pharmacokinetics and biodistribution of nanoparticles 79

glycol (PEG) with a macromolecular drug (e.g., a therapeutic protein) is an effectual strategy to increase the size of the drug and prevent its renal clearance (Duan and Li 2013).

7.2 Influence of plasticity of nanoparticles

The elasticity of nanoparticles can influence their interactions with cell membranes (Yi et al. 2011). At the time of cellular uptake via endocytosis, the adhesive interaction between a rigid particle and the cell membrane can force the membrane to deform and wrap around the particle, involving only a small increase in elastic energy, whereas a soft particle can initially spread along the membrane with a poor membrane deformation requiring a more abrupt increase in elastic energy for the full membrane wrapping of the particle. As a result, soft nanoparticles could overcome the hurdle of being internalized by macrophages, thus facilitating delivery of their cargoes to the target sites.

7.3 Influence of protein corona formed around nanoparticles

7.3.1 Opsonin-facilitated phagocytosis

Formation of a protein corona on the surface of a nanoparticle could alter its size and surface charge and thus influence its distribution profile. Some of the proteins that adsorb on the nanoparticles, such as complement factors, fibrinogen, and IgG (collectively known as opsonins), are recognized through their epitopes by the macrophages of MPS, facilitating the phagocytosis and the clearance of the nanoparticles as foreign materials from the body. Surface-immobilized IgG or complement C3 can trigger macrophage-aided phagocytosis of the nanoparticles with their subsequent accumulation in MPS tissues and concomitant clearance from blood (Karmali and Simberg 2011) (Figure 7.2).

7.3.2 Dysopsonin-enhanced blood circulation time

There are some proteins, known as dysopsonins, which, upon binding on the nanoparticle surface, could prolong the blood circulation time of the nanoparticles. Being a dysopsonin, albumin can prevent the opsonization of nanoparticles by associating with their surfaces, although the dysopsonin effect can be abolished by the proteins with higher affinities for the nanoparticles and the ability to replace the electrostatically bound albumin. Covalently linking albumin to the nanoparticles or drugs can block this replacement phenomenon (Figure 7.3). Commercially available Abraxane, which consists of paclitaxel, a conventional anticancer drug

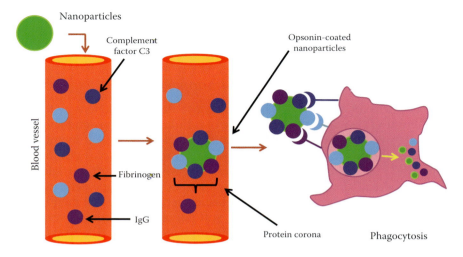

Figure 7.2 Protein corona around nanoparticle and opsonin-mediated phagocytosis.

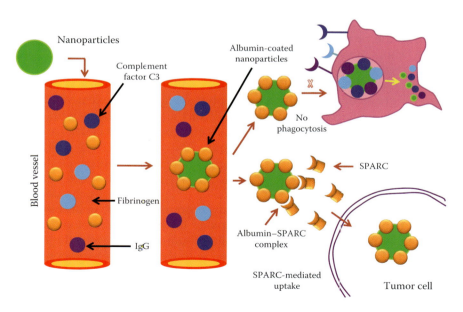

Figure 7.3 Prolonged blood circulation and tumor uptake of albumin-coated nanoparticles.

Chapter seven: Pharmacokinetics and biodistribution of nanoparticles 81

covalently bonded to albumin, greatly enhances the blood circulation time of the drug (Fu et al. 2009). In addition to the dysopsonin effect, albumin can enhance the tumor distribution and accumulation of the conjugated drug by recognizing and facilitating the uptake of Abraxane through the SPARC (secreted protein acidic and rich in cysteine), an albumin-binding protein highly expressed in tumor microenvironment (Desai et al. 2008).

7.3.3 Uptake by nonphagocytic cells

Apart from the existence of macrophage-specific proteins and dysopsonins in a protein corona, there are some other proteins such as apolipoprotein E (ApoE) that can be recognized by the cells other than the macrophages and Kupffer cells. Receptors for ApoE are expressed at the blood–brain barrier as well as in liver and spleen, and therefore, covalent attachment of ApoE to nanoparticles enables them to transport into the brain, liver, and spleen (Aggarwal et al. 2009).

7.4 Influence of charge and hydrophilicity

Hydrophilic nanoparticles adsorb less plasma proteins and therefore show prolonged half-lives with desirable pharmacokinetics by preventing opsonization in contrast to their cationic and anionic counterparts. A number of polymers, such as PEG, poly(ethylene oxide), block copolymers of poloxamers and poloxamine, dextran, and amphipathic polymers, are commonly used for producing a hydrophilic layer on the surface of both organic and inorganic nanoparticles. As the most convenient and biocompatible macromolecule, PEG received the utmost attention to block through steric hindrance the nonspecific protein interactions with the nanoparticles. The antiopsonization effect of PEG is highly dependent on its molecular weight, grafting density, and chain architecture. Indeed, with the increase of the polymer density and the consequential decrease in the space between each polymer chain, the configuration of the PEG molecule can be transformed from the "mushroom" to the "brush" state, the latter having more antiopsonization activity than the former (Figure 7.4). Additionally, the conformation and density of PEG could also influence the composition of protein corona. Despite the beneficial roles, there are concerns with high-molecular-weight PEG polymers that can be accumulated in the body as a result of nonbiodegradability of the original polymer and size-restricted renal clearance (Photos et al. 2003). Additionally, there are reports that PEG can elicit production of anti-PEG immunoglobulin M (IgM), which later binds to the PEG after the second dose of PEGylated nanoparticles and activates the complement system, thus leading to opsonization and liver accumulation of the PEGylated particles. This phenomenon, which ends up with a much shorter half-life of the second dose of

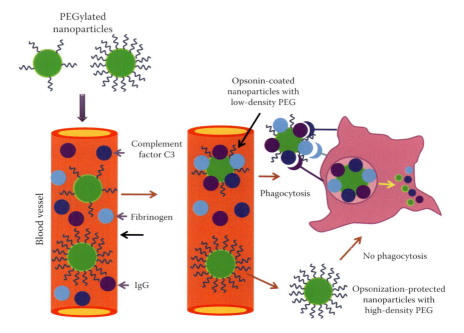

Figure 7.4 Anti-opsonization effect of PEG.

PEGylated nanoparticles compared to that of the first dose, is known as accelerated blood clearance (ABC) (Ishida and Kiwada 2008).

7.5 Influence of endogenous membrane coating

Endogenous membranes could be useful materials for coating nanoparticles in order to prevent their interactions with blood components. Thus, the nanoparticles coated with red blood cell (RBC) membrane were reported to increase their blood half-lives even more significantly than the PEG-coated nanoparticles. Since the erythrocyte naturally circulates in the blood, its membrane may not interact strongly with the blood components. Moreover, this approach is highly promising to avoid the ABC phenomenon that quickly removes the PEGylated nanoparticles in case of the second dose administration. However, the blood circulation time of RBC-coated particles is still much shorter than that of RBCs alone (Hu et al. 2011).

7.6 Influence of ligand coating

Modification of nanoparticle surfaces with hydrophilic polymers or endogenous membranes is useful in reducing the distribution of nanoparticles

Chapter seven: Pharmacokinetics and biodistribution of nanoparticles 83

to MPS tissues, but not beneficial in facilitating their uptake by the target cells. Attachment of specific ligands to the nanoparticles has no role in driving the delivery system to the target site but enables recognition by the target cells through ligand–receptor interaction and thus increases the cellular uptake as the nanoparticles are distributed to the target site in addition to other nontarget sites.

7.7 Influence of coating of CD47 as a "self" marker

CD47, an integrin-associated protein, acts as a key self-recognition marker for RBCs so as to enable them to escape the phagocytosis by macrophages for prolonged blood circulation, by interacting with the inhibitory receptor signal regulatory protein alpha (SIRP) on macrophage membranes. Recombinant CD47-coated polystyrene particles were shown to significantly inhibit the phagocytosis even after opsonization of the particles, suggesting that the CD47–SIRP interaction could be a potential approach for the development of macrophage-evading nanoparticles (Figure 7.5). Recently, polystyrene nanobeads with surface-attached minimal "Self" peptides (computationally designed from human CD47) were

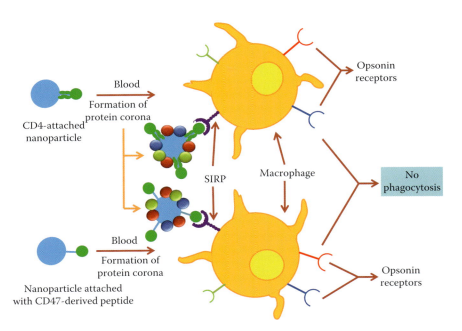

Figure 7.5 Inhibition of phagocytosis with CD47-bearing nanoparticles.

reported to delay macrophage-mediated clearance of the nanoparticles, promoting their persistent blood circulation and enhancing drug delivery to tumors (Rodriguez et al. 2013).

7.8 Extravasation from blood through vascular endothelium

After reaching the bloodstream, the nanoparticles whose target site of action lies beyond the vascular wall must cross the vascular endothelium that lines the blood vessels and controls the homeostasis of tissue fluids including plasma proteins and the transmigration of leukocytes between the blood and the interstitial space. Long circulating nanoparticles must extravasate from the blood across the vascular endothelium to reach the target cells by crossing the interstitium where high pressure may also interfere with their movement particularly in solid tumors. Other subsequent factors modulating the biodistribution include the cellular, endosomal, and nuclear membranes, depending on the target site.

7.8.1 Permeability of vascular endothelia

Blood capillaries are categorized into the following: continuous capillary, which is characterized by the presence of tight junctions between the adjacent endothelial cells and an uninterrupted basement membrane; fenestrated capillary, which possesses fenestrae (pores) or gaps in the endothelium supported by a continuous basement membrane; and, finally, discontinuous capillary, which is a thin-walled vessel containing much larger gaps than the fenestrae and either lacking a basement membrane as in the case of liver sinusoidal capillary or containing an interrupted basement membrane usually found in the vessels of spleen and bone marrow (Figure 7.6). Most of the endothelium in our body is continuous and found in muscle, skin, lung, connective tissue, and the nervous system. The tight intercellular junctions in the continuous epithelium of the brain are particularly restrictive, even hindering the flux of small molecules across the blood–brain barrier. Fenestrated endothelium with gaps of up to 100 nm is, on the other hand, located in exocrine glands, kidney, and intestinal mucosa. Discontinuous endothelium with pores of >100 nm readily permits entry of the nanoparticles into the interstitium from the blood circulation. In fact, the exclusive localization after systemic injection of large inorganic particles (>1 µm) in the liver, spleen, and bone marrow is the combined effect of the highest permeability of the discontinuous endothelium and the phagocytosis event of the MPS (Gentile et al. 2008).

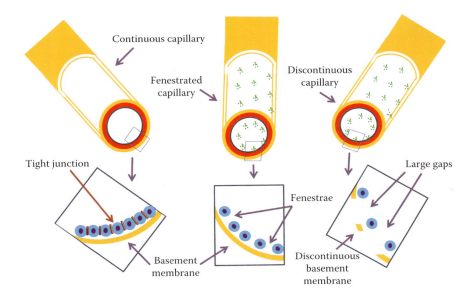

Figure 7.6 Different types of blood capillaries.

7.8.2 *Different routes of traffic across continuous endothelium*

The continuous endothelium or endothelial cell monolayer that provides a size-selective and semipermeable barrier between the blood plasma and interstitium allows passage of plasma proteins and solutes across via either the paracellular route through the tight junctions between the endothelial cell or the transcellular route through the endothelial cell membrane by vesicular trafficking, called transcytosis. While the paracellular route blocks passage of molecules larger than 3 nm, the transcellular path permits selective delivery of macromolecules across the endothelium (Komarova and Malik 2010). Nanoparticles are generally unable to use the paracellular transport except for some dendrimers capable of disrupting and opening the tight junctions to facilitate the transport. Transcellular transport is therefore a promising approach to facilitate the nanoparticle transport. The transcellular trafficking of albumin occurs sequentially via fission of caveolae-enriched plasma membrane macrodomains from the luminal surface of the endothelial cell, transport of the resultant caveolar vesicles toward the basal surface, and fusion of the vesicles with plasma membrane of the abluminal side discharging albumin through exocytosis. The endothelial cell surface-associated 60-kD glycoprotein, gp60, or albondin is believed to be important in albumin binding for initiating the transcellular albumin transport (Minshall et al. 2000) (Figure 7.7).

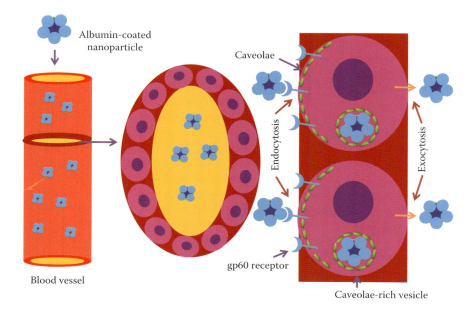

Figure 7.7 Transcytosis of albumin-coated nanoparticle mediated by gp60 receptor and caveolin-1.

7.8.3 Deregulated vascular endothelium

Diseases involving acute inflammation induce mediators such as vascular endothelial growth factors (VEGF or VEGF-A) that are responsible for changing the organization of the intercellular tight junction fenestrations and thus increasing vascular permeability of the endothelium for plasma protein extravasation. VEGF also contributes to the proliferation and migration of endothelial cells from preexisting vessels, leading to the formation of new blood vessels in a process called angiogenesis. Deregulated angiogenesis is often associated with many pathologies, such as cancer, rheumatoid arthritis, atherosclerosis, and psoriasis (Moros et al. 2013). In tumor, the new vasculature formed as a result of the imbalance between pro-angiogenic and anti-angiogenic factors is characterized by having wide fenestrations with increased vessel leakiness and heterogeneous hyperpermeability, thus potentially enabling escape of the nanoparticles of defined size range to the interstitium (Jain 2005). Indeed, the increase in vascular permeability along with the ineffective lymphatic drainage system accounts for the enhanced permeability and retention (EPR) effect and forms the basis for the passive targeting of nanoparticulate drugs to the tumor (Figure 7.8).

Chapter seven: Pharmacokinetics and biodistribution of nanoparticles   87

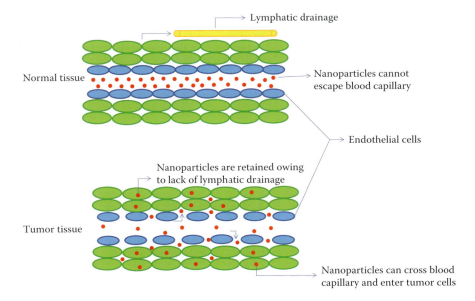

Figure 7.8 EPR effect in tumor tissue.

7.8.4 *Vascular endothelium as a target for drug delivery*

Vascular endothelium itself could be an important target for therapeutic interventions in various diseases including ischemia, atherosclerosis, inflammation, edema, oxidative stress, thrombosis, and cancer (Kim et al. 2014). Nanocarriers in blood circulation can specifically or nonspecifically adhere to the endothelial cells, releasing their drugs to the ECM of the endothelium. Nanoparticles can be coated with specific ligands in order to recognize the endothelial molecules, such as P- and E-selectins, vascular cell adhesion molecule-1, and intercellular adhesion molecule-1, that are exposed in the diseased areas (Chacko et al. 2011).

7.9 *Transport across the interstitium*

After extravasation, nanoparticles must cross the interstitium to reach the target cells by overcoming the physiological and physicochemical barriers of the interstitial space containing fibrous proteins and polysaccharides, as discussed earlier in Chapter 6. Collagen is considered to provide more resistance compared to glycosaminoglycan or hyaluronan to the diffusion of macromolecules, such as IgG, and therefore, collagenase was found more effective in improving macromolecule delivery than hyaluronidase. The lymphatic vessels at the center of a tumor are generally compressed and collapsed by the proliferating cancer cells while the functional vessels exist in the periphery to carry fluid, growth factors, and cancer cells. The inefficient drainage of fluid from the tumor center coupled with the leakage of vascular contents leads to the development of interstitial fluid pressure (IFP) or interstitial hypertension with the consequence of a net fluid flow from the high-pressure core to the tumor periphery, thus preventing effective penetration of the nanoparticulate drugs inside the solid mass or interior portions of the tumor and promoting their removal through the lymphatic vessels in the periphery (Jain 2010) (Figure 7.9). The potential strategies for lowering IFP mechanisms are often dependent on tumor type and dose and suffer from limitations of toxicity and lack of tumor selectivity. Some examples include applications of platelet-derived growth factor antagonist (anti-PDGF) to decrease contraction and interaction of stromal fibroblasts with ECM, hyaluronidase to degrade hyaluronan to reduce the physical resistance of ECM, bradykinin agonist to increase vascular surface area and pore size, angiogenesis inhibitors to produce transient vascular normalization, nicotinamide to reduce the heterogeneity of microregional perfusion, and tumor necrosis factor-alpha to destruct tumor vessel, improve vascular permeability, and thus increase the pressure gradient across the vessel (Li et al. 2012).

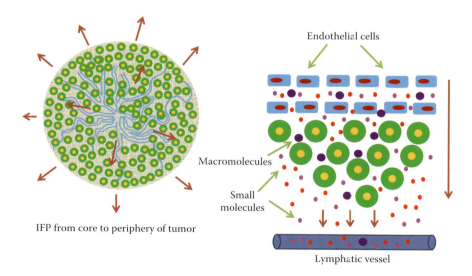

Figure 7.9 Effect of IFP on nanoparticle destination in tumor tissue.

7.10 Cellular uptake, metabolism, and excretion

Most of the nanoparticles are internalized by cells through endocytosis with its rate and mechanism dependent on cell type and varying with the size, shape, charge, aggregation state, and other surface characteristics (ligand arrangement and length) of the carriers. Liposomes or viral particles, however, enter cells either via endocytosis or by fusion with the cell membrane.

After cellular internalization, nanocarriers must release their bound drugs in order to make them therapeutically active and should themselves be eliminated from the body either via degradation as biodegradable nanoparticles or by excretion as nondegradable carriers. Inorganic nanoparticles, such as those composed of gold, silica, or titanium dioxide, or polymers, such as PEG or HPMA [N-(2-hydroxypropyl)methacrylamide] copolymer, are unlikely to be degraded by either target or nontarget cells and therefore they could remain in the body for an indefinite period in the absence of renal clearance, with its threshold for rapid excretion being approximately 5.5 nm in hydrodynamic diameter. Among the natural polymers, collagen or chitosan can undergo biodegradation while some synthetic polymers such as poly(aspartic acid) can be degraded by lysosomal enzymes (Markovsky et al. 2011). The nanoparticles of carbonate apatite and iron oxide are believed to be metabolized via dissolution in endosomes or lysosomes, respectively. The liver is another major route of excretion of the nanoparticles with >6 nm of diameter through the biliary excretion pathway.

chapter eight

Specific roles of nanoparticles in various steps of drug transport

8.1 Protection of nucleic acid– and protein-based drugs against degradation

Unlike traditional small drugs, macromolecular therapeutics, such as nucleic acids (plasmid DNA [pDNA], oligodeoxyribonucleotides, and small interfering RNAs [siRNAs]) and proteins, are quite sensitive to degradation by extreme pHs and hydrolytic enzymes (nucleases or proteases), resulting in loss of their activities during trafficking through the delivery routes particularly after oral administration and during blood circulation. The upper gastrointestinal (GI) tract (stomach) has acidic pH, whereas the lower part (small intestine) has alkaline pH. Moreover, there are various digestive enzymes in the upper and lower segments of the tract. As a result, oral delivery of nucleic acids and proteins has always resulted in massive degradation by both extreme pHs and enzymes. Nanoparticles can be employed to electrostatically complex with or encapsulate chemically or physically the macromolecular drugs in order to protect them against the degradation either in the GI tract or in the blood vessel.

8.1.1 Determinants of polyplex stability

Cationic polymers, such as poly(ethyleneimine) (PEI), the benchmark among those, are commonly used to complex with nucleic acids that carry negative charges owing to their phosphate backbone, forming polyplexes. The stability of polyplexes depends on the ratio of positive to negative charges (N/P), that is, the number of a cationic polymer to an anionic nucleic acid chain, the length of the polymer or nucleic acid, and the configuration of the polymer (linear or branched type). For instance, linear PEI (LPEI)/pDNA complexes were less stable than branched PEI (BPEI)/pDNA (Madani et al. 2011). For siRNA, which is much shorter than pDNA, the stability of siRNA/BPEI complexes was comparable to that of LPEI/pDNA while LPEI/siRNA showed less stability and inefficacy in gene knockdown. The best transfection and silencing efficiency achieved with LPEI/pDNA and BPEI/siRNA, respectively, was correlated with their similar dissociation behaviors in the presence of heparin, an anionic polymer.

91

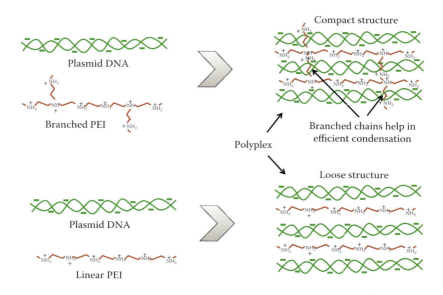

Figure 8.1 Effects of branched and linear cationic polymers on polyplex formation.

BPEI/pDNA and LPEI/siRNA, which were too stable and too unstable, respectively, did not serve as efficient carriers. siRNA, being a very small molecule, showed limited interactions with LPEI (Zhang et al. 2008). The stability of self-assembled polyplexes of siRNA could be improved by introducing cross-links between the incorporated cysteine groups within the polymer. The larger DNA molecule allows extensive electrostatic interactions with the cationic polymers conferring sufficient stability to the polyplexes without further cross-linking. For systemic administration, polyplex nanomicelles can be formed based on the self-organization of amphiphilic block copolymers composed of a polyethylene glycol (PEG) or any hydrophilic block and a cationic polymer segment, which forms the core by condensing with nucleic acid while the former, with its presence on the surface, prevents destabilization by hindering nonspecific interactions with blood components, such as nuclease and opsonins (Itaka and Kataoka 2011) (Figures 8.1 and 8.2).

8.1.2 Determinants of stability of lipoplex and other lipid-based complexes

The stability of lipoplexes, which consist of electrostatically associated cationic lipids and nucleic acids, is usually influenced depending on the initial conditions under which the lipoplexes are prepared. Thus, a liposomal

Chapter eight: Specific roles of nanoparticles in drug transport 93

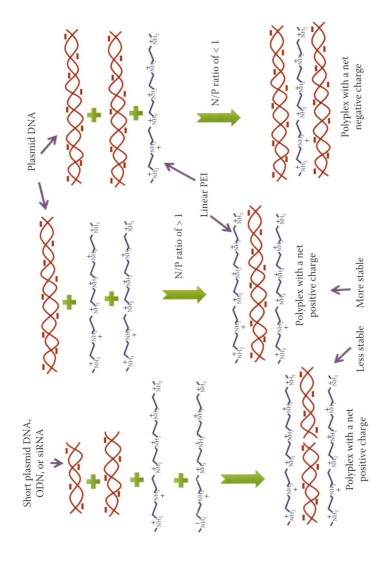

Figure 8.2 Effects of N/P ratio and nucleic acid length on polyplex formation.

formulation with highly condensed nucleic acid under low ionic strength solutions would be destabilized, dissociating the nucleic acid payloads, when exposed to physiological saline and serum (Madeira et al. 2008). The presence of PEG on the surface of lipoplexes provides a steric barrier at the bilayer surface against a variety of interactions with molecular and cellular components in blood and tissues. However, although the addition of PEGylated lipids to these formulations increases nuclei acid protection against degradation by serum and inhibits complement activation, it also hampers their transfection activities (Tros de Ilarduya et al. 2010).

As for lipopolyplexes that are formed by the addition of liposomes to preformed polyplexes, stability is mainly determined by the outer liposome shell independent of pDNA or siRNA and the cationic polymer forming the polyplexes. Since siRNA does not possess sufficient numbers of negative charges to form a sufficiently stable interelectrolyte complex, siRNA-containing lipoplex formation requires inclusion of a large polyanion such as calf thymus DNA or hyaluronic acid for optimum stability. In the case of the stable nucleic acid–lipid particle consisting of a lipid bilayer including a mixture of cationic and neutral lipids (cholesterol and fusogenic lipids) and a PEG-lipid coating, stability is mainly conferred by the outer liposomal shell (Morrissey et al. 2005).

8.2 Passive targeting to facilitate endothelial escape

In order to prevent the small-molecule drugs from crossing the blood capillaries immediately after reaching the circulation, macromolecules or nanoparticles with the ability to resist phagocytosis and renal clearance could be adsorbed to or conjugated with them for prolonging their circulation time in an attempt to pass them through the leaky vasculature of a tumor or an inflamed tissue. In a similar fashion, the macromolecular or nanoparticulate drugs could be selectively delivered to the target sites. Although small-molecule drugs passively cross the capillaries throughout the body, their retention time in tissues is usually no more than 10 minutes, apparently because of their fast diffusion rate through the interstitium, easy removal through the lymphatic system, and reentry into the circulation (along with the renal clearance), whereas the macromolecules or nanoparticles could retain in the tumor more than days to weeks because of the "lack of lymphatic drainage" of the EPR effect. Nitric oxide (NO), prostaglandins, and bradykinin were reported to facilitate the EPR effect in tumor by increasing vascular permeability as vasodilators. Another approach to the artificial augmentation of the EPR effect is through the elevation of systemic blood pressure by means of angiotensin II (AT-II) (Figure 8.3) or in combination with nitroglycerin that can be converted via nitrite to NO in hypoxic tumor tissue, thus markedly increasing both

Chapter eight: Specific roles of nanoparticles in drug transport 95

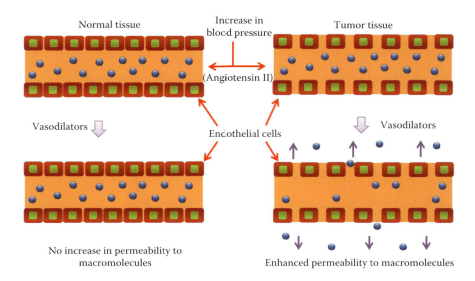

Figure 8.3 Artificial augmentation of the EPR effect.

drug delivery and therapeutic efficacy against highly refractory solid tumors (Maeda 2010). Biocompatibility and molecular size are two prime factors that regulate the passive targeting of nanoparticulate drugs to a tumor by taking advantages of the EPR phenomenon. Consequently, the surface properties and resultant size of protein corona-decorated particles have a notable influence on the outcome of passive targeting. The molecular size of the particles circulating in blood should be larger than 40 kDa, while their half-lives should be sufficiently high so as to exert the EPR effect. The luminal surface of blood vessels is negatively charged owing to the presence of sulfated and carboxylate sugar moieties, and therefore the particles with high positive charges can bind nonspecifically to the luminal surface with rapid clearance from the blood circulation (Maeda et al. 2013). The influences of particle surface properties on protein corona formation have been discussed in Chapter 6. In addition to the tumor and inflamed tissues, there are organs, such as liver, spleen, bone marrow, lymph nodes, exocrine glands, kidney, pancreas, and intestinal mucosa, that could be passively targeted depending on the extent of endothelial gaps of the respective tissue capillaries and fabricating biocompatible particles of optimized size ranges (Figure 8.4).

8.3 *Drug delivery via the lymphatic system*

The tight junctions between adjacent endothelial cells of the blood vessels with a continual basement membrane dictate the access for relatively

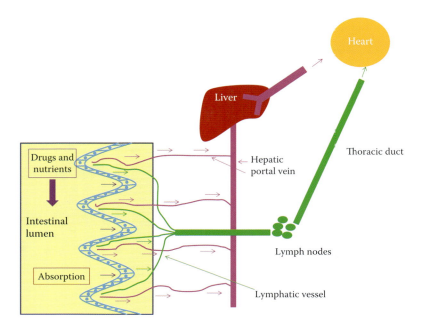

Figure 8.4 Gut-associated lymphatic vessels for therapeutic transport.

small lipophilic drugs transcellularly after administration through different routes other than the intravenous or intra-arterial injection. On the other hand, the adjacent lymphatic endothelial cells are arranged in a highly gapped and overlapped manner in the absence of tight junctions forming a porous wall, thus allowing absorption of hydrophilic nanoparticles across the lymphatic vessels, particularly from the interstitial spaces in the periphery and the gastrointestinal submucosa after subcutaneous and oral administrations, respectively. An absorption enhancer could further accelerate the absorption process by opening up the paracellular route (Kaminskas and Porter 2011).

Drug delivery through the lymphatic system can circumvent first-pass metabolism in the liver and enable targeting of drugs for the diseases that spread through the lymphatic system (e.g., certain types of cancer and human immunodeficiency virus). The lymphatic system plays an active role in dispersing infectious agents and metastatic cancer cells throughout the body with the replication of the human immunodeficiency virus (HIV) observed in macrophages of lymphoid tissue at all stages of the acquired immunodeficiency syndrome and the secondary tumors of many human cancers detected in the lymphatic system. The identification of specific markers including growth factors and receptors in the lymphatic endothelium, such as vascular endothelial growth factor (VEGF)-C, VEGF-D,

Chapter eight: Specific roles of nanoparticles in drug transport 97

VEGF-A, VEGF-D receptor (VEGFR-3), and Prox-1, would assist in specific drug targeting of the lymphatic system (McAllaster and Cohen 2011).

There are additional advantages of drug delivery via the intestinal lymphatic vessels. First, gut-associated lymphoid tissues consisting of either isolated or aggregated lymphoid follicles (Peyer's patches) provide an entry point for the drugs to the vessels. Secondly, intestinal chylomicron production is involved in delivery of lipophilic compounds into the lymphatic system (Sailor and Park 2012).

Subcutaneously injected nanoparticles that cannot penetrate the blood capillaries traverse the interstitium and enter the lymphatic capillaries. During the trafficking through the lymphatic system, a small percentage of the nanoparticles administered via any of the available routes are captured by regional lymph nodes while the rest reach the general circulation where they behave like the intravenously injected nanoparticles.

8.4 Targeting cell surface receptors and facilitated uptake

Active targeting involves covalent or noncovalent attachment of a targeting moiety (ligand) on the surface of nanoparticles in order to enable them to recognize the corresponding antigens or receptors present on particular cells. The strength of interactions between the nanoparticle surface groups and plasma membrane antigens/receptors depends on specificity as well as density of the ligands and the antigens/receptors present on a nanoparticle and a cell, respectively. Binding of an anchored ligand with the specific antigen or receptor is followed by internalization of the nanoformulation by the target cell. Although normal cells can be targeted on the basis of the approach, such as targeting of gastrointestinal tract epithelial cells with vaccine formulations in oral route, most of the attempts have been focused so far on the cells expressing disease-associated biomarkers, as in the case of cancer. The diverse moieties that have been investigated as targeting ligands include monoclonal antibodies (e.g., anti-Her2, anti-EGFR), carbohydrates (e.g., galactose), peptides (e.g., Arg–Gly–Asp or RGD), proteins (e.g., lectins, transferrin), vitamins (e.g., vitamin D), and aptamers (e.g., RNA aptamers against HIV glycoprotein).

8.4.1 Monoclonal antibody–mediated targeting

Engineered monoclonal antibodies have been used successfully not only as therapeutic drugs but also as targeting agents because of their abilities to bind with exceptional specificity to a wide range of target antigens (Figure 8.5). Indeed, antibodies have been employed for the delivery of active therapeutics including cytokines, such as L19–IL2 consisting of the cytokine interleukin-2 (IL-2) fused to a human single-chain Fv (scFv)

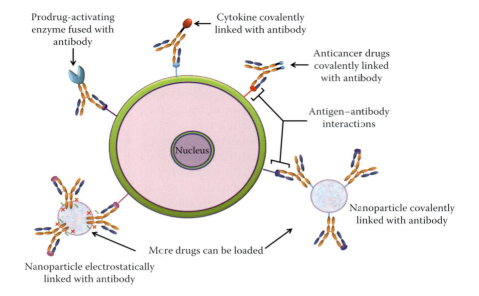

Figure 8.5 Antibodies as targeting agents for nanocarriers and small and macromolecular drugs.

antibody fragment for binding with the extra-domain B (ED-B) of fibronectin (L19) on tumor cells; prodrug activation enzymes, such as TAB2.5 having β-lactamase fused with a single-chain fragment (scFv) with specificity for a carbohydrate epitope that is overexpressed and exposed on the cell surface in a large number of solid malignancies (Alderson et al. 2006); and chemotherapy drugs, such as gemtuzumab consisting of a CD33-specific monoclonal antibody conjugated with calicheamicin, a potent antitumor antibiotic for the treatment of acute myeloid leukemia. In addition, conjugation of radioisotopes with targeting antibodies has been developed for radioimmunotherapy, such as Zevalin consisting of anti-CD20 monoclonal antibodies coupled to cytotoxic radioisotopes for the treatment of non-Hodgkin lymphoma.

Lipid- or nonlipid-based nanocarriers can be either conjugated or complexed with antibodies, thus circumventing some of the issues associated with direct drug conjugation with antibodies, such as the possible inactivation of the drug entity during the chemical process, the necessary inclusion of pH-labile or reducible linkers for release of the drug after being internalized into endosomal or lysosomal vesicles, and very low drug-to-antibody ratios (usually one drug molecule per antibody) limiting the concentration of the drug that can be targeted to the disease site. Antibody-based targeted therapeutics have frequently been directed to the cell membrane markers frequently upregulated or uniquely expressed on

Chapter eight: Specific roles of nanoparticles in drug transport   99

either the tumor or tumor-associated cells, such as EGFR, Her-2, VEGFR, prostate-specific membrane antigen (PSMA), and endocytic transmembrane receptors such as C-type lectins, for treating various types of cancer.

8.4.2 Carbohydrate-mediated targeting

Among the carbohydrates, mannose- and galactose-containing nanoparticles have been extensively used for targeted delivery of therapeutics.

Mannose receptors that are highly expressed in the cells of our immune system have been proven highly useful for improving the efficacy of vaccines. As such, mannosylated cationic liposomes were electrostatically adsorbed to melanoma-associated antigen-expressing pDNA (pUb-M) for intraperitoneal administration in order to develop a novel APC-targeted DNA vaccine against melanoma. The treatment induced significantly higher pUb-M gene expression into dendritic cells and macrophages than the unmodified lipoplex or naked DNA and strongly induced cytotoxic T lymphocyte (CTL) activity against the melanoma, thus inhibiting its growth and prolonging the survival compared with the unmodified liposomes or the standard method via intramuscular administration of naked pUb-M (Lu et al. 2007). Aiming to develop a vaccine against HIV, a single subcutaneous immunization with oligomannose-coated liposomes containing an immunodominant peptide (15 amino acids) of the envelope glycoprotein gp120 of HIV-1 was shown to induce a major histocompatibility complex class I–restricted CD8+ CTL response in mice, whereas noncoated liposomes did not (Fukasawa et al. 1998).

The hepatic uptake of intravenously administered mannosylated liposomes was preferentially mediated by liver nonparenchymal cells through mannose receptors. Thus, muramyl dipeptide (MDP), an immunomodulator when incorporated into the mannosylated liposomes, significantly reduced the number of metastatic colonies in the liver in contrast to free MDP or mannose-free MDP liposomes (Opanasopit et al. 2002).

On the other hand, the asialoglycoprotein receptor on mammalian hepatocytes provides a unique strategy for development of liver-specific carriers based on galactose-carrying liposomes, recombinant lipoproteins, and polymers. Both natural ligands, such as asialofetuin, and synthetic galactosylated or lactosylated ligands have shown significant efficacy in targeted drug delivery for acute liver injury (Wu et al. 2002).

8.4.3 Peptide-mediated targeting

The arginine–glycine–aspartic acid (RGD) motif is required for cell adhesion with ECM proteins, such as fibronectin, collagen, or vitronectin through cell surface integrins. Nanoparticles coated with RGD-containing peptides were used for targeted drug delivery to the integrin-overexpressing endothelial cells in tumor neovasculature or to the brain tumors overcoming the blood–brain barrier. Cationic lipid-based nanoparticles coupled to an integrin αvβ3-targeting ligand could also deliver mutant Raf genes selectively to angiogenic blood vessels in tumor-bearing mice, leading to regression of both primary and metastatic tumors.

Long-circulating liposomes with surface-bound PR_b, a peptide sequence that mimics the cell adhesion domain of fibronectin, were found to deliver the anticancer drug 5-fluorouracil specifically to the colon cancer cells that express the integrin α5β1, causing a greater cytotoxicity compared to the nontargeted nanoparticles (Garg et al. 2009).

Surface functionalization of siRNA-loaded PEGylated poly(propyleneimine) dendrimers by conjugating the synthetic analog of the luteinizing hormone–releasing hormone (LHRH) peptide to the distal end of PEG chains resulted in selective accumulation of the siRNA along with its gene silencing activity into the tumor of A549 human lung cancer cells that overexpress LHRH receptors (Taratula et al. 2009).

8.4.4 Aptamer-mediated targeting

Aptamers are nonimmunogenic short DNA or RNA oligonucleotides that can selectively bind antigens by secondary or tertiary structures (Figure 8.6). For example, the A10 2'-fluoropyrimidine RNA aptamers can fold into unique tertiary conformations that are capable of specific binding to the PSMA abundantly expressed on the surface of prostate cancer cells. Docetaxel-encapsulated PEGylated poly(lactic-co-glycolic acid) (PLGA) nanoparticles with A10 2'-fluoropyrimidine RNA aptamer covalently linked to the surface via an amido bond promoted docetaxel uptake by LNCaP prostate epithelial cells that express PSMA and produced greater cytotoxic effects compared to the nontargeted nanoparticles. Moreover, a single intratumor injection of the nanoformulation resulted in a complete tumor reduction in five of seven LNCaP xenograft nude mice (Farokhzad et al. 2006). In another approach, a fourth-generation polyamidoamine dendrimer with a conjugated single-stranded oligodeoxynucleotide that was subsequently hybridized with PSMA-specific A9 RNA aptamer was shown to specifically deliver doxorubicin to the prostate cancer cells both in vitro and in vivo (Lee et al. 2011) (Figure 8.7).

8.4.5 Transferrin receptor–mediated targeting

Transferrin receptor (TfR), which functions in cellular iron uptake through its interaction with transferrin, is upregulated on the surface of diverse cancer types. Therefore, incorporation of transferrin moieties or specific monoclonal antibodies (or their fragments) to the TfR onto the

Chapter eight: Specific roles of nanoparticles in drug transport   101

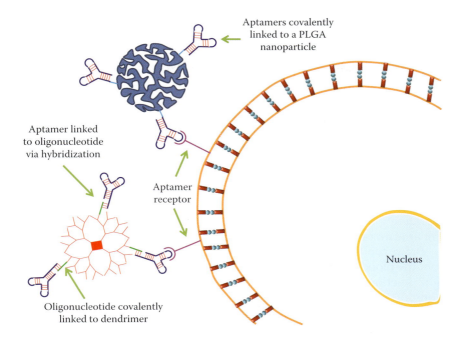

Figure 8.6 Aptamer-aided targeting.

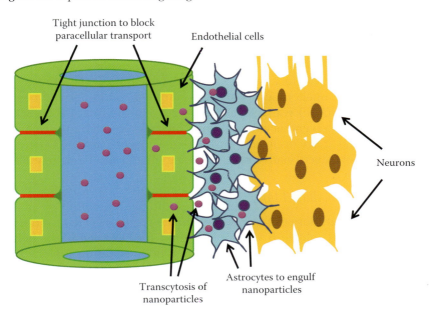

Figure 8.7 Blood–brain barrier to drug delivery.

surface of nanoparticles has been extensively used for targeted delivery of many different therapeutic agents with significant cytotoxic effects in various cancer cells in vitro and in vivo. As an example, paclitaxel-loaded PLGA nanoparticles with surface-bound transferrin promoted threefold greater uptake of the drug by human prostate PC3 cancer cells than the unmodified nanoparticles while a single intratumor injection of the functionalized nanoparticles led to the complete tumor regression in PC3 tumor-bearing mice. In another study, the PEG–TNF-α (tumor necrosis factor-alpha) conjugate with covalently attached transferrin delayed blood clearance and enhanced tumor targeting in comparison to the PEGylated TNF-α in Kunming mice bearing subcutaneous S-180 murine sarcomas (Jiang et al. 2007). On the other hand, intravenous administration of the TNF-α gene in complexation with the transferrin–PEI conjugate resulted in preferential expression of the cytokine within the tumors with significant antitumor effects in three different cancer models (Kursa et al. 2003).

Brain capillaries, which allow transport of only selective molecules from the bloodstream to the brain tissues owing to the existence of many tight junctions between the endothelial cells (Möschwitzer 2013), highly express TfR in order to mediate the delivery of iron to the brain. Intravenous injection of transferrin-anchored PEGylated albumin nanoparticles with the encapsulated antiviral drug azidothymidine demonstrated significant uptake of the drug by rat brain after 4 hours compared to the nontargeted PEG-albumin particles (Mishra et al. 2006). Since TfR could be saturated with endogenous circulating transferrin, a monoclonal antibody that will bind to an extracellular epitope other than the transferrin binding site of the TfR (thus preventing competition for binding sites between the drug targeting vector and natural ligand) should be a better option, such as OX26, a mouse monoclonal antibody against the rat TfR (Figure 8.8). As an example, for amelioration of stroke symptoms, brain-derived neurotrophic factor (BDNF) was coupled to OX26, and the final OX26–BDNF conjugate was intravenously injected into the rats subjected to permanent middle cerebral artery occlusion, with the consequence of a 243% increase in motor performance relative to BDNF alone (Zhang and Pardridge 2006).

8.4.6 Folate-mediated targeting

The majority of tumor cells overexpress folate receptors (FRs) for uptake of folate, a water-soluble B vitamin required for DNA synthesis before cell division. Thus, folate-decorated nanoparticles could be exploited for targeted delivery of molecular therapeutics to a large number of cancers. Folate moieties can be attached either by covalently binding to amino groups with appropriate linkers to carboxyl groups of the nanoparticle surface or by conjugating with the distal end of PEG chains (Figures 8.9 and 8.10).

Chapter eight: Specific roles of nanoparticles in drug transport   103

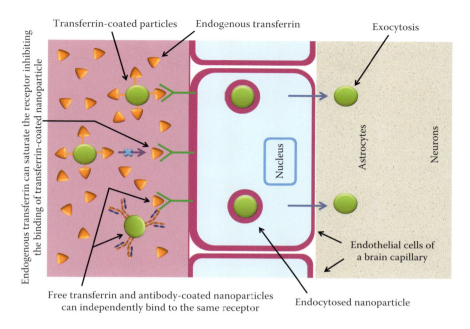

Figure 8.8 TfR-mediated transcytosis for overcoming the blood–brain barrier.

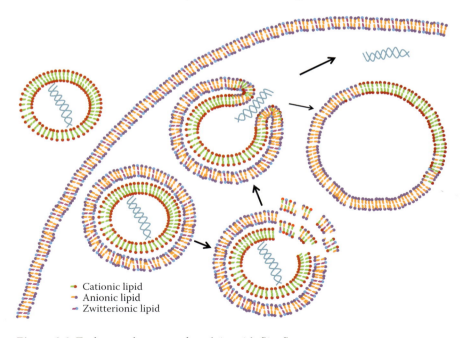

Figure 8.9 Endosomal escape of nucleic acid: flip-flop.

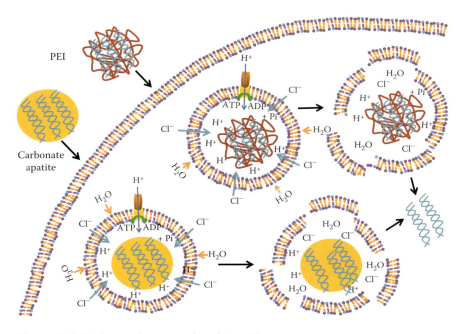

Figure 8.10 Endosomal escape of nucleic acid: proton sponge.

The FR-targeted doxorubicin-loaded liposomes (synthesized from folate–PEG–distearoyl phosphatidylethanolamine) were shown to enhance cytotoxicity relative to nontargeted liposome formulation in FR-β-expressing KG-1 human acute myelogenous leukemia cells, and the effect was enhanced by selective FR-β upregulation with all-*trans* retinoic acid (ATRA). Moreover, the FR-targeted nanoformulation showed increased antitumor activity compared to the nontargeted control in mouse ascites leukemia models while ATRA treatment further increased the effect (Pan et al. 2002). In a similar study, DOX-loaded folate-conjugated polymeric nanoparticles showed greater cytotoxicity to FR-expressing HeLa cells compared to other treated groups and better therapeutic efficacy in inhibiting tumor growth in the HeLa xenograft mouse model (Zhang et al. 2010).

8.5 Endosomal escape

The superior transfection efficacy of viral vectors over the nonviral particles is partly attributed to the ability of the former to escape the endosomes after cellular internalization. For example, in the case of adenoviruses, acidic pH–induced conformational changes on the viral capsid enable endosomal membrane lysis. Unless the nuclease- or proteinase-sensitive

Chapter eight: Specific roles of nanoparticles in drug transport 105

drugs are aimed for reaching lysosomal targets, the transport of these macromolecules from endosomes to lysosomes will lead to their massive degradation, reducing the therapeutic efficacy. There are two major strategies undertaken for nonviral vectors to facilitate their endosomal release, such as inclusion of fusogenic lipids or peptides to enable fusion with endosomal membrane and pH buffering of endosomes, which is known as the "proton sponge" effect.

8.5.1 Fusogenic lipids or peptides

In the case of cationic liposomes or lipid-based carriers, endosomal escape is promoted by the interactions between positive charges of the liposomes and anionic phospholipids of the endosomal membrane, causing some of them to displace in a "flip-flop" mechanism and membrane fusion with the resultant release of the liposomal contents into the cytosol (Uhumwangho and Okor 2005). The inclusion of neutral helper lipids, such as 1,2-dioleoyl-*sn*-glycero-3-phosphatidylethanolamine, in liposome formulation can accelerate the fusion of lipid layers by switching from a lamellar to a hexagonal phase as the pH drops in the endosome. The presence of other components, such as PEG in the liposome surface, could impair the phase transition, accompanied by an incomplete endosomal escape. Another mechanism for destabilization of the endosomal membrane using fusogenic peptides (which are short sequences of mainly basic amino acid residues) has, indeed, been inspired by the viral strategy of escaping endosomes. The change in pH inside the endosomes triggers the conformational changes of these peptides, enabling them to insert into and fuse with the endosomal membrane and releasing the therapeutic contents into the cytosol. For instance, as a model, a fusogenic peptide sequence within the hemagglutinin protein of the influenza virus has been used to design and synthesize amphipathic synthetic peptides, such as GALA and KALA, for endosomal membrane destabilization (Canton and Battaglia 2012).

8.5.2 Endosomal buffering

With an excess of uncharged amines and thus a high buffering capacity, nonviral vectors, particularly PEI, could neutralize the endosomal acidic pH by absorbing the protons that are pumped inside with the help of V-ATPase present in the endosomal membrane. The substantial accumulation of protons forces an influx of counterions, such as chloride into endosomes, causing osmotic swelling and, finally, rupturing the endosomal membrane. Among the other efficient nonviral vectors, pH-sensitive inorganic nanoparticles of carbonate apatite or Mg-substituted hydroxyapatite (CaP) particles, which are dissolved at endosomal acidic pH, have been proposed to facilitate the endosomal escape of nucleic acids by the same proton sponge effect (Chowdhury 2007; Canton and Battaglia 2012; Parhi and Suresh 2012).

8.6   Nuclear targeting

pDNA carrying gene(s) of interest must enter the nucleus in order to be transcribed into mRNAs, which subsequently transport to the cytoplasm for translating into desirable proteins. Short hairpin RNAs, which exert long-term effects on gene silencing compared to siRNAs, are also similarly synthesized in the nucleus from plasmid-based vector before being transported to the cytoplasm. Some chemotherapy drugs, such as cisplatin, can passively translocate to the nucleus for binding their targets. While dividing cells, such as cancer cells, permit entry of pDNA-based macromolecular drugs into the nucleus during mitosis when the nuclear membrane is disrupted, nondividing cells (which include majority of our cells) strictly restrict trafficking of molecules larger than 40–45 kDa through the pore complexes of the nuclear membrane, unless the molecules possess nuclear localization signals. Small carriers, such as 2.4-nm gold nanoparticles with cell-penetrating peptide, 2-and 6-nm tiopronin-coated gold nanoparticles, or polymer–drug conjugates that are <10 nm, could cross the cell nucleus without nuclear targeting sequences. However, 15-nm-sized tiopronin-coated gold nanoparticles were only found in the cytoplasm (Chowdhury 2009).

Nanoparticles that are unable to overcome the nuclear barrier could be modified with specific signals such as nuclear localization sequences (NLSs), which were originally derived from the Simian virus 40 large T antigen (Figure 8.11). Thus, 16-nm gold nanoparticles functionalized with NLS were shown to enter the nucleus after cellular uptake. By coating gold nanoparticles with nucleoplasmin, which contains two NLSs and then with the receptor for NLS, it was possible to let the macromolecules up to 39 nm cross the nuclear membrane. However, even though a nanocarrier with bound DNA can successfully reach the cell cytoplasm after cellular uptake (and endosomal escape in case of endocytosis) and finally cross the nuclear membrane, the DNA must be dissociated or released from the carrier for transcription in the nucleus. If the nanoparticles are degraded in endosomes (e.g., carbonate apatite) or fused with the endosomal membrane (e.g., liposome), the plasmid eventually released in cytosol will be prone to nuclease-mediated degradation during its waiting period until the breakdown of the nuclear membrane in the case of dividing cells or will try to passively diffuse across the nuclear pores depending on its size in both dividing or nondividing cells. In such a case, attachment of an NLS peptide to the nanocarriers will not confer any advantage to the nuclear transport of the DNA (Chowdhury 2009).

Chapter eight: Specific roles of nanoparticles in drug transport   107

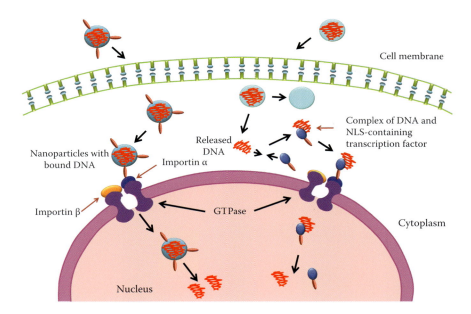

Figure 8.11 Nuclear targeting approaches.

One fascinating approach to the nuclear targeting of pDNA is to covalently link it to an NLS peptide with the help of a suitable cross-linking agent. In addition, streptavidin-conjugated NLSs can be coupled to biotinylated DNA while biologically stable peptide nucleic acid (PNA) can be used to couple NLSs to DNA by precisely hybridizing PNA to its target DNA sequence without interrupting transcription and translation. Another strategy is to incorporate (clone) specific DNA sequences, such as a 366-bp DNA containing the simian virus 40 origin of replication and early promoter, into the pDNA by recombinant DNA technology (Figure 8.11). Since a number of transcription factors that possess NLS sequence so as to enter the nucleus after activation or synthesis bind to the regulatory sequence (promoter/enhancer) of a gene during transcription, a pDNA carrying such a sequence would be able to bind to a specific NLS-containing transcription factor before the nuclear transport using the nuclear transport machinery (Chowdhury 2009).

chapter nine

Nanotechnology approaches to modulate transport, release, and bioavailability of classical and emerging therapeutics

9.1 Controlled release and bioavailability of oral nanoformulations

Controlled release of drugs from nanoformulations in the gastrointestinal (GI) tract offers several advantages, such as reducing high total dose, dosing frequency, and GI side effects along with the improvement of patient compliance. Nevertheless, there are concerns with oral controlled-release systems particularly to stop the pharmacological action of a drug in case of serious adverse effects or unwanted intolerance and maintain the reproducibility of drug action as influenced by the rate of gastric emptying and stability of the formulations. Alongside controlling drug release, nanoparticles can also help the intestinal absorption of the associated drugs through either the lymphatic vessels of Peyer's patches (Figure 8.4) or the portal veins after transport via the transcellular route (transcytosis) (Figure 9.1) or paracellular pathway (by interrupting the tight junctions) in the epithelium. Small particles ranging from 50 to 100 nm and the large particles are more likely to be taken up, respectively, by the enterocytes and the M cells (of the Peyer's patches) via transcytosis. The factors that affect the particle uptake in the intestine include the particle size, surface charge or hydrophobicity, amount or concentration, and existence of surface ligands having affinity to the specific receptors expressed on the epithelium (Ahmad et al. 2012, 2014).

9.1.1 Pharmaceutical techniques for controlling drug release

After oral administration, the stability of macromolecular drugs can be affected by the acidic and alkaline pHs, hydrolytic enzymes (such as pepsin, trypsin, chymotrypsin, carboxypeptidase, amylase, lactase, and lipase) and bile salts in addition to the drug residence time, the

110　Nanotherapeutics

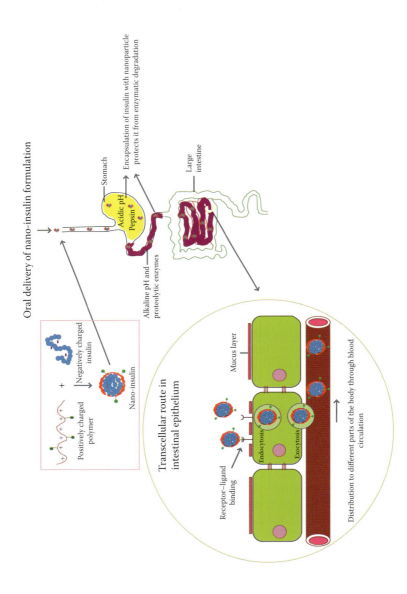

Figure 9.1 The intestinal absorption of drug-loaded nanoparticles via the transcellular route.

Chapter nine: Nanotechnology approaches to therapeutics   111

volume of fluid available in the GI tract, and its motility (Figure 1.1). Absorption, distribution, and elimination of these drugs during gut transit are determined by the tunable physicochemical properties of the nanocarrier, such as carrier chemical composition, size, interface forces, morphology, surface decoration, surface charge, and hydrophile–lipophile balance.

9.1.1.1   Enhancement of drug stability against digestive enzymes and bile salts

Generally all types of nanocarriers can protect the surface-immobilized or encapsulated drugs against the attack of metabolizing enzymes that have very limited capacity to bind their substrates (i.e., drugs) embedded on the surface or penetrate the interior of the nanocarrier systems (Nguyen et al. 2011).

The integrity of liposomes is, however, controversial because of the possible degradation of these nanocarriers by the acidity of the stomach, bile salt, and lipase. Liposomes can also be destabilized in the GI tract if H^+ ions from the gastric medium diffuse in their inner aqueous phase or the bile salt monomers penetrate into their lipid bilayers disrupting the vesicular structure. Interestingly, an increase in cholesterol quantities or addition of saturated phospholipids was shown to improve the rigidity of liposomes, thus decreasing the possibility of their enzymatic degradation. In addition, coating the surface of liposomes with polyethylene glycol (PEG) or the sugar portion of mucin protected them against bile salt destabilization and increased the stability in the GI tract. While uncoated liposomes released 20%–50% of insulin in the presence of bile salts, distearoylphosphoethanolamine-PEG– and mucin-coated liposomes released approximately 2% and 10% of insulin, respectively, within 6 h. The presence of poloxamer or PEG on the surface of nanoparticles of polylactide (PLA), poly(lactic-co-glycolic acid) (PLGA), and palmitic acid (PA) as well as the solid lipid nanoparticles (SLNs) could sterically prevent the cleavage of the ester bonds of PLA, PLGA, and PA polymers in an acidic environment or by the enzymes and the degradation of SLNs by the pancreatic lipase. The burst release of therapeutic proteins, such as insulin release, could be alleviated in a similar approach by modifying PLGA with β-CD or hydroxypropyl methylcellulose phthalate, increasing insulin bioavailability in diabetic rats (Ahmad et al. 2012).

Furthermore, various polymeric excipients, such as the chitosan–EDTA (ethylenediaminetetraacetic acid) conjugate, are capable of inhibiting the enzymes either present on the surface (e.g., aminopeptidase N) or inside the intestinal epithelial cells (e.g., cytochromes P450). The chitosan–EDTA conjugate acts by binding to Zn^{2+} ions that are essential for the enzyme activity (Ahmad et al. 2012).

9.1.1.2 Promoting drug retention and uptake at the intestinal epithelium

Nanocarriers ensuring an intimate contact with the intestinal mucosa could minimize the drug degradation by reducing the overall time of exposure to the enzymes and increasing the probability of absorption via either the paracellular or the transcellular route (Figure 9.1). Thus, oral administration of calcitonin, a polypeptide hormone to reduce blood calcium (Ca^{2+}), using the liposome carriers coated with a mucoadhesive polymer such as chitosan resulted in a significant decrease in blood Ca^{2+} level compared to the uncoated liposomes in rats. In a similar approach, modification of the PLGA surface with chitosans resulted in greater bioadhesion and improved pharmacological activity of oral insulin compared to subcutaneous insulin. Being cationic in nature, chitosan exhibits electrostatic attraction with the mucus, which is negatively charged owing to the presence of sialic acid groups. Furthermore, chitosan-based nanocarriers have the ability to cause reversible opening of tight junctions, thus enhancing the paracellular transport. Although mucoadhesion prolongs the residence time of nanocarriers at absorption sites, they can hinder their diffusion through the mucus layer toward the epithelium (Ahmad et al. 2012).

9.1.2 Strategies for drug release from nanoparticles

Three strategies for drug release from nanoparticles could be designed on the basis of the relative constant transit time in the digestive tract, the pH, and the enzyme environment.

9.1.2.1 Time-dependent strategy

This strategy is based on transit time in the GI tract of the nanoformulations from which the drug is released in a time-regulated manner depending on the rate of breakdown or collapse of the outer layer because of either swelling or the pH-responsive dissolution. The release time profile can be modulated by changing the enteric coating and the concentration or thickness of the layer. Enteric-coated solid dosage forms are used to protect the active drug substances against the acidic gastric juice, making them available after a time delay in the intestine (Figure 9.2). As an example, after oral administration in rats with induced diabetes, enteric-coated capsules containing the pH-responsive nanoparticles of chitosan and poly(γ-glutamic acid) with loaded exendin-4 (a potent insulinotropic peptide) were found intact in the stomach but dissolved in the proximal segment of the small intestine, thus releasing the drug with a maximum plasma concentration at 5 h after the treatment and stimulating the insulin secretion with a prolonged glucose-lowering effect. On the contrary, no detectable plasma exendin-4 was found after treatment with the capsule filled with the empty nanoparticles blended with free exendin-4, suggesting that although the

Chapter nine: Nanotechnology approaches to therapeutics 113

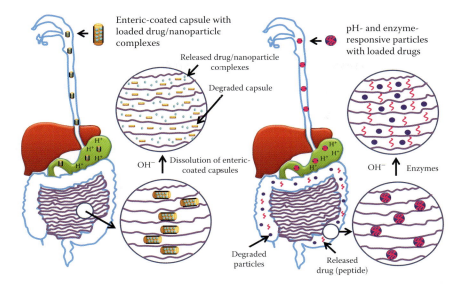

Figure 9.2 Time-, pH-, and enzyme-dependent drug release in the intestine.

capsule prevented degradation of the free exendin-4 in the acidic stomach, it was finally poorly absorbed in the intestine, whereas the pH-sensitive nanoparticles mediated adhesion and infiltration of the loaded exendin-4 into the local mucus layer and delivery in proximity to the epithelium for absorption of the free drug or the loaded one via either the paracellular or the transcellular route (Nguyen et al. 2011) (Figure 9.2).

9.1.2.2 pH- and enzyme-dependent strategy

The GI tract presents a wide range of pH values, 2–3 in the stomach, 5–6 in the small intestine, and 7 in the colon, which can be reduced in active ulcerative colitis. pH-sensitive polymers having acidic or alkaline groups can be harnessed to develop nanoparticles for pH-dependent drug release in the GI tract. For example, several polysaccharides, such as chitosan or alginate, have been used for controlled delivery purposes by virtue of their pH-dependent stability. Thus, pH-sensitive hydrogels of chitosan that swell under acidic conditions owing to protonation of their free amino groups have been used for sustained drug delivery in the stomach via the oral route. On the other hand, alginate is known to shrink at low pH of gastric environment, preventing release of the encapsulated drugs in the stomach (Chen et al. 2004). Besides, these polysaccharides can be degraded by the enzymes produced by colonic bacteria (Figure 9.2). Thus, in order to deliver an anti-inflammatory tripeptide, Lys-Pro-Val (KPV), to the colon and assess its therapeutic efficacy in a mouse model of colitis, a polysaccharide gel composed of alginate and chitosan was used to encapsulate KPV-loaded hydrophilic

PLA nanoparticles. When exposed to a gastric solution of pH 1, 2, or 3 for 24 hours, these hydrogel beads remained stable but completely collapsed upon exposure to a solution of intestinal pH 5. The release of nanoparticles from the gel (which could collapse at the colonic pH or be degraded by the colonic enzymes) and thus the release of KPV from the nanoparticles after oral delivery took place mostly in the colonic lumen compared with other parts of the GI tract such as the stomach and small intestine, thus reducing the severity of intestinal inflammation compared with the hydrogel with free KPV at an equivalent concentration (Laroui et al. 2010).

9.2 Sustained release and bioavailability of ocular drugs

Formulation of a topical ocular delivery system with consistent bioavailability and minimal adverse effects is still a challenging goal for pharmaceutical industries. Controlled-release ocular drug delivery systems have the advantages of decreasing the dose regimen and the washout effects, thereby reducing the systemic side effects as well as improving patient compliance. The formulation approaches for sustained release of ocular or ophthalmic drugs mainly include ophthalmic inserts, polymeric gels, and colloidal carriers. Posterior segment ocular diseases, such as age-related macular degeneration (AMD), cytomegalovirus (CMV) retinitis, and diabetic retinopathy, are the most prevalent causes of visual impairment in industrial countries. Nonetheless, the ocular drug market is dominated by anterior segment drugs, such as the eye drop formulations of antibiotics and anti-inflammatory agents for local administration. The potential treatment approaches for posterior segment diseases include intravitreal injections of the antisense and aptamer drugs and the monoclonal antibodies. In addition to the local delivery, the ocular tissues can be reached by the drugs after the systemic administration. While the corneal and conjunctival epithelia act as the barriers to the local drug administration, the blood–aqueous barrier limits the transport of drugs from the systemic blood to the anterior chamber, and the blood–retina barrier restricts the drug diffusion from the blood to the retina (Del Amo and Urtti 2008) (Figure 9.3).

9.2.1 Different routes for controlled release of ophthalmic drugs

9.2.1.1 Corneal route

After topical administration of eye drops that are used only for the treatment of the anterior segment disorders (such as the diseases found in the cornea, conjunctiva, sclera, iris, ciliary body, and aqueous humor), only less than 5% of the dose is absorbed into the eye across the cornea to the anterior chamber while most of it is absorbed to the systemic blood circulation via the conjunctival and nasal blood vessels. While only moderately

Chapter nine: Nanotechnology approaches to therapeutics   115

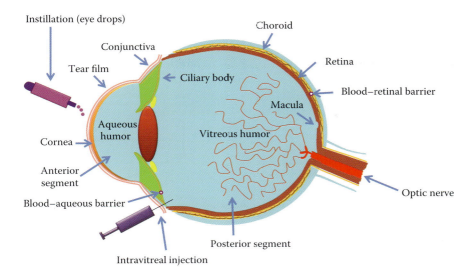

Figure 9.3 Ocular routes of delivery and barriers.

lipophilic drugs are allowed to pass through the cornea, the penetration rate of macromolecule drugs remains too low to exert a therapeutic effect (Del Amo and Urtti 2008; Kim et al. 2014).

9.2.1.2 Noncorneal routes

The noncorneal route via conjunctiva and sclera, which avoids the counterflow of aqueous humor and the lens barrier, could be a potential option for the administration of macromolecular drugs to the retina and vitreous. Although the sclera is permeable even to macromolecules, choroidal blood flow and retinal pigment epithelium (RPE) are the major barriers to the drug penetration (Figure 9.3). Intravitreal administration by injecting into the vitreous cavity can be performed to provide adequate drug concentrations in the posterior segment, provided that designed drug formulations can maintain the therapeutic drug concentrations over a prolonged period, minimizing the number of injections. Currently, many macromolecular drugs including those for the treatment of neovascular AMD are applied to treat choroidal and retinal diseases via intravitreal injections (Del Amo and Urtti 2008) (Figure 9.4).

9.2.2   *Strategies for controlled release and enhanced bioavailability of ophthalmic drugs*

Intraocular administration of the drug-containing polymeric implants by placing them intravitreally (posterior to the lens and anterior to the retina)

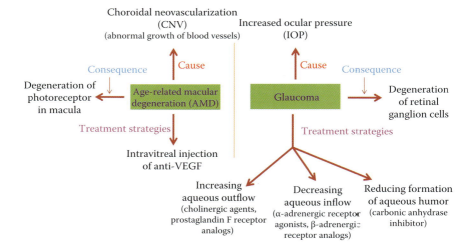

Figure 9.4 Ocular diseases: causes and current treatment strategies.

through a minor surgery provides controlled releases of the drug with prolonged activity, thus bypassing the blood–ocular barriers and avoiding the side effects associated with frequent systemic and intravitreal injections. Although nonbiodegradable implants usually provide more accurate control with longer release periods for the drug than the biodegradable ones, the former require the surgical removal of the implant with the associated risks. Another promising strategy is intravitreal injection (a less invasive procedure than the surgical implantation) of drug-encapsulated microparticles or nanoparticles to facilitate sustained drug release for weeks or even months. However, microparticles and nanoparticles may sink to the lower part of the vitreal cavity and cause clouding in the vitreous, respectively.

9.2.2.1 Micelles

A polymeric micelle of PEG-*b*-P(Lys) encapsulating a negatively charged dendritic photosensitizer that is capable of preventing aggregation of its core sensitizer (porphyrin) and thus inducing a highly effective photochemical reaction was tested in photodynamic therapy of exudative AMD. The highly selective accumulation of the formulation on choroidal neovascularization (CNV) lesions after its tail vein injection resulted in a remarkably efficacious CNV occlusion with minimal unfavorable phototoxicity (Ideta et al. 2005).

9.2.2.2 Liposomes

Two commercially available liposome products that are sprayed on the eyelids to improve the stability of the lipid layer of the tear film and

Chapter nine: Nanotechnology approaches to therapeutics 117

treatment of dry eye symptoms include ClaryMist (Savant, UK) and Ocusoft (Ocusoft, USA). Liposome technology has been used to develop light-induced systems for the retinal diseases. For example, verteporfin (Visudyne1, Novartis Pharmaceuticals, USA), the only ocular liposomal drug currently available for clinical use, works as a photosensitizer for photodynamic therapy to treat CNV and AMD. After intravenous infusion of the formulation, a nonthermal red laser is applied to the retina to activate verteporfin, which causes local damage to the neovascular endothelium, thus resulting in occlusion of the targeted abnormal vessels. Since photodynamic therapy itself induces an increased local production of vascular endothelial growth factor (VEGF) and potential reappearance of the choroidal neovessels, patients need repeated treatments with Visudyne1. The potential application of liposome technology to treat ocular diseases such as CMV retinitis was shown by intravitreal delivery of a phosphodiester oligonucleotide encapsulated within sterically stabilized liposomes resulting in sustained release of the protected small interfering RNA (siRNA) into the vitreous and the retina–choroid with a reduced distribution to nontarget tissues (sclera, lens) (Bochot et al. 2002).

9.2.2.3 Nanosuspensions

Formulation of poorly water-soluble ophthalmic drugs such as nanosuspensions will enhance the ocular absorption and bioavailability of the drugs. For instance, the nanosuspension of methylprednisolone acetate (MPA), an anti-inflammatory glucocorticoid, formulated using a copolymer showed localized and controlled ocular anti-inflammatory activity more significantly than the microsuspension of MPA in rabbits with endotoxin-induced uveitis. Hence, the copolymer nanosuspension of the MPA is a potential nanoformulation for the prevention of inflammatory symptoms in ocular diseases (Adibkia et al. 2007).

9.2.2.4 Polymeric nanoparticles

PLAs, chitosan, gelatin, alginate, and albumin were studied for efficient drug delivery to the ocular tissues. For example, PLA-based nanoparticles encapsulating two flourochromes (Rhodamine 6G and Nile red) showed a transretinal movement with a subsequent localization in the RPE cells after a single intravitreous injection, suggesting the feasibility of targeting the posterior segment of the eye with the nanoparticle formulation (Bourges et al. 2003). In another study, after topical instillation into the eyes of rabbits, chitosan nanoemulsion carrying indomethacin, a nonsteroidal anti-inflammatory drug, showed therapeutic concentration of the drug in the cornea and aqueous humor with a clearer healing effect on corneal chemical ulcer and inhibition of polymorphonuclear leukocytic infiltration (Badawi et al. 2008).

9.2.2.5 Solid lipid nanoparticles

SLNs have been evaluated as carriers for topical ocular delivery. For instance, SLNs loaded with tobramycin, an aminoglycoside antibiotic, was administered topically to rabbits with a significantly higher drug bioavailability in the aqueous humor than an equal dose of the drug delivered by standard commercial eye drops apparently because of longer retention of the former on the corneal surface and in the conjunctival sac (Cavalli et al. 2002).

9.2.2.6 Dendrimers

The unique architecture of polyamidoamine (PAMAM) dendrimers makes them potential carriers for ophthalmic drugs. PAMAM dendrimers were used to complex with puerarin, a therapeutic agent for cataracta glauca and instilled onto the center of the right and left corneas of rabbits with eventually longer ocular residence times compared with puerarin eye drops (Yao et al. 2010). In another study, PAMAM dendrimers were shown to improve the bioavailability of pilocarpine nitrate and tropicamide to a small but statistically significant extent as a result of prolonged ocular residence time. In addition to the size and molecular weight, the charge and molecular geometry of dendrimers also influence the residence time (Vandamme and Brobeck 2005).

9.2.2.7 Niosomes

Vesicles consisting of one or more surfactant bilayers enclosing aqueous spaces are called nonionic surfactant vesicles or niosomes. Niosomes offer several advantages over liposomes particularly with respect to chemical stability and cost. They have been explored as ocular vehicles for a wide range of therapeutic classes including anticholinergic (e.g., cyclopentolate), antiglaucomic (e.g., acetazolamide and timolol maleate), and antibiotic (e.g., gentamicin) with minimal signs of ocular irritation. For instance, different formulations of niosomes were investigated to incorporate acetazolamide, which has a limited aqueous solubility and poor corneal permeation. Although positively charged niosomes were found with higher efficiency of drug entrapment and good corneal permeability than their neutral and negatively charged counterparts, they demonstrated significant corneal toxicity in rabbits. A bioadhesive niosomal formulation was, however, associated with a much lesser toxicity than the positively charged niosomes while possessing the similar intraocular pressure–lowering capacity compared to the latter (Aggarwal et al. 2004). In a similar study, chitosan-coated (bioadhesive) niosomes with loaded timolol maleate, a β-blocker, demonstrated improved ocular absorption of the drug with high residence time in aqueous humor after instillation in rabbits (Kaur et al. 2010).

9.3 Sustained release and bioavailability of dermal drugs

Skin diseases are mainly caused by numerous infectious pathogens (bacteria, fungi, or viruses) or inflammatory situations and the treatment strategies depend on the type of pathogens involved and the integrity of skin layers and structures. Chronic inflammatory skin diseases (e.g., psoriasis, atopic dermatitis [AD], and allergic contact dermatitis) are associated with infiltration of inflammatory T cells with increased production of cytokines in the lesions. Skin tumors that are basically derived from hair follicles represent another concerning issue. Controlled and directed delivery can overcome the limitations with regard to patient compliance and therapeutic efficacy and minimize the side effects resulting from unspecific delivery and systemic exposure (Gupta et al. 2012).

The majority of conventional topical drug formulations are intended for a local rather than a systemic therapeutic action. Traditional topical treatments of the skin rely on the application of drug-carrying ointments or creams on the outer surface of the skin. Skin penetration of drugs from these systems is usually very low with high variations. On the other hand, transdermal patches are placed on the skin usually for systemic delivery of a wide variety of pharmaceuticals (Figure 9.5). Thus, nicotine patch releases nicotine to help quit the habit of tobacco smoking while other skin patches are designed to deliver estrogen (to prevent osteoporosis after menopause), nitroglycerin (for angina), hormonal contraceptives, antidepressants, and even pain killers. All of the drugs presently administered across skin via patches are of low molecular weight (<500 Da) and high lipophilicity. In contrast, transdermal delivery of large hydrophilic drugs is extremely limited. Although penetration enhancers can increase the transport rate through the skin layers, they also induce unwanted irritative or even toxic side effects. Transdermal transport of

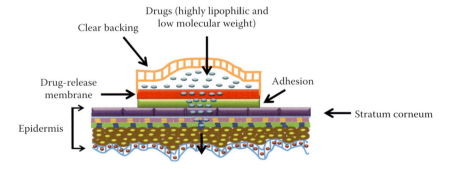

Figure 9.5 A dermal patch.

synthetic molecules and small macromolecules (<10 kDa) was significantly increased using electroporation. A combination of electroporation and chemical enhancement methods was applied for transdermal transport of larger macromolecules, such as heparin, insulin, vaccines, oligonucleotides, DNA, and microparticles (Prausnitz et al. 2004).

9.3.1 Roles of nanoparticles in sustained release

In order to target specific cells in the skin tissues, nanocarriers have the potential ability to improve penetration across the stratum corneum, which is traditionally bypassed by intravenous, subcutaneous, or intramuscular injection (Figure 9.6). Since skin acts as a negatively charged membrane, positively charged nanoparticles would strongly interact with the skin surface with better permeability and demonstrate prolonged pharmacological activity owing to sustained release of the entrapped drugs.

9.3.1.1 Liposomes

Vesicular carriers could facilitate the transport of drug and increase its concentration in various skin layers where they serve as a regional depot or reservoir, reducing the dosing frequency and the systemic side effects commonly associated with conventional topical formulations (Manosroi et al.

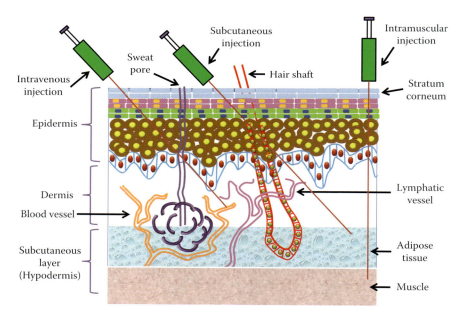

Figure 9.6 Barriers to dermal delivery and traditional ways of bypassing the obstacles.

Chapter nine: Nanotechnology approaches to therapeutics 121

2004). The first vesicular topical product is a liposomal gel of the antifungal drug econazole (Pevaryl; Cilag AG, Schaffhausen, Switzerland). In order to curtail the adverse effects of conventional formulations of fluconazole, a first-choice antifungal drug in clinically stable patients, fluconazole-loaded liposomal gel was tested via topical administration in experimentally induced cutaneous candidiasis in rats with findings of better antifungal activity owing to localized drug-depot formation and subsequent controlled release of the drug. Another example is ultraflexible liposomes (comprising phospholipids and an edge activator) with loaded miconazole, a widely used antifungal agent with limitations similar to other traditional topical formulations in treating deep-seated fungal infections. An in vivo study based on topical administration of the formulation in a rat cutaneous candidiasis model showed better antifungal activity as compared to traditional liposomes, apparently attributed to the facilitated entry into the tough barrier consisting of the subcutaneous layer (Pandit et al. 2014).

9.3.1.2 SLNs and nanostructured lipid nanoparticles

While SLNs are composed solely of a solid lipid, nanostructured lipid nanoparticles (NLCs) are formed from a combination of solid and liquid lipids. Fluconazole-loaded SLNs with optimized lipidic core were reported to promote the accumulation of the embedded drug into the upper skin layers after topical delivery, facilitating its long-term sustained release at the site of fungal infection in rats (challenged with *Candida albicans*) with improved skin tolerability and antifungal activity (Gupta et al. 2013). In a subsequent study, the same research group showed that NLCs mediated a higher level of localization with sustained release of the same drug compared to SLNs, although both of the lipidic formulations offered significantly high therapeutic efficacy with resultant low fungal burden in the skin (Gupta and Vyas 2012).

For topical application of the local anesthetic agent lidocaine (LID), both SLNs and NLCs were formulated into hydrogels and compared with a marketed LID formulation (Xylocaine gel). The in vivo efficacy test that was evaluated on guinea pigs showed that the LID SLN gel and the LID NLC gel resulted in fivefold and sixfold increase in duration of anesthesia, respectively, compared to that of the Xylocaine gel (Pathak and Nagarsenker 2009).

9.3.1.3 Polymeric nanoparticles

Polymeric nanocarriers are excellent candidates for transdermal delivery because of their high drug entrapment efficiency, controlled drug release rates, and reduced enzymatic degradation. Among the biodegradable polymers, mucoadhesive chitosan could enhance transepidermal penetration through disruption of intercellular tight junctions. To provide anti-inflammatory and antioxidant benefits in the treatment of AD, chitosan nanoparticles were used for percutaneous codelivery of hydroxytyrosol (HT), a potent antioxidant, and hydrocortisone (HC), an anti-inflammatory drug, thus revealing that HC–HT-NPs significantly accumulated both drugs in the skin layers and alleviated the signs and symptoms of dermatosis in an NC/Nga mouse model of AD (Hussain et al. 2013). Topical application of pDNA-loaded chitosan nanoparticles in another study showed sustained expression of a transgene in rat skin (Özbaş-Turan and Akbuğa 2011).

9.3.1.4 Lipid–polymer hybrid nanoparticles

A biodegradable lipid–polymer hybrid nanoparticle system composed of a cyclic head group–containing cationic lipid shell and a negatively charged hydrophobic PLGA core was employed in a self-assembly process to entrap an anti-inflammatory drug, capsaicin (Cap), into the core and electrostatically complex with siRNA against TNF-α (siTNFα [TNF being tumor necrosis factor]) onto the shell so as to treat difficult skin inflammatory conditions such as psoriasis and dermatitis. Although both Cap and anti–TNF-α siRNA have independent mechanistic actions, they worked in concert (synergism) in reducing the cutaneous inflammation after topical delivery in a psoriatic plaque–like model using the hybrid nanoparticles that seemingly enhanced the penetration as well as the release of the siTNFα and Cap in the subcutaneous layer of the skin (Desai et al. 2013). In another study, diflucortolone valerate (DFV), a potent corticosteroidal molecule commonly used in the treatment of psoriasis and AD, was loaded into lecithin/chitosan nanoparticles and was later incorporated into a chitosan gel. The topical application of the formulation led to significant accumulation and retention of the drug in the stratum corneum and epidermis of the rat skin with substantial inhibition of induced edema compared with a commercial cream of DFV (Özcan et al. 2013).

9.4 Sustained release and bioavailability of pulmonary drugs

Delivery of inhalation medications to the respiratory tract is nowadays commonly recommended for the first-line treatment of asthma, chronic obstructive pulmonary disease, and respiratory infection. While local delivery to the upper airways maximizes the concentration of a therapeutic agent in the lung tissue and thus reducing its systemic exposure, the drug formulation delivered to the lower respiratory tract promotes systemic absorption. Despite the availability of various formulations for treating respiratory illnesses, the short duration of their actions with the need to deliver them several times a day necessitates the development of a controlled-release system. Thus, prolonging the presence of 2-agonists for binding to the 2-adrenergic receptors would reduce the exacerbation of asthma during sleep. The small molecular drugs that have been

Chapter nine: Nanotechnology approaches to therapeutics   *123*

investigated in local controlled-release formulations in the lungs include antivirals, antibacterials, antifungals, cytotoxic agents, and immunosuppressives in addition to the macromolecular drugs, such as recombinant human DNase for the treatment of cystic fibrosis, alpha-1 antitrypsin as a potential treatment of emphysema, and DNA vaccines for mucosal immunity. The large therapeutic molecules designed for systemic absorption from the lower respiratory tract including insulin, calcitonin, interferons, parathyroid hormone, vaccines, gene therapy, human growth hormone, leuprolide, and potent gonadotropin-releasing hormone receptor would also benefit from a controlled-release mechanism. Among the small-molecule drugs, morphine would be advantageous for postoperative or cancer patients' pain management (Salama et al. 2009).

9.4.1   Challenges of controlled release from nanoformulations

9.4.1.1   Mucociliary escalator

The rapid mucociliary clearance system, which is a micrometer-thick viscoelastic mucus blanket on top of the ciliated epithelium in the upper respiratory tract, can trap micro- and nanoparticles and excrete them to the larynx of the GI tract along with the moving mucus, thus posing a serious limitation to any inhalable controlled-release device (Figure 9.7).

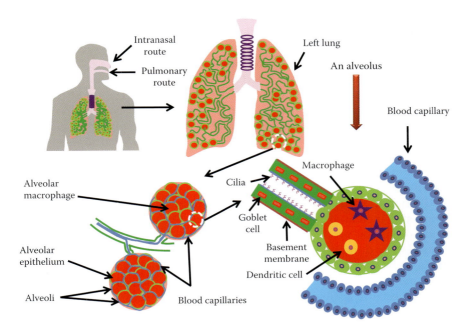

Figure 9.7 Barriers to the pulmonary route.

A substantial increase in diffusion rate of nanocarriers in mucus (muco-penetration) or minimizing their mucoadhesion could help them escape the mucociliary clearance (Kirch et al. 2012).

9.4.1.2 Transport to blood or the lymphatic system

Nanoparticles that reach the lower respiratory tract might be absorbed through the epithelium into the bloodstream or the lymphatic system, thus hampering the controlled release of drugs in that region (Labiris and Dolovich 2003) (Figure 9.7).

9.4.1.3 Clearance via alveolar macrophages or dendritic cells

Once the nanoparticles reach the lower respiratory tract, they could be engulfed by alveolar macrophages or dendritic cells resulting in their clearance via the mucociliary escalator or migration to lymph nodes, respectively (Mühlfeld et al. 2008) (Figure 9.7). PEGylation of nanoparticles or macromolecular drugs would enhance their retention time in the lungs by inhibiting their cellular uptake.

9.4.2 Controlled release and bioavailability of pulmonary drugs

9.4.2.1 Liposomes and lipid–core micelles

Liposomes are well tolerated by the lungs as a result of dissolution in the respiratory fluid of their constituent lipids, which then serve as pulmonary surfactants. In order to produce vasodilation in pulmonary arterial hypertension (PAH), a disorder of the pulmonary circulation, a Rho-kinase inhibitor (fasudil) was encapsulated in liposomal vesicles before being subjected to intratracheal instillation in a rat model of PAH. The fasudil-loaded liposomes reduced pulmonary arterial pressure in a controlled-release manner. Moreover, an approximately 10-fold increase in plasma half-life of the drug was observed when liposomal fasudil was administered as aerosols compared to intravenous fasudil (Gupta et al. 2013). Similarly, cell-penetrating and lung-homing peptide-conjugated PEG–distearoyl phospho-ethanolamine micelles were used to deliver fasudil through intratracheal instillation in a rat model of PAH, resulting in its prolonged retention in lungs and increased half-life in circulation. Overall, the micelle formulation of fasudil produced approximately two- and threefold increase in vasodilatory duration and lung targeting, respectively, as compared with the liposomal formulation (Gupta et al. 2014). In another study, liposomes were employed for pulmonary delivery of ciprofloxacin, a potent and broad-spectrum antibiotic, for assessing the controlled release and therapeutic efficacy of the drug against intracellular infection. Aerosol administration of the nanoformulations by jet nebulization resulted in significantly higher drug levels and prolonged drug retention in the lower respiratory

Chapter nine: Nanotechnology approaches to therapeutics 125

tract compared to the free drug, as well as complete protection to mice against a pulmonary lethal infection of *Francisella tularensis* (Wong et al. 2003). For systemic absorption of insulin through the pulmonary route, delivery of aerosolized insulin-encapsulated liposomes led to an increase in drug retention time in the lungs and a reduction in plasma glucose level in diabetic mice (Huang and Wang 2006).

9.4.2.2 Solid lipid nanoparticles

SLNs have shown increasing potential as an efficient and nontoxic mediator for pulmonary delivery and controlled release of insulin. For example, insulin-encapsulated SLNs after intrapulmonary administration in diabetic rats were found homogeneously distributed in the lung alveoli. Plasma insulin concentration increased dramatically after the administration, reaching a maximum value within 4 hours with a concomitant reduction in fasting plasma glucose level. Moreover, in the presence of SLNs, the pharmacological bioavailability of insulin in lungs was remarkably enhanced (Liu et al. 2008). A similar report showed that intratracheal instillation of insulin-loaded SLNs in diabetic rats led to a prolonged hypoglycemic effect with an enhanced pharmacological bioavailability in the lungs (Bi et al. 2009).

9.4.2.3 Polymeric nanoparticles

Mucoadhesive PLGA nanospheres were developed by surface modification with chitosan for pulmonary delivery of elcatonin, a polypeptide hormone of a calcitonin derivative that acts to reduce blood calcium level, via the trachea of guinea pigs using a nebulizer. After the pulmonary administration, the nanospheres were more slowly eliminated from the lungs than unmodified ones because of the adherence of the former to the bronchial mucus and lung tissue, promoted sustained drug release at the adherence site, and thereby enabled reduction of blood calcium levels to 80% of the initial calcium concentration with a prolonged pharmacological action (Yamamoto et al. 2005). In a very similar study, pulmonary delivery of calcitonin-loaded poly(methyl vinyl ether maleic acid) nanoparticles in rats using a microsprayer device enhanced and prolonged the hypocalcemic effect of the drug (Varshosaz et al. 2014). In order to reduce the drug dosage frequency and improve patient compliance in tuberculosis chemotherapy, lectin-functionalized PLG nanoparticles encapsulating antitubercular drugs isoniazid, rifampicin, and pyrazinamide were administered to guinea pigs through the aerosol route (as well as the oral route). The coated nanoparticles prolonged the half-lives of the three drugs in blood compared to the uncoated formulations and more substantially than the free drugs, while retaining the antibacterial activity in the lungs and spleens of *Mycobacterium tuberculosis*–infected guinea pigs (Sharma et al. 2004). In another study, a PAMAM dendrimer with conjugated methylprednisolone, an important

126 *Nanotherapeutics*

corticosteroid used in inhalation-based therapy of asthma-associated lung inflammation, was shown to enhance the drug's ability to decrease allergen-induced inflammation probably by improving drug residence time in the lung after airway delivery in a pulmonary inflammatory murine model (Inapagolla et al. 2010).

9.5 Intracellular and extracellular transport vehicles

9.5.1 Gene, siRNA, and ODN

Intracellular delivery of therapeutic genes or antisense DNA or RNA sequences or "gene therapy" is a potential revolutionary approach paving the way for treating critical human diseases at the genetic level. Since nucleic acids are vulnerable to degradation, or susceptible to renal clearance particularly in the case of siRNA or ODN, they often require nanocarriers for being safely carried to their designated extracellular or intracellular destinations. Despite possessing advantages and disadvantages as discussed earlier, both viral and nonviral vectors have been extensively explored as nucleic acid carriers through many preclinical and clinical trials. However, no gene therapeutics have so far been approved by the US Food and Drug Administration (FDA) except for Glybera used in Europe.

9.5.1.1 Retroviral vectors

Among the retroviruses, only lentiviruses can replicate in nondividing cells, making them attractive for transducing human cells, most of which are nondividing. Being a member of retroviruses, they can stably integrate their genomes into the host chromosome, which is both advantageous and disadvantageous by enabling sustained gene expression and increasing the risk of proto-oncogene activation with cancer development. Targetable lentiviral vectors can be produced by fusing a ligand protein or antibody to viral glycoproteins through genetic engineering. Thus, the lentiviral vector with the MLV Env glycoprotein engineered to display anti-CD3 single-chain antibody could efficiently transduce primary lymphocytes. Apart from the specific uptake after receptor–ligand engagement, lentiviral vectors with a tissue-specific promoter (such as neuron-specific, dendritic cell–specific, tumor angiogenesis–specific, or hepatocyte-specific promoters) can be used to target specific cell type. Integration-deficient lentiviral vectors can reduce insertional mutagenesis and promote short-term transgene expressing in nondividing cells. Preclinical animal studies involving lentiviral vectors were carried out for correction of β-thalassemia and sickle cell anemia, hemophilia B, Parkinson's disease, cystic fibrosis, and spinal muscular atrophy. The lentiviral genome harboring short hairpin RNA (shRNA) to target the rev and tat mRNAs of HIV-1, a nucleolar-localizing TAR RNA decoy,

Chapter nine: Nanotechnology approaches to therapeutics 127

and an anti-CCR5 ribozyme-expressing cassette were used to successfully inhibit HIV-1 infection (Sakuma et al. 2012).

9.5.1.2 Adenoviral vectors

The adenoviral vector is the most clinically used vector, with approximately 25% of all gene therapy trials currently based on this. Both dividing and nondividing cells can be efficiently transduced with the vector. Adenovirus binds to most cells via interaction between the virus fiber "knob" and cellular coxsackievirus–adenovirus receptor. After cellular uptake and translocation to the nucleus, the vector genome remains episomal, and therefore, when delivered into relatively quiescent tissues, such as the liver, muscle, or brain, stable production of therapeutic proteins (e.g., coagulation factors, $\alpha 1$ antitrypsin, or erythropoietin) was noticed throughout the lifetime of a mouse (Wang et al. 2010). Despite high levels of gene transfer into many tissues after systemic delivery, the vector can induce severe toxicity that stems from an immediate innate immune response and a secondary antigen-dependent response. Because of deletions of additional viral genes, the second- and third-generation adenoviral vectors are less toxic than the first-generation vectors. Among the gene therapy trials (>400) that have been or are being conducted with human adenoviral vectors, the majority are for the treatment of a variety of cancers. Quite a large number of different transgenes have been incorporated into the adenoviral genome for clinical trials, such as ornithine transcarbamylase, cystic fibrosis transmembrane conductance regulator (CFTR), VEGF, *Escherichia coli* cytosine deaminase, HIV proteins (gag, pol, nef), p53, thymidine kinase, granulocyte macrophage colony-stimulating factor, human interleukin-12 (IL-2), human interferon, human fibroblast growth factor 4, heat shock protein 70, TNF, prostate-specific antigen and malaria circumsporozite protein, Her-2, and hepatitis C virus nonstructural proteins, among others (Wold and Toth 2013).

9.5.1.3 Adeno-associated viral vectors

Adeno-associated virus (AAV), one of the most attractive gene therapy vectors, infects primates and is nonpathogenic. The viruses usually bind to heparan sulfate proteoglycans on the cell membrane before cellular internalization. Like adenoviruses, they can induce strong immune responses, infect both dividing and nondividing cells, and almost always remain episomal in the nucleus. The short genome size (4 kb) of the vector presents a limitation in carrying long therapeutic genes. Clinical trials using AAV vectors have yielded promising results. For example, in clinical trials for familial lipoprotein lipase (LPL) deficiency, intramuscular injection of an AAV1 vector encoding the gain-of-function LPLS447X variant resulted in persistent gene expression and thus sustained decreases in the incidence of pancreatitis, a disease in which the pancreas becomes inflamed. On the basis of these outcomes and the safety profile, the product known as "alipogene tiparvovec" (Glybera) received marketing approval in 2012 in the European Union (the first approved gene therapy in Western nations). In another example, administration of an AAV1 vector containing ATP2A2, which encodes sarcoplasmic/endoplasmic reticulum calcium ATPase 2, was shown to improve various key outcomes in patients with advanced heart failure (Kotterman and Schaffer 2014). AAVs can also be employed to deliver either a shRNA or an artificial miRNA. However, there are currently no AAV-mediated RNAi vectors in clinical development. Among the preclinical trials, one example is intravenous injection of a recombinant AAV expressing three shRNAs in order to treat hepatitis C virus infection, thus resulting in sustained expression of the shRNAs.

The common route harnessed by different viral vectors to produce therapeutic proteins and shRNAs is shown in Figure 9.8.

9.5.1.4 Nonviral vectors

Although ~70% of gene therapy clinical trials have been carried out so far with viral vectors, they have severe limitations particularly in relation to carcinogenesis and immunogenicity. Nonviral vectors are currently being sought for their increased safety profile and capacity to deliver larger genetic payloads despite the issues pertaining to their much lower efficacy

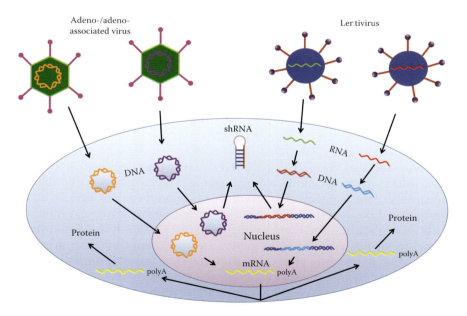

Figure 9.8 Viral vectors for generation of therapeutic proteins and shRNAs in target cells.

Chapter nine: Nanotechnology approaches to therapeutics 129

record in comparison to the former. Many nonviral physical methods of therapeutic DNA delivery, such as gene gun, electroporation, hydrodynamic delivery, sonoporation, and magnetofection, are generally less applicable to systemic delivery in humans than in small experimental animals. As a consequence, a range of synthetic delivery vectors that are relatively easier to synthesize than viral vectors have been developed. On the other hand, although the nuclease susceptibility and immunogenicity of unmodified siRNA can be prevented by modifying the chemical structure of the siRNA strands through replacement of the 2′-OH group of ribose with -O-methyl or 2′-fluoro groups, incorporation of locked or unlocked nucleic acids, or substitution of phosphorothioate linkages in place of phosphodiester bonds, the backbone modifications might lead to inappropriate strand selection by RISC (RNA-induced silencing complex) and partial hybridization to nontarget mRNAs, thus resulting in nonspecific gene silencing and consequential off-target effects. Hence, the most convenient approach to prevent the degradation and, additionally, escape the renal clearance of siRNA as well as block the siRNA-triggered immune recognition is to complex the siRNA with the nanoparticles (Yin et al. 2014).

9.5.1.4.1 Strategies for long-term and tissue-specific expression of nonviral DNA Plasmids that are routinely used as expression vectors carrying the gene(s) of interest in nonviral gene delivery study remain episomal without integrating with the host chromosome after nuclear translocation, reducing the risk of insertional mutagenesis in contrast to the retroviral vectors. The choice of a regulatory sequence of a desirable gene, that is, enhancer or promoter, has a great impact on both the intensity and the duration of transgene expression. Although viral enhancers and promoters derived from CMV, respiratory syncytial virus, and simian virus 40 are frequently used to induce high levels of gene expression in a range of mammalian cell and tissue types, the expression is often transient, limiting the overall therapeutic potential. Constitutive mammalian promoters, such as the human ubiquitin C and the eukaryotic translation elongation factor 1 alpha 1 promoters, can be harnessed to promote the persistent expression. In addition, numerous cis-acting sequences including various polyadenylation signal introns and scaffold/matrix attachment regions have been shown to increase the level and the duration of transgene expression. Moreover, DNA size and topology can also affect the gene expression profile with the small covalently closed circular plasmids leading to greater levels of transgene expression than the large or linearized plasmids and the compact DNA vectors that lack a bacterial backbone (minicircles) maintaining superior levels and duration of gene expression relative to the full-length DNA plasmids. Another strategy to prolong the expression is via the transgene integration into the chromosome utilizing a variety of transposition systems, such as recombinases phiC, PiggyBac, and Sleeping

Beauty. On the other hand, tissue-specific promoters or enhancers, such as the alpha-fetoprotein enhancer or albumin promoter, which regulates gene expression exclusively in the liver, can assist in minimizing the unwanted transgene expression in other tissues, thus increasing the overall delivery efficacy and minimizing the off-target effects (Yin et al. 2014).

9.5.1.4.2 Lipid-based DNA/siRNA vectors Cationic liposomes have traditionally been the most commonly used nonviral delivery systems for plasmid DNA. Various cholesterol or PEG-modified cationic liposomal formulations are being tested clinically, including DOTAP [1,2-bis(oleoyloxy)-3-(trimethylammonio)propane]–cholesterol for the delivery of the fus1 tumor suppressor gene and GL67A–DOPE (1,2-dioleoyl-*sn*-glycero-3-phosphoethanolamine)–DMPE (1,2-dimyristoyl-*sn*-glycero-3-phosphoethanolamine)–PEG for the delivery of the CFTR (pGM169) gene in patients with non–small-cell lung cancer and cystic fibrosis, respectively (Yin et al. 2014). Despite their ability of efficiently condensing siRNA and silencing target genes in vitro, cationic liposomes have demonstrated limited success for in vivo gene downregulation (Ozpolat et al. 2014). Nanoliposomes based on neutral DOPC (1,2-dioleoyl-*sn*-glycero-3-phosphatidylcholine) were shown to deliver siRNA targeting either EphA2, FAK, neuropilin-2, IL-8, TMRRS/ERG, EF2K, or Bcl-2 with remarkable antitumor efficacy in orthotopic and subcutaneous xenograft tumor models of various cancers after intravenous administration. These neutral liposomes were able to deliver siRNA in vivo into tumor cells 10-fold more effectively than the cationic liposomes (DOTAP), without any induction of pro-inflammatory cytokines and reactive oxygen species (Ozpolat et al. 2014). In addition, several formulations of stable nucleic acid–lipid particles are currently in clinical trials for siRNA-assisted knockdown of target genes. For example, for the treatment of hypercholesterolemia, ALN-PCS02 (Alnylam Pharmaceuticals), which targets the proprotein convertase subtilisin/kexin type 9 (*PCSK9*) transcript, has been reported to substantially reduce expression of their target genes and subsequently of low-density lipoprotein cholesterol in a Phase I trial without demonstration of serious adverse effects (Yin et al. 2014).

9.5.1.4.3 Polymeric DNA/siRNA vectors Cationic polymers constitute another attractive class of nonviral DNA vectors partly because of their chemical diversity and potential for functionalization. Although poly(L-lysine) (PLL) was one of the earliest cationic polymers investigated as a DNA vector, in the absence of a lysosomal disruption agent, such as chloroquine, generally it shows very poor transfection activity, since, at physiological pH, all of its amine groups tend to be positively charged and therefore possess low endosomal buffering capacity. PEI and its variants, on the other hand, are relatively more efficient in intracellular transgene delivery even in the absence of chloroquine, since PEI having a high charge

Chapter nine: Nanotechnology approaches to therapeutics 131

density at reduced pH values could aid in condensation and endosomal escape of DNA. However, the transfection efficiency and cytotoxicity of PEI strongly depend on its structural properties, such as molecular weight and linear versus branched forms. To reduce the substantial cytotoxicity of PEI, a range of modifications have been investigated including block copolymers of PEG and PEI for improved stability and biocompatibility, degradable disulfide cross-linked PEIs for reduced toxicity, and alkylated PEI for increased potency (Yin et al. 2014). In mice, intravenous injection of PEI–DNA polyplexes was reported to facilitate transgene delivery to the lungs. In humans, PEI has been studied for local gene therapy of various cancers, such as bladder, ovarian, and pancreatic cancers; multiple myeloma; B-cell lymphoma; and pancreatic ductal adenocarcinoma. A PEG–PEI–cholesterol lipopolymer is under clinical investigation for immunotherapy of ovarian and colorectal cancers through forced expression of the cytokine IL-12. To address issues of efficacy and toxicity associated with PLL and PEI, numerous other polymers are currently being evaluated preclinically for DNA delivery. For example, cyclodextrin polymer (CDP)–siRNA formulation has been evaluated in several therapeutically relevant animal models as well as in Phase I clinical trials in which the ribonucleotide reductase M2 mRNA was targeted in patients with solid cancers with a consequential reduction in target mRNA levels (Yin et al. 2014).

The roles of intracellular delivery of DNA and siRNA with the help of nonviral vectors in gene expression and knockdown have been illustrated in Figure 9.9.

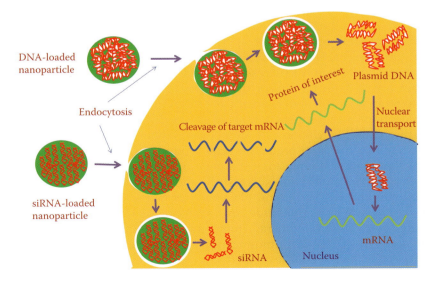

Figure 9.9 Nanoparticle-mediated intracellular delivery of DNA and siRNA.

9.5.2 Synthetic peptides and recombinant proteins

Advances in molecular biology research have led to an increased understanding of the roles of peptides and proteins in clinical therapy while the recombinant DNA technology has conferred large-scale manufacturing capability, thereby providing us immense opportunities to increase the range of biopharmaceutical-based nanotherapeutics, such as proteins, synthetic vaccines, and monoclonal antibody for addressing various critical diseases.

9.5.2.1 Current obstacles to protein delivery

Despite the revolutionary progress in the development of biopharmaceuticals, effective and convenient delivery of this rapidly expanding class of therapeutic agents in the body remains to be a major challenge particularly because of their poor permeability through biological membranes, large molecular size, wide tissue distribution, potential immunogenicity, and short plasma half-lives attributed to enzymatic degradation, aggregation, and renal clearance, thus necessitating frequent dosing and finally resulting in increased cost and patient noncompliance. The most common routes for delivery of therapeutic proteins clinically are intravenous, intramuscular, and subcutaneous (Ahmad et al. 2014). In recent years, tremendous efforts are being made to improve delivery of these therapeutic agents.

9.5.2.2 Major approaches to overcome the shortcomings of protein delivery

9.5.2.2.1 Chemical modifications of proteins and peptides

9.5.2.2.1.1 Conjugation with polymers Conjugation with polymers represents an attractive approach to enhance the blood circulation time (i.e., half-life) and reduce the immunogenicity of peptides and polypeptides. The most successful approach is the attachment of PEG to either one or both termini of the peptides or proteins. A PEG mass of 40 to 50 kDa is sufficient to increase the size of a small peptide through conjugation so as to prevent its renal clearance via glomerular filtration. Furthermore, PEGylation can protect the peptide from proteolytic degradation by forming a protective shell around it and lower its immunogenicity by sterically masking the potential antigenic sites (Figure 9.10). Conjugation with PEG was successfully shown to increase the half-life of calcitonin and GLP-1. Several PEGylated protein products are currently marketed, such as peginesatide, an erythropoiesis-stimulating agent, and many are under development. Among the other polymers, polysialic acid can reduce the potential immunogenicity by forming a glycocalyx around the peptide and XTEN (extended recombinant polypeptide) has been applied to several peptide drugs including exenatide, a GLP-1 agonist showing a long terminal half-life of the potential drug in clinical trials.

Chapter nine: Nanotechnology approaches to therapeutics   133

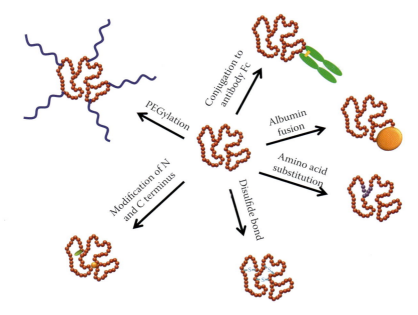

Figure 9.10 Chemical modifications of proteins and peptides.

However, although conjugation with polymers is often associated with a decrease in therapeutic activity of a peptide or protein, the increased half-life can have a far greater influence on the overall clinical efficacy than the decreased potency (Diao and Meibohm 2013).

9.5.2.2.1.2 Conjugation to antibody Fc portion   Since, by interacting with the neonatal Fc receptor (FcRn), the Fc portion of a human antibody causes endosomal recirculation of the entire molecule and protects it against metabolism and elimination, conjugation of a peptide to the antibody Fc portion is another tool to extend the half-life of the peptide while retaining its biological action (Figure 9.10). For example, romiplostim, which was approved by the FDA for chronic idiopathic thrombocytopenic purpura, is a thrombopoietin (TPO) mimetic peptibody comprising two human immunoglobulin IgG1 Fc domains, each of which is covalently linked to a peptide chain that contains two TPO receptor-binding peptides (Diao and Meibohm 2013).

9.5.2.2.1.3 Albumin fusion   With an unusually long half-life (approximately 19 days), wide distribution, and negligible immunogenic potential, albumin can enhance the half-lives of short-lived peptides (Figure 9.10). Two examples of peptide–albumin fusion proteins currently in clinical trials are CJC-1134 and albiglutide, which are, respectively, a recombinant exenatide–albumin fusion protein and a dipeptidyl peptidase-4-resistant GLP-1 dimer fused to human albumin. The half-lives in humans are 8 and 6–7 days for CJC-1134 and albiglutide, respectively, which are considerably longer than other marketed GLP-1 analogs, such as exenatide and liraglutide, with half-lives of 2.5 and 13 hours, respectively (Diao and Meibohm 2013).

9.5.2.2.1.4 Amino acid substitution One of the most widely used strategies to increase the stability of therapeutic peptides in blood is the substitution of amino acids that make the peptides susceptible to proteolytic enzymes, with unnatural or D-amino acids that are three-dimensional mirror images of natural L-amino acids (Figure 9.10). Examples of such modified therapeutic peptides include octreotide, GnRH antagonists and agonists, ipamorelin, GLP-1 analogs (exenatide and liraglutide), and D-peptides (rotigaptide and PIE12-trimer) that solely consist of D-amino acids. Octreotide, a cyclic octamer synthetic analog of natural somatostatin, which retains the critical Phe–Trp–Lys–Thr portion of somatostatin with the tryptophan (Trp) residue in the D configuration, has greater potency than somatostatin and a prolonged half-life (113 minutes) compared to the native somatostatin having a very short half-life (2–3 minutes (Diao and Meibohm 2013). Synthetic decapeptide GnRH antagonists (cetrorelix, degarelix, and abarelix) and GnRH agonists (triptorelin, leuprorelin, buserelin, goserelin, and nafarelin) all contain unnatural amino acids and show prolonged half-lives compared to GnRH. Among the other therapeutic peptides with amino acid replacement and consequential extended half-lives, ipamorelin is a selective growth hormone–releasing pentapeptide, exenatide is a synthetic version of exendin-4 (a naturally occurring 39–amino acid peptide) that mimics the pharmacological action of GLP-1, liraglutide is a homolog of native GLP-1 with an additional fatty acid chain that allows the peptide to bind to albumin, rotigaptide is a D-peptide analog of stable gap junction conduction-enhancing antiarrhythmic peptide analog, and the PIE12-trimer consists of three D-peptides (PIE12) linked together in order to block the HIV gp41 pocket critical for HIV entry into the cell. Amino acid substitution has also been successfully applied to produce long-acting insulin with full biological activity, such as insulin glargine, which has poor solubility in physiological pH but high solubility in pH 4.0. As a result, on subcutaneous injection, the insulin glargine solution precipitates, thus producing a relatively constant 24-hour concentration profile (Diao and Meibohm 2013; Kumar et al. 2006).

9.5.2.2.1 5 Modification of peptide terminus Chemical modification of the amino (N) or carboxyl (C) terminus through N-acetylation and C-amidation can reduce the susceptibility of a therapeutic peptide to exopeptidase-mediated proteolysis (Figure 9.10). For example, the

Chapter nine: Nanotechnology approaches to therapeutics 135

FDA-approved tesamorelin, which consists of a synthetic 44–amino acid sequence of human growth hormone–releasing hormone (GHRH) with a hexenoyl moiety attached to the tyrosine residue at the N-terminus, is used for the reduction of excess abdominal fat in HIV-infected patients with lipo-dystrophy. Tesamorelin was found to be resistant to dipeptidyl peptidase-4 degradation with its half-life (1 hour) in healthy human subjects being much longer than that of natural GHRH (6.8 minutes) (Diao and Meibohm 2013).

9.5.2.2.1.6 Disulfide-rich peptides Disulfide linkages can affect the folding and structural stabilization of therapeutic peptides (Figure 9.10). Currently, there are several disulfide-rich peptides in the market, such as lepirudin, ziconotide, calcitonin, octreotide, and eptifibatide. Linaclotide, a 14–amino acid peptide recently approved by the FDA and the European Medicines Agency for irritable bowel syndrome with constipation, has three disulfide bridges, which make it stable enough to elicit a significant, dose-dependent increase in GI transit rates and subsequently work as a guanylate cyclase C agonist after oral administration in rats (Diao and Meibohm 2013).

9.5.2.2.1.7 Lipidization By linking the carboxylic group of a fatty acid to the amine group of the N-terminal residue of a peptide through a stable amide bond, the stability of the peptide can be effectively increased (Figure 9.10), although such modification often compromises the bioactivity by reducing the affinity of the peptide for receptor binding owing to an alteration of its conformation. Octreotide, an octapeptide that mimics natural somatostatin pharmacologically, was chemically modified using reversible aqueous lipidization technology to increase its plasma half-life for greater therapeutic potential in the treatment of liver cancers, such as hepatocellular carcinoma. Several other proteins such as soma-tostatin analog, desmopressin, and salmon calcitonin were lipidized to improve their biological half-lives (Kumar et al. 2006).

9.5.2.2.2 Controlled delivery of proteins from injectable formulations An injectable protein delivery system could provide sustained release of a therapeutic protein over a period of days, weeks, or even months to enhance therapeutic action and patient compliance. The challenges that should be addressed to develop a clinically viable controlled-release device for protein drugs include optimum drug loading, maintaining the integrity of the drug during processing and storage, choosing the right polymer with acceptable safety profile, and, finally, controlling the drug release profile without a burst effect.

9.5.2.2.2.1 Polymeric microspheres Many of the parenteral controlled-release products of proteins available in the market are based on polymeric microspheres of 10 to 100 μm in diameter for either intramuscular or subcutaneous administration. There are several methods for microencapsulating proteins or peptides, such as emulsification followed by solvent evaporation/cross-linking, emulsion polymerization, spray drying, supercritical fluid technology, ProLease, PolyShell, and electrospray. The release of protein- or peptide-based drugs from the microspheres depends on the type of polymers used in microsphere preparation, physicochemical properties of drugs, interactions between polymers and drugs, drug concentrations, and the encapsulation process. The major limitations with the microsphere formulation of proteins/peptides include sensitivity of the drugs toward denaturation, aggregation, oxidation, and cleavage, in addition to inadequate drug loading in the microspheres. Lupron Depot (leuprolide acetate), the first depot introduced into the market, can be applied subcutaneously or intramuscularly as a GnRH agonist for the treatment of severe endometriosis and advanced prostate cancer (Kumar et al. 2006).

9.5.2.2.2.2 Injectable in situ gel forming solution Polymeric depot forming or gel forming solution technologies have several unique advantages over microspheres. Atrigel, a depot forming delivery system, consists of biodegradable polymers (dissolved in biocompatible organic solvent, such as *N*-methyl-2-pyrrolidone) that convert into depot after intramuscular injection. Atrigel technology has been used in the production of Eligard, a formulation of leuprolide that acts as a luteinizing hormone–releasing hormone agonist for the treatment of hormone-responsive cancers. On the other hand, Ascentra is a temperature-sensitive polymer depot technology utilizing ReGel, which is a thermosensitive, biodegradable triblock copolymer composed of PLGA and PEG and formulated exclusively with water, making it more suitable than Atrigel in protein formulation and delivery. Immediately upon injection and in response to body temperature, an insoluble gel depot is formed at the injection site, enabling long-term release of the associated therapeutic agents. ReGel has been evaluated for the delivery of few proteins, such as hGH, IL-2, and insulin (Kumar et al. 2006).

9.5.2.2.2.3 Lipid-based protein delivery systems Conventional liposomal technologies were found to be unsuitable for encapsulating a large amount of water-soluble peptides or proteins. To overcome the hurdle, a multivesicular lipid-based drug delivery system, known as DepoFoam technology, has been reported for fabrication of sustained-release formulations for therapeutic proteins and peptides with high loading capacity. DepoFoam formulations of insulin, leuprolide, enkephalin, and octreotide have been developed. Pharmacodynamic studies in rats showed a sustained therapeutic effect of the formulations over a

Chapter nine: Nanotechnology approaches to therapeutics   137

prolonged period (Ye et al. 2000). The first clinically available formulation contains the antineoplastic agent cytarabine (DepoCyt) for the treatment of malignant lymphomatous meningitis. Intrathecal injection of DepoCyt was reliably shown to result in the sustained release of cytarabine at cytotoxic concentrations in cerebrospinal fluid over a period of at least 2 weeks. The second available formulation contains morphine (DepoDur) for single epidural injection in the treatment of postoperative pain (Angst and Drover 2006).

9.5.3   Conventional small drugs

Nanoparticles as carriers of small-molecule drugs can dramatically influence the pharmacokinetics of the drugs by preventing their fast removal from the blood vessels across the endothelial cell layer (via transcellular or paracellular route) into the interstitial space and tissues or through the glomerular filtration into the urine. In addition, nanocarriers depending on their surface-bound ligands or size distributions can selectively deliver the small drugs to the diseased tissues through active or passive targeting, respectively. The intravenous route is therefore highly expected to deliver larger doses of drugs with nanoparticles to the target organ or tissue of a patient.

9.5.3.1   Liposomes

PEG-modified liposomes demonstrate a reduced recognition by the MPS, which increases their circulation time and accumulation particularly in solid tumors resulting in a higher antitumor activity and reduced toxicity. There are several liposome formulations approved in the United States and the European Union for intravenous injection. Liposomal amphotericin B (AmB; AmBisome) was the first FDA-approved liposomal product for the treatment of visceral leishmaniasis, a chronic fatal parasitic disease of the viscera. Doxil, a parenteral liposome-based drug formulation, was designed taking advantage of the enhanced permeability and retention (EPR) effect. Unlike Doxil, Myocet is a non-PEGylated liposomal formulation of doxorubicin (DOX). Because of the variations in primary target, liposome size, and lipid composition, Doxil and Myocet have completely different pharmacokinetic and toxicological profiles (Wacker 2013). There are many other similar anticancer liposomal formulations that are currently in preclinical or clinical phases (Noble et al. 2014). For example, modified nanoliposomes containing irinotecan (CPT-11), a topoisomerase I inhibitor, were reported to promote tumor accumulation of the drug and enhance its antitumor effects in an intracranial U87MG brain tumor model (Merisko-Liversidge and Liversidge 2011). PEGylated liposomes could also selectively transport anti-inflammatory drugs to the inflamed site in rheumatoid arthritis (RA) because of their long blood circulation time along with the EPR effect. However, no liposomal formulations for the treatment of RA have yet to be approved. On the basis of the preclinical studies, liposomal formulations of methotrexate and glucocorticoids appear to be promising candidates for the treatment of RA (van den Hoven et al. 2011).

9.5.3.2 Nanoparticle albumin-bound technology

The nanoparticle albumin-bound (nab) platform utilizes albumin as a therapeutic carrier for the delivery of hydrophobic chemotherapeutics. Albumin naturally interacts with hydrophobic molecules through reversible noncovalent binding and mediates transcytosis after binding with the glycoprotein (gp60) receptor, inspiring the development of the nab technology that converts insoluble drugs into injectable nanoparticle formulation using human albumin. The first commercial product based on the nab platform was the 130-nm nab-paclitaxel (Abraxane R) for the treatment of breast cancer. Several other nab-based chemotherapeutics are currently in clinical evaluations.

9.5.3.3 Polymeric nanoparticles

Many of the nanocarriers in clinical and preclinical investigations of small anticancer drugs are polymeric nanoparticles, most of which are formulated via self-assembly of block copolymers consisting of a hydrophobic block to form the core with hydrophobic drugs and a hydrophilic segment forming the shell. The pharmacokinetics study of the water-insoluble anticancer drugs, such as taxol or cisplatin, incorporated into the polymeric micelle core showed increased stabilization, prolonged blood circulation time, reduced side effects, and EPR-enhanced accumulation of the drugs in tumor tissues. The conjugation of drugs with the core or drug incorporation into the cross-linked core of a micelle could further improve the distribution of the drugs to the desirable site (tumor) (Wang et al. 2013). Mixed micellar nanoparticles consisting of AmB and polystyrene-block-polyethylene oxide (PS-block-PEO) were investigated for the potential treatment of fungal diseases in patients who are at risk of renal dysfunction (Han et al. 2007). Nephrotoxicity was markedly reduced when AmB was intravenously administered to rats as mixed micellar nanoparticles with PS-block-PEO without altering the antifungal activity of AmB.

9.5.4 Combinations of drugs and nucleic acids

Considering the heterogeneity of a disease such as cancer with complex networks that control tumor growth, progression, and metastasis, combined delivery of traditional small drugs and nucleic acid–based therapeutics (DNA/siRNA) with the help of a versatile nanocarrier offers a potential approach for achieving the best treatment outcomes (Figure 9.11).

Chapter nine: Nanotechnology approaches to therapeutics 139

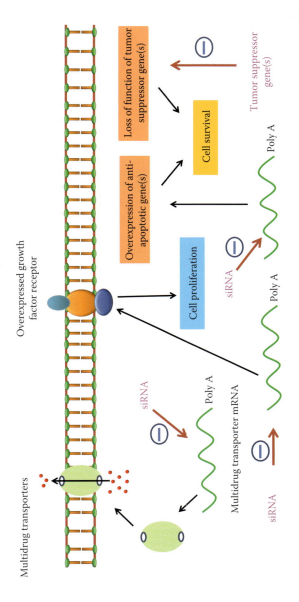

Figure 9.11 Nanoparticle-facilitated delivery of oncogene-specific siRNAs and tumor suppressor genes sensitizes cancer cells to anticancer drugs.

9.5.4.1 Combining chemotherapy with siRNAs to overcome multidrug resistance

Resistance to chemotherapy drugs is a major cause of treatment failure and relapse of many tumors. Despite the effort of administering multiple drugs in combination to overcome the chemoresistance, the cancer cells often develop resistance to other drugs as well. Nanoparticle-mediated delivery of both gene silencing tools (siRNAs) to suppress the expression of efflux transporters, such as P-glycoprotein (P-gp) (also known as MDR1), MRP1, MRP2, or BCRP, and chemotherapeutic agents presents a logical solution to overcome the multidrug resistance. Thus, successful inhibition of P-gp expression by gene silencing with siRNA was shown to dramatically increase the accumulation of chemotherapy drugs in tumors, resulting in improved antitumor efficiency (Figure 9.11). Since the efflux proteins have important physiological roles in healthy tissues, delivery of siRNA or ODN should be precisely targeted to the tumor tissue in order to avoid the undesirable pharmacological activity in those healthy tissues. For example, since the Notch signaling pathway is a key regulator for inducing tumor resistance to cisplatin in ovarian cancer, either a Notch-pathway inhibitor or a Notch 3-siRNA was demonstrated to sensitize the tumors to platinum therapy, making the Notch pathway a therapeutic target for treating platinum-resistant relapses (Li et al. 2013).

9.5.4.2 Combining chemotherapy with genes or siRNAs that promote apoptosis

Cancer cells can escape programmed cell death, that is, apoptosis, even after undergoing fatal DNA damage. This is because a balanced network of pro-apoptotic and anti-apoptotic genes that regulates apoptosis often becomes dysfunctional during tumor progression as a result of multiple gene mutations. Thus, delivery of a functional gene, such as tumor suppressor genes (p53 or retinoblastoma protein), TNF-related apoptosis-inducing ligand (TRAIL), or TNF-α, to reverse or retain the apoptotic functionality of the mutated cells together with apoptosis-inducing chemotherapy is a possible strategy to effectively kill the cancerous cells (Figure 9.11). For example, systemic delivery of the TNF-α gene using epidermal growth factor receptor–targeting polyplexes synergistically enhanced DOX activity by increasing the accumulation of liposomal DOX and significantly delayed tumor growth in subcutaneous murine neuroblastoma as well as liver metastases of human LS174T colon carcinoma (Su et al. 2012). In another report, intratumoral delivery of the TRAIL gene by AAV-2 was demonstrated to exert synergistic antitumor effects together with cisplatin in an established head and neck squamous cell carcinoma mouse model (Jiang et al. 2011). Other potential approaches rely on silencing of anti-apoptotic genes such as survivin, Bcl-2, or BAX and providing

Chapter nine: Nanotechnology approaches to therapeutics   *141*

apoptosis-inducing drugs. Thus, a combined treatment of PEG-coated anti–Bcl-2 siRNA lipoplex and a prodrug of 5-FU showed superior tumor growth suppression in a human colorectal cancer xenograft model (Li et al. 2013).

9.5.4.3   Combining drugs with nucleic acids with antiangiogenic functions

Tumor progression and growth requires an adequate supply of oxygen and nutrients through blood circulation. Hence, inhibiting angiogenesis with specific siRNA targeting VEGF is a promising therapeutic option for cancer. Indeed, downregulation of VEGF with siRNA or the anti-VEGF antibody drug bevacizumab together with chemotherapy drugs has been shown to suppress tumor growth and prolong survival (Glade-Bender et al. 2003).

Furthermore, codelivery of VEGF siRNA and DOX using micelles composed of PEI grafted with stearic acid significantly suppressed the tumor growth in a mouse model of human hepatocarcinoma (Huang et al. 2011). In a recent study, codelivery of siRNA and paclitaxel via somatostatin receptor-targeting core–shell nanoparticles basically composed of a PEGylated cationic lipid shell containing hydrophobic paclitaxel and a ternary complex core of siRNA, chondroitin sulfate, and protamine resulted in significant drug distribution in tumor tissues and inhibition of tumor growth in the BALB/c mice bearing in situ MCF-7 tumor (Feng et al. 2014).

chapter ten

Nanotechnology in the development of innovative treatment strategies

10.1 Gene therapy

Gene therapy can be defined as the introduction of protein-coding or non-coding nucleic acids to provide the missing protein function(s), correct the abnormal function(s), or silence the overexpression of protein(s), thereby reversing pathological processes (Kay 2011).

10.1.1 Transgene expression

If the exogenous nucleic acid is a functional gene, it is always inserted using recombinant DNA technology either in the viral genome or in the plasmid DNA commonly used as a vector in nonviral delivery. Although gene addition is the most common of the techniques attempted in current preclinical and clinical studies, gene correction is a promising tool that uses zinc finger nucleases and DNA recombination technologies to correct a mutation (e.g., in the case of a genetic disease) or create a mutation (e.g., in C–C chemokine receptor type 5 [CCR5]), thus making the cells resistant to human immunodeficiency virus (HIV) infection. Long-term gene expression is desirable in order to avoid the repeated administration of gene therapeutics. After nuclear translocation, genomes of Ad, adeno-associated virus (AAV), and herpes simplex virus 1 and plasmid DNA remain episomal without being integrated into the chromosomal DNA. One of the strategies to achieve transgene integration in the target cells is based on a variety of class II transposable elements that allow a therapeutically important DNA sequence to integrate into the host chromosome. Another approach relies on the use of bacteriophage recombinases, enabling more site-selective integration than the DNA transposons (Boudes 2014; Kay 2011).

10.1.2 Knockdown of endogenous gene expression

Antigene technology based on small interfering RNA (siRNA), microRNA (miRNA), ribozyme, and oligodeoxyribonucleotide (ODN) can be used to silence expression of a disease-causing gene or an overexpressed gene typically found in cancer. While these antigene molecules are chemically synthesized (and sometimes further modified to improve their stability against nuclease-mediated degradation), short hairpin RNA (shRNA) or pre-miRNA (via pri-miRNA) can alternatively be produced from a plasmid-based vector in the nucleus and further processed into the cytoplasm through cleavage by Dicer into siRNA or mature miRNA, respectively. The mechanisms of shRNA- and siRNA-mediated gene knockdown are depicted in Figure 10.1. While a siRNA breaks a specific mRNA and prevents its translation, a mature miRNA targets and cleaves a set of mRNAs, via the formation of RNA-induced silencing complex (RISC). Ribozyme, a self-cleaving RNA, has two flanking sequences required for specific binding of its target RNA molecule and a catalytic core responsible for cleavage of the target. Antisense ODNs in the form of single-stranded DNA as the basic structure silence gene expression through blockage of translation, RNA transport, or splicing. The resultant RNA–DNA hybrid can be cleaved by RNase H. The efficacy of the antigene technology is determined by the target binding sequence, nonspecific interactions with proteins and half-life of the antigene molecule, the secondary and tertiary structure, and, thus, the accessibility to the binding site of the target RNA (Scanlon 2004).

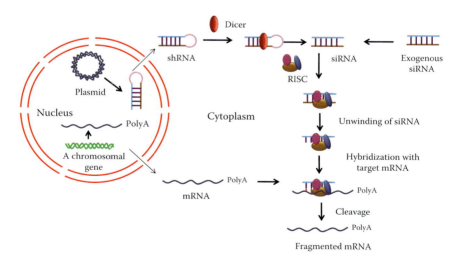

Figure 10.1 Cellular actions of shRNA and siRNA on gene knockdown.

Chapter ten: Nanotechnology in innovative treatment strategies 145

10.1.2.1 Chemical modifications of antigene molecules

A diverse range of chemical modifications have been employed to improve the binding affinity of the antigene molecules toward RNA targets, their nuclease stability, and immunostimulatory properties (Deleavey and Damha 2012).

10.1.2.1.1 Internucleotide linkage modifications

10.1.2.1.1.1 Phosphorothioate linkage The phosphorothioate (PS) linkage in which sulfur substitutes for one nonbridging phosphate oxygen confers significant resistance to the nuclease-mediated degradation without affecting the RNase H–mediated mRNA cleavage and therefore is widely used despite the reduced binding affinity for a complementary strand and some degree of toxicity observed for the modified nucleic acid molecule. PS linkages can also enhance the affinity of the antigene molecules for binding with serum albumin, leading to their improved pharmacokinetics and prolonged circulation time. Fomivirsen, a Food and Drug Administration (FDA)–approved 21-bp ODN-based antiviral drug, carries PS linkages. PS linkages have also been incorporated into siRNAs without significant loss of activity.

10.1.2.1.1.2 N3′ phosphoramidate linkages The N3′ phosphoramidate linkages replace 3′-OH groups for 3′-amine functionality, providing favorable binding affinity for target sequences and high nuclease resistance but diminishing the ability to activate RNase H–mediated cleavage. These linkages allow the substitution of a sulfur for a nonbridging oxygen in the internucleotide linkage, thus mimicking the PS linkage modification with enhanced target binding affinity.

10.1.2.1.1.3 Boranophosphate internucleotide linkages Substitution of one nonbridging oxygen with a borano (-BH3) group results in a boranophosphate internucleotide linkage. siRNAs with such linkages are capable of triggering potent gene silencing and possess improved nuclease stability. Boranophosphate-modified ODNs can be potent triggers for RNase H–mediated cleavage of a target RNA.

10.1.2.1.1.4. Phosphonoacetate linkages The phosphonoacetate (PACE) linkages substitute an acetic acid group for a nonbridging oxygen in the internucleotide linkage, thus retaining the negative charge character of unmodified DNA and enhancing the nuclease resistance. ODNs with PACE modifications are able to activate RNase H activity with slightly reduced binding affinity for RNA complement strands.

10.1.2.1.2 Sugar modifications ODNs that are modified at the 2′-position of the furanose sugar improve their affinity for the target mRNA, enhancing metabolic stability and improving pharmacokinetic and toxicological properties.

10.1.2.1.2.1 2'-O-Me sugar The nucleoside analog carrying 2'-O-Me sugar is one of the most widely used modifications that are well tolerated in siRNAs as well and have been shown to reduce immunostimulatory properties of the siRNAs.

10.1.2.1.2.2 2'-F-RNA The 2'-F-RNA presents another popular sugar modification in antigene molecules and is very well tolerated in siRNA in both the guide and passenger strands. Although 2'-F-RNA phosphodiester linkages are not nuclease resistant, chimeric 2'-F-RNA/DNA PS oligonucleotides are highly nuclease resistant in addition to exhibiting enhanced binding to the RNA target. Both 2'-O-Me and 2'-F-RNA modifications can be found in the FDA-approved aptamer Macugen.

10.1.2.1.2.3 2'-O-MOE A methoxyethyl modification at the 2'-OH (2'-O-MOE) can also be applied to antisense oligonucleotide constructs to improve their target binding affinity and nuclease stability.

10.1.2.1.2.4 Locked nucleic acid Locked nucleic acid (LNA), a chemically modified RNA analog in which a methylene bridge joins the 2'-OH to the C4' and forms a bicyclic nucleoside, can form a highly stable duplex with the target RNA, improving nuclease resistance of the antigene molecules and additionally reducing siRNA immunostimulation.

10.1.2.1.3 Morpholino phosphoramidates Antisense oligonucleotides with morpholino phosphoramidates, the uncharged substitutes of the internucleotide phosphodiester linkages and the furanose sugars, are resistant to nuclease degradation and possess similar or even improved binding affinity for the target RNA, although they lack the ability to trigger RNase H.

10.1.2.1.4 Peptide nucleic acid Peptide nucleic acid (PNA) has a neutral charge backbone composed of N-(2-aminoethyl)glycine, exhibiting significant nuclease and protease resistance and high binding affinity for its target nucleic acid owing to the absence of electrostatic repulsion between the PNA and the target strand. Although the structure of PNA substitutes both the internucleotide linkages and sugars for a peptide-based backbone, the nucleobases remain unchanged, allowing complementary base pairing with the target. However, PNAs are unable to activate RNase H.

10.1.2.1.5 Nucleobase modification A major goal in chemically modifying the nucleobases of antisense oligonucleotides is to enhance

Chapter ten: Nanotechnology in innovative treatment strategies 147

the duplex stability while maintaining or improving the native complementary base pairing with the target RNA via hydrogen bonding. For example, 5-bromo-uracil and 5-iodo-uracil as substitutes of uracil, and 2,6-diaminopurine in place of adenine, can be used to stabilize the adenine–uracil base pairing in the duplexes formed between the antigene molecule and target RNA. Other forms of modified nucleobases lacking oxygen, nitrogen, or classical hydrogen bond donating groups have been used to improve the performance of siRNAs.

10.1.3 Inhibition of endogenous miRNA functions

Since miRNAs play significant roles in regulating numerous biological processes from cell differentiation and development to apoptosis with roughly 60% of the protein-coding genes regulated by the hundreds of miRNAs, anti-miRNAs have been designed as a perfect complement to the miRNAs to sterically block the RISC loading of the latter, thereby preventing the ability of the miRNAs to cleave their target mRNAs. The most widely used anti-miRNA modifications are LNA, 2'-F-RNA, 2'-O-Me, PNA, and morpholinos.

10.1.4 Gene therapy applications

The nucleic acids or vectors can be administered in vivo via different routes (e.g., intravenous) with or without the help of a nanocarrier or ex vivo by genetically modifying the autologous cells derived from a patient and reintroducing them in the patient's body. Ex vivo gene therapy was attempted for the monogenic diseases of blood cells, such as sickle cell disease or β-thalassemia (Boudes 2014; Kay 2011).

10.1.4.1 Gene therapy of hereditary diseases

10.1.4.1.1 X-linked severe combined immune deficiency X-linked severe combined immune deficiency (X-SCID) is a combined cellular and humoral immunodeficiency that is caused by mutations in interleukin 2 receptor, gamma (IL2RG) gene leading to almost complete absence of T and natural killer (NK) lymphocytes and nonfunctional B lymphocytes. X-SCID is fatal in the first 2 years of life without reconstitution of the immune system via bone marrow transplantation or gene therapy. In 2000, autologous transplantation of retrovirally transduced bone marrow cells expressing the IL2RG gene resulted in a functional immune system in children with X-SCID, although 5 out of the 20 patients subsequently developed leukemia attributed to an integration of the proviral retrovirus and the eventual activation of the LIM domain only 2 (LMO2) proto-oncogene in those hematopoietic cell–derived cells.

10.1.4.1.2 Hemophilia B A mutation in factor IX gene leads to the deficiency of factor IX clotting activity and consequentially prolonged bleeding after an injury. In a preclinical trial of hemophilia B, recombinant AAV vectors were systemically administered for liver-based delivery and expression of hepatic coagulation factor IX gene to restore the function of the factor. A single dose of AAV2 vectors carrying the factor IX gene led to a long-term correction of hemophilia B in mice and dogs, although preexisting antibodies against the capsid of the AAV2 serotype or a CD8+ T-cell response to the capsid led to short-term expression of factor IX, thereby preventing a successful outcome in a patient with hemophilia. Recently, the peripheral infusion of an AAV8 vector expressing a codon-optimized human factor IX transgene in six patients with severe hemophilia B resulted in a long-term expression of factor IX, which was sufficient to improve the bleeding phenotype.

10.1.4.1.3 X-linked adrenoleukodystrophy X-linked adrenoleukodystrophy (X-ALD) is caused by mutations in the *ABCD1* gene encoding the peroxisomal membrane protein ALDP, which is involved in the transmembrane transport of very long-chain fatty acids. Thus, boys with a deficiency in ALD protein develop a severe lipid storage disorder with brain demyelination. The first successful use of a lentiviral vector was reported for transferring the *ABCD1* gene in the autologous hematopoietic stem cells (HSCs) from the patients with X-ALD with the final outcome apparently comparable to the allogeneic hematopoietic cell transplantation.

10.1.4.1.4 β-thalassemia A genetic deficiency in the synthesis of β-globin chains causes the reduced production of hemoglobin in β-thalassemia. A lentiviral vector was used to transfer the β-globin gene into the HSCs of an 18-year-old patient with β-thalassemia, resulting in 10% of the patient's blood cells that has the normal hemoglobin as a result of the β-globin gene integration without any insertional mutagenesis.

10.1.4.2 Gene therapy for cancer

Cancer, a leading cause of human deaths worldwide, is caused by uncontrolled cell division and differentiation owing to mutational and epigenetic changes leading to overexpression or suppression of cellular genes. siRNAs, shRNAs, and ODNs have been extensively used to silence anti-apoptotic genes, such as bcl-2; pro-angiogenic growth factor genes, such as vascular endothelial growth factor and basic fibroblast growth factor; and multidrug transporter genes, such as MDR1 and MRP1 genes in many preclinical studies to inhibit tumor cell growth, angiogenesis, metastasis, and chemoresistance. The therapeutic genes that have been investigated for cancer treatment are those of p53; thymidine kinase in combination with gancyclovir as a prodrug; tumor necrosis factor (TNF)–related

Chapter ten: Nanotechnology in innovative treatment strategies 149

apoptosis-inducing ligand (TRAIL); p21; pro-inflammatory cytokines, such as granulocyte macrophage colony-stimulating factor (GM-CSF) or fms-like tyrosine kinase 3 receptor ligand; angiotensin II type-2 receptor; interleukin-2 (IL-2) and IL-12; interferon (IFN)-α, -β, and -γ; esophageal cancer–related gene 2 protein; Fas ligand; and inducible Caspase-9 (Bakhtiar et al. 2014; Chowdhury 2011).

10.1.4.3 Gene therapy for cardiovascular diseases

Development and progression of atherosclerosis and hypertension lead to congestive heart failure, myocardial infarction, peripheral vascular disease, and kidney damage. The renin–angiotensin system (RAS), which plays an important role in atherosclerosis and hypertension, has become an important target for drug development in treating the cardiovascular diseases. In RAS, angiotensinogen (AGT) is the precursor for synthesis of angiotensin II (AngII) in liver, which in turn increases the blood pressure (BP) and accelerates the progression of atherosclerosis, accompanied by the generation of a larger amount of superoxides and reactive oxygen species (ROS) and the impaired endothelium-dependent dilation owing to a reduction in nitric oxide production. Therefore, silencing of AGT expression via nanoparticle-assisted delivery of the shRNA targeting the AGT mRNA transcript was studied to markedly reduce the protein expression of AGT and eventually that of AngII. Interestingly, along with the decline in BP, the atherosclerotic lesions were found to be markedly attenuated in AGT shRNA-treated rats. In addition, overexpression of vasodilators, such as kallikrein, adrenomedullin, atrial natriuretic peptide, and endothelial nitric oxide synthase either by naked DNA or by using viral delivery methods, was demonstrated to result in an effective lowering of high BP and attenuation of the pathophysiology in different experimental models of hypertension (Raizada and Der Sarkissian 2006). ALN-PCS02 (Alnylam Pharmaceuticals) is an RNAi therapeutic currently under clinical trial for the treatment of hypercholesterolemia, by targeting the *PCSK9* gene to increase low-density lipoprotein (LDL) receptor levels in the liver and thereby lowering LDL cholesterol levels (Kubowicz et al. 2013).

10.1.4.4 Gene therapy for neurodegenerative diseases

Neurodegenerative diseases, such as Huntington's disease (HD), Alzheimer's disease (AD), and Parkinson's disease (PD), are characterized by the progressive loss of neurons and the gradual appearance of disabling neurological symptoms. The goal of the current therapeutics is to improve the symptoms rather than to prevent the disease progress. Selective suppression of the disease-causing genes is a promising approach for the treatment of such neurodegenerative diseases. Thus, AAV-mediated delivery of shRNA targeting mutant human huntingtin in the striatum and cerebellum of HD mice caused a significant pathological and behavioral improvement with a reduction in the size and number of neuronal inclusions, whereas lentiviral delivery of shRNA targeting BACE1 to the hippocampi significantly reduced Aβ production, amyloid plaques, and neuronal death, resulting in improved learning and memory in a transgenic mouse model of AD. For the treatment of PD, silencing of the expression of α-synuclein, a principal protein component of Lewy bodies (the pathological hallmark of PD), has been proposed as a potential therapeutic option (Deng et al. 2014).

10.1.4.5 Gene therapy for HIV

HIV of the retroviral family, which is responsible for acquired immune deficiency syndrome, binds to and destroys CD4+ T lymphocytes, thus decreasing and weakening the immune system and paving the way for opportunistic infections and tumor development. Despite being highly active, the recently formulated antiretroviral therapy leads to toxicity and generation of drug resistance variants of the virus. RNAi-mediated inhibition of HIV-encoded RNAs is still challenging because of the mutation in the genome by which the virus can escape from being targeted. Therefore, targeting the cellular components involved in the viral infection process, such as CD4, the "primary receptor" by which HIV enters the cells, could be an alternative strategy. Indeed, repression of CD4 expression through RNAi was demonstrated to inhibit viral entry, syncytia formation, and viral load. The first clinical trial (NCT00569985; 04047) was based on a lentiviral vector (rHIV7-shI-TAR-CCR5RZ) encoding three anti-HIV RNA genes, which was used to transduce autologous, CD34-positive hematopoietic progenitor cells. The three RNA products include an shRNA targeted to an exon of the HIV-1 genes tat/rev (shI) to destroy the viral mRNA, a decoy for the HIV TAT-activated RNA to antagonize the viral transactivation, and a ribozyme targeting the host T-cell CCR5 cytokine receptor (CCR5RZ) (a coreceptor) to block viral entry (Deng et al. 2014).

10.1.4.6 Gene therapy for viral hepatitis

Hepatitis C virus (HCV) infection, a major cause of chronic liver disease, is currently treated by a combination of pegylated IFN (IFN-α) and ribavirin with associated adverse effects, high cost, and poor response rate, necessitating the development of additional agents that could act specifically. A single-stranded HCV genome, which also functions as a messenger RNA, is an attractive target for potential RNAi-based therapeutics. Thus, siRNAs or shRNAs could be designed to directly target the 5′ UTR and 3′ UTR, the most conserved regions of the HCV RNA, which signifies their functional importance in the viral life cycle; NS3 serine protease, which is involved in HCV polyprotein maturation; NS3 helicase, whose primary function is to unwind the viral genomic RNA during replication; and NS5A, a pleiotropic protein with key roles in both viral RNA

Chapter ten: Nanotechnology in innovative treatment strategies 151

replication and modulation of the physiology of the host cell (Motavaf et al. 2012).

10.2 Protein- and DNA-based prophylactic vaccines

Despite the remarkable success of vaccination in preventing infectious diseases, effective vaccines have yet to be developed for many important diseases, such as HIV, malaria, and tuberculosis (Moyle and Toth 2013). One of the traditional approaches to vaccine development was to use the attenuated whole microorganisms incapable of causing diseases but capable of eliciting long-lasting immunity. However, these modified microorganisms are potentially harmful as they could revert to the pathogenic forms. Moreover, the more attenuated a pathogen becomes, the less immunogenic it is (Bachmann and Jennings 2010).

10.2.1 Subunit vaccines

A subunit vaccine consists of the only microbial components with protein/peptide and carbohydrate antigens that are required to produce an appropriate immune response, thus significantly improving vaccine safety. Although subunit vaccines can be produced with a high level of lot-to-lot consistency, the ability to generate specific immune responses, and the flexibility for nonrefrigerated transport and storage in freeze-dried forms, their ability to stimulate potent immune responses is much weaker than the attenuated whole microorganism-based vaccines, therefore necessitating the incorporation of potent immunostimulatory compounds, known as adjuvants, and the administration of the vaccines at multiple points in a patient's life (boosters) so as to ensure long-term protective immunity. Proteins and peptides that serve as the source of Th and CTL epitopes as well as many B-cell epitopes are a major component of many subunit vaccines. Both the recombinant DNA technology and the various synthetic techniques confer batch-to-batch reproducibility of the proteins and peptides, respectively. Protein folding is required if the antibody response is expected for the surface antigens of the folded proteins, whereas in other cases, denatured protein may elicit more potent B- and T-cell responses. Inclusion of adjuvants, such as aluminum salts (alum), is usually required to strengthen the overall immune response.

10.2.2 DNA vaccines

The DNA vaccine involves a plasmid vector carrying one or more functional genes in order to generate within APCs the antigenic peptides that associate with the major histocompatibility complex (MHC) class I or II molecules, eliciting both cellular and humoral immune responses. The traditional vaccines, however, mostly induce only humoral response. Like proteins in subunit vaccines, DNA molecules have short half-lives because of the action of endonucleases in extracellular (blood) and intracellular (endosomes or lysosomes) compartments. In addition, there are other barriers associated with the intracellular processing of plasmid DNA, such as inefficient cellular uptake and nuclear translocation. One of the approaches to improve DNA vaccine immunogenicity is the inclusion of additional plasmids or additional inserts in the same plasmid, encoding molecular adjuvants.

10.2.3 Vaccine delivery

In order to prevent rapid degradation and renal clearance, the amino acid sequence of proteins/peptides could be modified or suitable nanocarriers could be employed for delivery of these antigenic molecules. For example, the influenza virus vaccine Inflexal (Crucell) incorporates the influenza virus protein hemagglutinin (HA) into liposomes, whereas the hepatitis B virus (HBV) vaccine Engerix b (GlaxoSmithKline biologicals) consists of virus-like particles (VLPs) and liposomes (Bachmann and Jennings 2010). The enzymatic cleavage of DNA can also be protected by electrostatically complexing with cationic lipid or polymers. Combined delivery of peptides/proteins (or DNA) and adjuvants using nanoparticles could further boost the immune responses. Targeted delivery could even direct the vaccines to specific cells (e.g., APCs) of the immune system for efficient uptake and processing of the antigenic peptides/proteins (or DNA). The rational development of such vaccines thus requires an understanding of the immune responses in order to select the appropriate components as the constituents of a vaccine. Thus, codelivery of plasmids encoding cytokines, chemokines, or costimulatory molecules could augment the immune responses.

10.2.3.1 Administration site

Intramuscular administration is most commonly chosen for vaccine delivery. Although alternative delivery routes are highly sought for some local applications as well as for easier administration of the vaccines, involvement of nanoparticle carriers or other physical methods are often required to achieve the goal. For example, antigens that are inefficient in crossing the nasal mucosa upon nasal vaccination could be loaded into the mucoadhesive nanocarriers for retention and subsequent transport across the barrier. As for dermal vaccination, several techniques that allowed the vaccine to cross the stratum corneum of the skin include microneedles, skin disruption methods, and jet injectors. Similarly, oral vaccine formulations should be able to resist the low pH and enzymatic destruction in the digestive tract.

10.2.3.2 Physicochemical characteristics of nanoparticles

10.2.3.2.1 Size, shape, charge, and polarity Delivery vehicles could be designed to form the final vaccine particles of defined size and shape for their proper biodistribution and efficient internalization by APCs. Usually, particles in the range of 20–200 nm preferably enter the lymphatic system before being taken up by APCs, whereas those smaller than 10 nm are inefficiently internalized by those cells, and the larger ones that do not efficiently enter the lymph capillaries are carried into the lymph by specialized cells, such as dendritic cells (DCs). The effects of particle size, shape, hydrophilicity/hydrophobicity, and surface charge on binding of potential therapeutic agents and subsequent distribution of the particles have been discussed earlier in Chapters 6 and 7. Additionally, there are many adjuvants that contain large hydrophobic sites, including those recently licensed, such as MPLA (monophosphoryl lipid A)-plus-alum and the oil-in-water emulsions MF59 and AS03.

10.2.3.2.2 Surface organization Surfaces of viruses and bacterial structures (e.g., flagella) typically contain a few proteins, each of which is present in multiple copies as well as in an ordered manner for efficiently triggering the immune system. Thus, the nanoparticles fabricated with 60 peptide epitopes displayed per particle, with spaces of 5–10 nm, were shown to optimally induce antibody responses. In addition, the accessibility of different epitopes as determined by the spacing in between the epitopes and the conformation of the antigen is another prerequisite for eliciting a robust immune response. An influenza nanoparticle vaccine based on ferritin–hemagglutinin (HA) fusion proteins was designed by genetically fusing viral HA to ferritin, a protein that naturally forms nanoparticles composed of 24 identical polypeptides. Immunization with the self-assembled nanoparticles in which HA was inserted at the interface of adjacent subunits, thereby forming eight trimeric viral spikes on the surface, led to the induction of hemagglutination inhibition antibody titers 10-fold higher than those from the licensed inactivated vaccine (Kanekiyo et al. 2013).

10.2.3.2.3 Cellular targeting After uptake by APCs, antigens are entrapped by endosomal–lysosomal compartments and degraded there into peptides that are subsequently loaded onto MHC class II molecules and transported to the cell surface for stimulation of CD4+ T helper cells. Finally, the stimulated CD4+ T cells interact with and activate B cells for generation of antibody-producing plasma cells and memory B cells (Bachmann and Jennings 2010) (Figure 10.2). Although macrophages, some endothelial cells, and B cells can function as APCs, the most professional APCs are the DCs with subsets that differ in their surface receptors, such as Toll-like receptors (TLRs) and C-type lectin receptors. Targeting

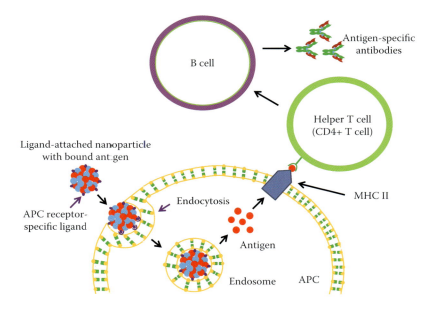

Figure 10.2 Nanoparticles for vaccine delivery.

of those receptors and other cell surface molecules either individually or in combination has potential applications in vaccine development. Thus, the nanoparticles with the coated antibodies recognizing DC-SIGN, a DC-specific adhesion receptor, and the encapsulated TLR ligands strongly increased immune responses at low doses of TLRs and with reduced toxicity.

10.2.3.3 Nanoparticles as adjuvants and vaccine carriers

10.2.3.3.1 Inorganic nanoparticles Inorganic nanoparticles can be used as both adjuvants and delivery vehicles for the antigens. Among them, aluminum-based nanoparticles (AlNPs) are commonly used in tetanus, diphtheria, and influenza type B vaccines to facilitate strong antigen-specific immune responses (Smith et al. 2015). The rod-like AlNPs were shown to stimulate a stronger DC response than the spherical counterparts with the consequence of a more robust immune response against a codelivered antigen (Sun et al. 2013). High affinity of gold nanoparticles (AuNPs) for the sulfur groups present in peptide epitopes makes conjugation of peptide antigens to the particles very straightforward, thus presenting the AuNPs as another potential candidate for vaccine delivery. Oral administration of chitosan-modified AuNPs with loaded tetanus toxoid was demonstrated to more effectively stimulate an immune response than tetanus toxoid alone (Barhate et al. 2014).

Chapter ten: Nanotechnology in innovative treatment strategies 155

10.2.3.3.2 Polymeric nanoparticles Nanoparticles based on degradable synthetic polyanhydrides were fabricated to encapsulate pneumococcal surface protein A (PspA), a virulence factor found on the surface of all *Streptococcus pneumoniae* strains, and subsequently administered subcutaneously to mice, which resulted in induction of a high-titer and high-avidity anti-PspA antibody response (Haughney et al. 2013).

10.2.3.3.3 Liposomes Modifying liposomes with lectin binding mannose on their surface and MPLA adjuvant enabled the nanovaccine system to target DCs, thereby facilitating enhanced antigen presentation to T cells against a model antigen (Wang et al. 2014). In another study, MPLA-modified liposomes carrying a fusion protein of multiple tuberculosis epitopes induced strong T cell–mediated immune responses in mice (Orr et al. 2013). Recently, liposomes (virosomes) displaying viral proteins, such as influenza-derived HA and neuraminidase on their surface with the ability to actively fuse with target cells (Felnerova et al. 2004), have been utilized for tetanus and hepatitis B vaccine development and vaccine-based protection against influenza-infected host cells (Monto et al. 2009).

10.2.3.3.4 Virus-like particles VLPs that resemble the traditional structure of viruses can be formed from either one or a combination of viral proteins (e.g., complete envelope or capsid) without involvement of any viral genetic material, making them incapable of replicating. The first VLP-based human vaccine, which has been administered widely, confers protection against hepatitis B (Smith et al. 2015). VLPs have recently been utilized to deliver multiple human papillomavirus (HPV) vaccines (including those marketed as Gardasil and Cervarix), triggering a highly cross-protective antibody response and thus effectively protecting against a vast majority of HPV serotypes (Bissett et al. 2014). Cytotoxic T cells play a key role in clearing infected cells and controlling pathogen load during chronic infection. By targeting the cytotoxic T cell response that plays a key role in clearing infected cells and controlling pathogen load during chronic infection, VLPs can be very effectively utilized in vaccine development (Smith et al. 2015).

10.2.3.3.5 Micelles Protein or peptide antigens can be covalently linked to the hydrophilic exterior of a micelle. Adjuvant-loaded polymer–lipid micelles with an attached model peptide antigen were shown to traffic to the lymph node, enhancing T-cell priming and functioning as an anticancer vaccine (Liu et al. 2014). In another approach, hydrophilic peptide antigens are covalently attached to hydrophobic moieties (i.e., lipids or aliphatic hydrocarbons) leading to self-assembly into micelles. Such peptide amphiphile micelles have been shown to induce strong antitumor

156 Nanotherapeutics

cytotoxic T-cell responses and antibody responses against *Streptococcus pyogenes* in mice (Black et al. 2010).

10.3 Immunotherapy

There are a number of causes for cancer development including viral infection (Epstein–Barr virus, HBV, and HPV), bacterial infection (*Helicobacter pylori*), carcinogens, and genetic abnormalities. Our immune system can recognize and eliminate those infections and the transformed or precancer cells, thus preventing tumor occurrence in a process called immune surveillance. Tumor-specific antigens expressed on tumor cells enable the immune cells to recognize and destroy the cancer and precancer cells, with the help of IFN-γ, IFN-α/β, perforin, NKG2D, and TRAIL. However, the tumor can escape the immune surveillance via multiple mechanisms, such as loss of MHC-I, loss of adhesion molecules, generation of regulatory T (Treg) lymphocyte, expansion of myeloid-derived suppressor cells, immunosuppression, blocking of NKG2D-mediated activation, and apoptosis induction of antitumor effector cells (Sheng and Huang 2011).

10.3.1 Strategies of cancer immunotherapy

Nanomedicine offers two strategies to kill the cancer cells based on the knowledge of immune surveillance mechanisms: nonspecific immune activation and tumor-specific immune activation. These two strategies can also be combined for effective elimination of tumor cells by increasing tumor antigen presentation and inducing specific CTL activity.

10.3.1.1 Nonspecific immune activation

This strategy involves treatment with recombinant cytokines, IFNs, or TLR agonist in order to activate the immune system.

10.3.1.1.1 Cytokines IL-2 promotes proliferation and activation of effector immune cells. For example, IL-2 has been shown to activate cytokine-induced killer cells that possess antitumor activity. In addition, systemic administration of IL-2 was found to suppress tumor growth with clinical efficacy observed in malignant melanoma and renal carcinoma. However, IL-2 therapy causes significant dose-related morbidity with noticeable toxicity observed in most organ systems, including heart, lungs, kidneys, and the central nervous system. Among the other cytokines, IL-21 activates CD4+ T cells, CD8+ T cells, NK cells, and B cells and suppresses Treg cells. Treatment with IL-21 was demonstrated to enhance the production of IFN-γ, IL-2, TNF-α, GM-CSF, IL-1β, and IL-6 by activating T cells. Moreover, the combination of IL-21 and anti-DR5 antibody elicited the tumor-specific CTL activity, suppressed TRAIL-sensitive

Chapter ten: Nanotechnology in innovative treatment strategies 157

tumor metastases, and enhanced memory responses to tumor rechallenge (Smyth et al. 2006). Administration of IL-21 alone resulted in antitumor activity in patients with metastatic melanoma and renal cell carcinoma activity in a phase I clinical trial study. IL-18 is another immunostimulatory cytokine with the capacity to augment antitumor therapy through induction of IFN-γ, IL-2, TNF-α, GM-CSF, and IL-1α; activation of effector T cells; and enhancement of NK cell–mediated cytotoxicity. IL-18 was also shown to promote protection against tumor challenges in mice (Micallef et al. 1997).

10.3.1.1.2 Interferons IFN-α and IFN-β demonstrate antitumor potency and enhance NK-cell activity. Clinical trials of these IFNs revealed their efficacy in the treatment of leukemia, melanoma, and renal-cell carcinoma. A phase I clinical trial based on adenovirus-mediated IFN-β gene delivery showed strong antitumor immune responses in malignant pleural mesothelioma and metastatic pleural effusions (Sterman et al. 2007). On the other hand, IFN-γ, which is secreted by NK cells and effector T cells, trigger apoptosis and antigen presentation in cancer cells and exert antiangiogenic effects.

10.3.1.1.3 TLR agonist TLRs induce DC maturation, stimulate proliferation of CD4+ and CD8+ T cells, and modulate the suppressive function of Treg cells. Administration of TLR7 or TLR9 agonists was shown to enhance the activity of cancer vaccines in patients with malignancies.

10.3.1.1.4 Nanoparticles in nonspecific immune activation The short half-lives of the cytokine-based nonspecific immune agents in the circulation and their systemic toxicities necessitate development of nanoparticles to carry the genes of cytokines specifically to the tumor tissues. For example, nanoparticles have been applied for improving the immune reactions by delivering GM-CSF genes or siRNA to inhibit the expression of immune suppression genes in tumor microenvironments, such as transforming growth factor β (Schneider et al. 2008).

10.3.1.2 Tumor-specific immune activation
This strategy enables the immune cells to recognize tumor cells specifically by activating the adaptive immune system.

10.3.1.2.1 Antibody-dependent cell-mediated cytotoxicity Antibody-dependent cell-mediated cytotoxicity (ADCC) is the killing of Ab-coated target cells by cytotoxic effector cells, such as NK cells (although macrophages, DCs, neutrophils, and eosinophils can also be involved), when the Fc region of the target cell-bound Ab interacts with an Fc receptor on the effector cells, resulting in the production of pro-inflammatory cytokines, such as IFN-γ, and release of cytotoxic compounds, such as perforin and granzyme, thereby killing the target cells. Since MHC loss is one of the mechanisms facilitating tumor escape from the immune surveillance, bispecific antibodies (bAbs) have been designed to act as adaptors linking an effector cell type with a tumor target. Thus, one end of a bAb interacts with a tumor-associated antigen, such as CD19 on B-cell lymphomas, ERBB2 on breast cancer, or CD30 on Hodgkin's lymphoma, while the other end binds to a surface marker on an effector cell, such as CD3 on a cytotoxic T cell (Tc cell) or CD16 on an NK cell, leading to tumor-specific immune responses and inducing apoptosis through ADCC (Figure 10.3). Among the bAbs currently in clinical trials, the bispecific tandem single-chain Fv molecule (MT103) directed against CD19 (tumor antigen of non-Hodgkin lymphoma) and CD3 (T cell) was found very potent in destroying CD19-expressing tumor cells in vitro and in vivo (Rogers et al. 2014).

10.3.1.2.2 Complement-dependent cytotoxicity In complement-dependent cytotoxicity (CDC), monoclonal antibodies (mAbs) with affinities to both C1q, a complement component, and a specific antigen on target cells are designed with the goal of increasing C1q accumulation on the target cells to trigger the complement cascade, leading to the formation of the membrane attack complex (C5b to C9) at the cell surface and, finally, lysis of the target cells (Figure 10.3). One example is ofatumumab, an FDA-approved anti-CD20 mAb that enhanced CDC-mediated killing

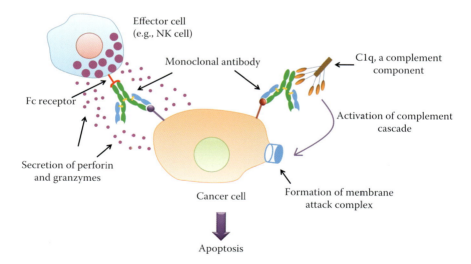

Figure 10.3 Antibody-dependent cellular cytotoxicity (ADCC) and complement-directed cytotoxicity (CDC).

Chapter ten: Nanotechnology in innovative treatment strategies 159

of cancerous B cells in chronic lymphocytic leukemia. Another FDA-approved anti-CD20 mAb to promote CDC in B-cell non-Hodgkin lymphoma is rituximab, whose epitope is distinct from that of ofatumumab. However, complement activation can promote both CDC and ADCC processes that act synergistically. Thus, complement activation by mAb can also serve to recruit the effector cells through the release of chemotactic components (C3a and C5a) (Rogers et al. 2014).

10.3.1.2.3 Antigen-loaded DCs for induction of T-cell immune responses Immature DCs take up antigens through phagocytosis, micropinocytosis, and receptor- and lectin-mediated endocytosis, and after encountering inflammatory mediators, such as TNF-α, they start to mature, upregulating the expression of MHC, and travel to the lymph nodes where they display the surface-bound antigen through MHC molecules to T cells. Thus, antigen loading of DCs ex vivo could be an attractive approach for cancer immunotherapy. The loading can be accomplished by pulsing autologous DCs with antigenic peptide or protein and tumor lysate, fusing the DCs with irradiated tumor cells, or transfecting the DCs with tumor antigen-encoding DNA or RNA with the help of viral or nonviral nanoparticles. Antigen-loaded mature DCs are finally injected into the patient to induce T-cell immune responses against the tumor (Sheng and Huang 2011).

10.3.1.2.4 Monoclonal Ab or siRNA to enhance tumor recognition by NK cells NK cells usually express at least one inhibitory killer cell immunoglobulin-like receptor (KIR) that is specific for a self-MHC class I allele. When NK cells with a KIR specific for a particular MHC class I allele bind through other interactions with the target cells lacking this allele, the NK cell will kill the target cells. Therefore, one exciting strategy is to block the KIRs with mAbs or silence the expression of the inhibitory receptors by specific siRNAs, thereby augmenting recognition of the tumor cells and their killing by NK cells (Ljunggren and Malmberg 2007).

10.4 Photodynamic therapy

Photodynamic therapy (PDT), an emerging therapeutic approach to treat infections, cancer, and other diseases by removing unwanted tissues, involves the use of light-sensitive molecules called photosensitizers (PS), which, after absorption of light (photon) of a specific wavelength, reach an excited state and thus undergo reaction with ambient oxygen, resulting in the production of potent ROS and, finally, leading to ROS-mediated cell death (Figure 10.4). By applying the light to the region of unwanted cells, PDT could be made highly selective.

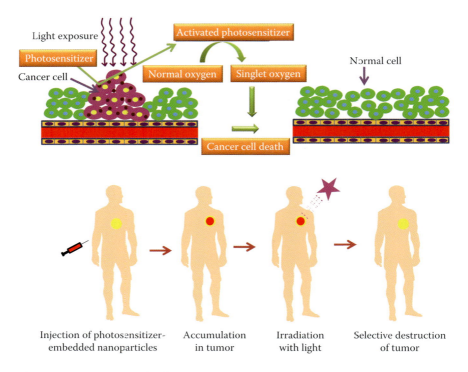

Figure 10.4 Nanoparticles for PDT.

10.4.1 Nanoscale drug delivery systems in PDT

Most highly effective PS are very hydrophobic, forming inactive aggregates in an aqueous environment and thus ending up with loss of their efficiency, since only the monomeric forms are photoactive. Moreover, PS should be delivered in therapeutic concentrations to the target cells alone to maximize their performance and minimize the undesirable side effects on healthy tissues. To overcome the hurdles, various strategies were undertaken to complex the PS with nanoparticles, for example, by embedding in polymeric nanoparticles and covalently or noncovalently binding to the surface of nanoparticles. In addition, there are cases where the nanoparticles themselves act as the PS. The following nanoparticles are most commonly used in PDT (Gupta et al. 2013).

10.4.1.1 Lipid-based carriers

10.4.1.1.1 Liposome-encapsulated verteporfin Verteporfin, an FDA-approved photosensitizer of benzoporphyrin derivative, is widely used in

Chapter ten: Nanotechnology in innovative treatment strategies 161

a liposomal formulation (Visudyne) to destroy neovasculature in the eye secondary to disease such as wet AMD (age-related macular degeneration). The intravenous administration of the liposomal Verteporfin, which is then selectively accumulated in neovascular spots of the eye, is followed by the application of nonthermal red light (689 ± 3 nm wavelength) after a 15-minute interval (Christie and Kompella 2008).

10.4.1.1.2 Liposomal mTHPC (foslip) A liposomal formulation of meso-tetra(hydroxyphenyl)chlorin (mTHPC–Foscan1), an active photosensitizer capable of significant and deep destruction of the irradiated tissue at relatively small drug and light doses, was examined for PDT in a mouse model of local recurrence of breast cancer.

Intratumoral delivery of the lipid formulation followed by irradiation at 652 nm showed progressive incorporation of mTHPC into tumor tissue with a maximum at 24 hours postinjection, which perfectly correlated with the best therapeutic effect at that time point (D'Hallewin et al. 2008).

10.4.1.1.3 Liposomal photofrin Photofrin, a photosensitizing agent, was entrapped in multilammelar liposomes. Intravenous administration of the formulation into mice with a human gastric cancer xenograft resulted in its enhanced tumor uptake, and subsequent irradiation at 630 nm demonstrated a significant volume of necrotic tumor tissue, compared to the free photosensitizer (Igarashi et al. 2003).

10.4.1.1.4 Liposomal AlClPc Cationic liposomes were employed to entrap aluminum chloride phthalocyanine (AlClPc) as a photosensitizing agent, and the resultant formulation was applied in PDT against cariogenic bacteria in a clinical study involving volunteer patients with dental caries lesions. Application of the AlClPc-cationic liposome formulation in the cavity allowing interactions of the cationic complexes with anionic bacterial surface and subsequently irradiation of the area with a red laser (660 nm) led to significant reduction (82%) of bacterial load in the treated cavity (Longo et al. 2012).

10.4.1.1.5 Ce6 nanolipoplexes Chlorine6 (Ce6), a hydrophilic negatively charged photosensitizer derived from natural chlorophyll *a*, was electrostatically complexed with a cationic liposome. Intravenous administration of the nanolipoplexes resulted in their preferential distribution and retention in the tumor in a murine squamous cell carcinoma model, and localized illumination (with 660 nm LED light) of the mice resulted in the formation of thick scabs over the tumor regions with complete ablation of the tumors after the spontaneous scab detachment (Shim et al. 2011).

10.4.1.2 Polymeric carriers

10.4.1.2.1 TPP-loaded PEG–PE   A poorly soluble photosensitizer, meso-tetraphenylporphine (TPP), was solubilized by encapsulating into the polymeric micelles composed of polyethylene glycol–phosphatidyl ethanolamine conjugate (PEG–PE) with surface-anchored tumor-specific monoclonal 2C5 antibody (mAb 2C5). One day after the mice with murine Lewis lung carcinoma were given an intravenous injection of TPP-loaded mAb 2C5–PEG–PE, the tumor sites were illuminated with light (630 nm), leading to the maximum level of tumor growth inhibition as a consequence of enhanced tumor accumulation of the drug-loaded immunomicelles (Roby et al. 2007).

10.4.1.2.2 PpIX-conjugated glycol chitosan   Hydrophobic protoporphyrin IX (PpIX) photosensitizers were conjugated to the glycol chitosan (GC) polymer to form stable nanoparticles of amphiphilic PpIX–GC wherein the PpIX molecules formed hydrophobic inner cores with the hydrophilic GC polymer as the outer shell. Intravenous administration of the PpIX–GC nanoparticles into the tail vein of mice with human colon adenocarcinoma (HT-29) and subsequent irradiation at 633 nm showed severe cell death at the irradiation site, thus significantly suppressing the tumor growth (Lee et al. 2011).

10.4.1.2.3 DPc–polymeric micelle complex   Dendrimer phthalocyanine (DPc), an anionic photosensitizer, was used to electrostatically complex with PEG–poly(L-lysine) block copolymer (PEG–PLL) to form DPc/PEG–PLL micelle. The antitumor activity of the DPc-loaded micelle was tested in the mice bearing human lung adenocarcinoma through irradiation with a diode laser 1 day after intravenous administration. A significantly higher PDT efficacy with much less toxicity was observed than clinically used Photofrin, apparently owing to the effective accumulation of the DPc-encapsulated micelle based on the EPR effect and its enhanced photocytotoxicity (Nishiyama et al. 2009).

10.4.1.2.4 ZnPcF16-incorporated PEG–PLA nanoparticles   Hexadecafluoro zinc phthalocyanine (ZnPcF16), a photosensitizer for PDT, was formulated in PEG-coated PLA nanoparticles. Intravenous injection of the formulation in EMT-6 mammary tumor-bearing mice and exposure of the tumors to red light (650–700 nm) at 24, 48, and 72 h postinjection provided improved tumor response compared to conventional Cremophor EL (CRM) emulsions. The best tumor response observed through irradiation 24 hours after the administration reflects the highest tumor-to-blood ratio of the loaded photosensitizer at that time point (Allémann et al. 1996).

Chapter ten: Nanotechnology in innovative treatment strategies 163

10.4.1.2.5 Verteporfin–PMBN–antibody A complex consisting of verteporfin (a photosensitizer), PMBN (a water-soluble amphiphilic phospholipid polymer), and an anti-EGFR antibody was designed for selective delivery of the photosensitizer into EGFR-expressing tumors exploiting the passive targeting through the EPR effect as well as the active targeting via receptor–ligand interactions. Exposing the A431 tumor-bearing mice to a diode laser light (at 640 nm) 1 hour after the intravenous injection of verteporfin–PMBN–antibody led to a significantly greater decrease in tumor size with subsequently higher mice survival rate compared to verteporfin–PMBN (Kameyama et al. 2011).

10.4.1.3 Fullerene-based nanomaterials

Fullerenes, a class of closed-cage, third allotropic form of carbon materials with at least 60 atoms of carbon, are usually surface modified with functional groups to make them soluble and able to interact with biological systems.

10.4.1.3.1 C60–PEG To enhance the photodynamic effect of C60, the most abundant form of fullerenes, water-insoluble C60 was chemically modified with PEG, making it soluble in water and enlarging its molecular size. After intravenous injection into the mice carrying a tumor mass of Meth-A fibrosarcoma cells, the C60–PEG conjugate exhibited higher accumulation and more prolonged retention in the tumor tissue than in normal tissue and, upon exposure to visible light (400–505 nm), suppressed the tumor growth (through induction of necrosis) more strongly than Photofrin. The antitumor effect of the conjugate was found to increase with increasing the irradiation power and the C60 dose (Tabata et al. 1997).

10.4.1.3.2 C60–HA Hyaluronic acid (HA), a naturally occurring linear polysaccharide of a repeating disaccharide unit (β-1,4-D-glucuronic acid–β-1,3-N-acetyl-D-glucosamine), was grafted onto C60 generating a water-soluble macromolecular photosensitizer (HA–C60) with excellent biocompatibility. Four hours after the mice bearing murine sarcoma S180 tumor were administered with HA–C60 via tail vein, irradiation of the tumor region with 532 nm laser showed a strong antitumor efficacy compared to the treated tumor without laser irradiation (Zhang et al. 2015).

10.4.1.3.3 C60–5-ALA 5-Aminolevulinic acid (5-ALA), which can be metabolized to an endogenous photosensitizer called PpIX, was conjugated to C60, resulting in the formation of monodisperse aggregates of C60–5-ALA. After intravenous injection of C60–5-ALA nanoparticles in tumor-bearing mice, selective PpIX formation was observed in the tumor.

Moreover, the tumor growth was successfully suppressed after the tumor region was irradiated with 630 nm laser (Li et al. 2014).

10.5 Image-guided therapy

Active and passive targeting of nanoparticles has become a promising tool for simultaneous delivery of drugs and various imaging contrast agents such as paramagnetic metal ions, superparamagnetic iron oxide nanoparticles, near-infrared (NIR) probes, and radionuclides for image-guided therapy with the help of currently accessible imaging techniques including magnetic resonance imaging (MRI), optical imaging, ultrasonography, positron emission tomography, computed tomography, and single photon emission computed tomography. These nanobased imaging and therapeutic agents that form the basis of theranostic nanomedicines enable an enhancement of the signal-to-noise ratio in the targeted tissue compared to the nontargeted ones, facilitating to detect even small lesions that are undetectable with traditional methods, and allow to monitor in real time the biodistribution of the nanocarriers and the disease progression, paving the way for the early treatment of critical diseases such as cancer and neurological and cardiovascular diseases (Mura and Couvreur 2012).

10.5.1 Personalized medicine

The noninvasive monitoring of drug accumulation at the target site would enable the screening of patients responding positively to the treatment based on the high accumulation of the nanomedicine from others who are not responding and would thus need a different therapeutic option. The drug accumulation in healthy tissues would also help in assessing the risk of development of side effects. In other words, determining the level of the drug accumulated at the target site could enable the prediction of the effectiveness of a treatment or could provide a justification for the failure in certain diseases like cancer. For example, a high variability of responses among patients after administration of macromolecular drugs, such as Doxil (doxorubicin-loaded nanoparticles) and Abraxane (albumin-bound nanoparticles form of paclitaxel), which are expected to passively accumulate in the tumor as a result of the EPR effect, is mainly attributed to the uncertainty of the EPR effect depending on the type of cancers, which differ in fibrosis and hypovascularization, or the density and structural integrity of the tumor neovasculature within the same type of tumor. In fact, a minimal EPR effect is apparently responsible for the lack of response in the treatment of solid tumors (e.g., pancreatic adenocarcinoma). A conceptual sketch of personalized nanomedicine on the basis of nanotherapeutics is given in Figure 10.5.

Chapter ten: Nanotechnology in innovative treatment strategies   165

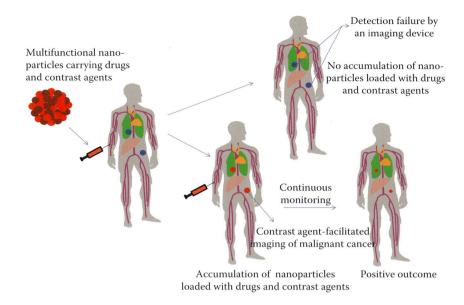

Figure 10.5 Personalized nanomedicine based on nanotheranostics.

10.5.2   Nanotheranostics in cancer

10.5.2.1   MRI and chemotherapy

MRI signal is produced as a result of interactions of protons with each other and the surrounding molecules under the influence of an applied magnetic field with the different T1 (longitudinal) and T2 (transversal) proton relaxation times in different tissues generating the endogenous contrast to detect the target tissue(s). Exogenous contrast agents act to increase the contrast by shortening the T1 or the T2 relaxation time. Nanoparticles carrying paramagnetic metal ions, such as Gd^{3+} (approved by the FDA), Mn^{2+}, or Mn^{3+}, are used as T1 contrast agents, while superparamagnetic iron oxide nanoparticles are the most popular T2 contrast agents. Multifunctional drug nanocarriers with enriched iron oxide nanoparticles or paramagnetic metal ions are attractive candidates for developing theranostic nanodevices for personalized nanomedicine.

10.5.2.1.1   Polymeric micelles   Among the several kinds of micelle-based approaches, one design employed a poly(ethylene glycol)-*b*-poly(glutamic acid) block copolymer that spontaneously formed micelles with a mean diameter of ~30 nm encapsulating both Gd^{3+} and the anticancer drug DACHPt, an oxaliplatin derivative in their core. In an orthotopic human pancreatic (BxPC3) xenograft model, when the Gd-DTPA–DACHPt-loaded micelles were intravenously injected, they were found to highly accumulate in the tumor area (up to seven times) compared to the unbound Gd-DTPA, leading to a strong enhancement of the MRI contrast signal and a significant reduction in the tumor size as a result of drug release from the complex (Kaida et al. 2010).

10.5.2.1.2 Liposomes Neural cell adhesion molecule (NCAM)–targeted liposomes encapsulating both a lipophilic gadolinium derivative (Gd-DOTA monoamide) and an anticancer agent (doxorubicin) were investigated in a severe immunodeficient mouse model of human Kaposi's sarcoma expressing NCAMs (Grange et al. 2010). Intravenous administration of the NCAM-targeted liposomes induced a significant reduction of tumor mass and vascularization with respect to the untargeted liposomes, while allowing the MRI visualization of drug delivery in the tumor region. In another approach, after intravenous administration of thermosensitive liposomes coencapsulating Gd^{3+} and doxorubicin (HaT-DOX-Gd) in a mouse model of mammary carcinoma (EMT-6) (Tagami et al. 2011), animals were imaged by MR in order to follow the release and the uptake of doxorubicin in locally heated tumors. A quantitative correlation was observed between the drug accumulation and the variation of the MR T1 relaxation time, along with a significant dose-dependent inhibition of the tumor growth.

10.5.2.1.3 Multifunctional magneto-polymeric nanohybrids Multifunctional magneto-polymeric nanohybrids (MMPNs) were prepared by simultaneously encapsulating the hydrophobic magnetic nanocrystals and doxorubicin with an amphiphilic block copolymer as a stabilizer using a nanoemulsion method and subsequently conjugating with an anti-HER antibody by utilizing the carboxyl group present on the surface of the MMPNs (Yang et al. 2007). The effectiveness of these nanohybrids has been evaluated for theranostic purposes on a subcutaneous model of breast cancer through imaging of the animals by MR at different time points after intravenous injection of the antibody-modified MMPNs. With an increase in the tumor contrast enhancement, there was a significant therapeutic activity compared to controls apparently attributed to active tumor targeting and the eventual cell internalization by receptor-mediated endocytosis.

10.5.2.2 MRI and chemo/photothermal therapy

Photothermal therapy (PTT) uses NIR light to excite photoabsorber molecules, converting the energy of the incident light to heat energy and thus inducing apoptotic cell death. PTT can increase the temperature between 45°C and 300°C and exert its effect at sufficient depth as well as with precision and minimal damage to the surrounding healthy tissues. An

Chapter ten: Nanotechnology in innovative treatment strategies   167

example of a multifunctional system for simultaneous imaging and therapy is based on gold nanorod (GNR)–decorated complex nanoparticles of paclitaxel-PLGA/iron oxide/quantum dots (Qds) (paclitaxel-loaded PLGA nanoparticles are conjugated to iron oxide nanoparticles and Qds) (Cheng et al. 2010). The GNRs have the capacity of converting NIR light to heat to achieve photothermal ablation of tumor tissue and release the entrapped drug via destruction of PLGA nanoparticles. Intratumoral injection of the multifunctional nanoparticles in A549 lung carcinoma–bearing mice followed by laser irradiation showed a progressive reduction of the tumor volume. The combination of photothermal destruction and chemotherapy was more effective in tumor regression than photothermal destruction or chemotherapy alone. Moreover, because of the presence of iron oxide nanoparticles, these multifunctional nanoparticles could also serve as a contrast agent for MRI with the T2-weighted imaging intensity found to be substantially darkened with increasing iron ion concentrations.

10.5.2.3   MRI and PDT

As discussed earlier, PDT utilizes photosensitizing agents that, upon exposure to light of a particular wavelength, produce ROS (e.g., singlet oxygen), killing the target cancer cells. Like PTT, PDT is also accompanied by minimal risks of damaging nearby health tissues because of the possibility of focusing the light to the target area. Thus, combined delivery of a photosensitizer and an imaging agent could allow image-guided treatment of cancer and monitoring the therapeutic progress. Multifunctional polyacrylamide-based nanoparticles consisting of a surface-localized F3 peptide (with affinity to nucleolin expressed on angiogenic endothelial cells within tumor vasculature) and encapsulating both iron oxide nanoparticles and Photofrin, a photosensitive hematoporphyrin agent, have been tested for the simultaneous imaging and treatment of brain tumor (Reddy et al. 2006). Significant MRI contrast enhancement was achieved and improvement in survival rate after PDT was noticed in intracranial rat 9L gliomas after intravenous administration of the F3 peptide–carrying multifunctional nanoparticles. In another approach, Gd^{3+}-chelated fullerene C60–PEG conjugate was shown to maintain an enhanced MRI signal at the tumor tissue for a prolonged period after intravenous administration and with subsequent light irradiation (400–505 nm) exerted significant antitumor effects depending on the light exposure time (Liu et al. 2007).

10.5.2.4   Optical imaging and chemotherapy

Optical imaging is based on excitation of a fluorophore at a certain wavelength of light and subsequent emission of light from the fluorophore at a different wavelength, enabling visualization of biodistribution of the fluorophore or fluorophore-containing molecules or particles. The technique is thus highly dependent on the optical properties of tissue constituents at different wavelengths, with the penetration depth currently representing the major limitation owing to the strong scattering properties of soft tissues. Scattering, however, decreases in the NIR region (700–900 nm), and therefore, recently developed fluorophores that emit/absorb in this region penetrate deeper in the tissues. A combination of NIR dyes and therapeutic molecules can potentially be used to develop nanotheranostics. In one approach, chitosan-based theragnostic nanoparticles were developed by labeling the nanoparticles with Cy5.5, a near-infrared fluorescent (NIRF) dye and loading them with paclitaxel. After intravenous delivery of the theragnostic nanoparticles in mice bearing SCC7 murine squamous carcinoma tumors, NIR fluorescence images exhibited significantly increased tumor uptake of the nanoparticles by virtue of the EPR effect and a progressive inhibition of tumor growth. Fluorescence signal intensity was found to correlate to the amount of the nanoparticles administered and the frequency of injection. Deep tissue imaging can also be carried out using NIR Qds, which are highly attractive fluorescent nanocrystals because of their large excitation spectrum and the possibility of modulating the emission as a function of their crystal size and chemical composition (Mura and Couvreur 2012).

10.5.2.5 Optical imaging and PDT

Photosensitizing agents can be designed not only to generate ROS for PDT but also to induce the emission of strong fluorescence signals for imaging, through irradiation. pH-sensitive block copolymer micelles encapsulating the photosensitizer PpIX were designed for the purpose of both optical imaging and PDT (Koo et al. 2010). Twenty-four hours after intravenous injection of the multifunctional nanoparticles in SCC-7 squamous cell carcinoma–bearing mice, irradiation with a 670-nm pulsed laser diode produced a visible strong fluorescent signal in the tumor that was distinguishable from the surrounding healthy tissues, while irradiation with a 633-nm laser led to the reduction of the tumor volume owing to the production of the cytotoxic oxygen singlet.

10.5.2.6 Optical imaging and PTT

Porphysomes are the lipid nanoparticles self-assembled in an aqueous buffer from phospholipid–porphyrin conjugates that exhibit a liposome-like structure where the hydrophobic chromophore (porphyrin) was positioned in place of an alkyl side chain, maintaining an amphiphilic structure (Lovell et al. 2011). Two days after intravenous injection of the porphysomes, high fluorescence intensity was observed in the KB tumors of xenograft-bearing mice, while laser irradiation at 660 nm 1 day after the systemic administration was found to induce the photothermal tumor ablation.

Chapter ten: Nanotechnology in innovative treatment strategies 169

10.5.2.7 Optical imaging and PTT/PDT

The GNR–photosensitizer complex is a promising multifunctional nano-medicine for NIR fluorescence imaging and cancer therapy through PTT/PDT, with the distance between the GNRs and the photosensitizers exploited to control the fluorescence emission and singlet oxygen generation. For instance, GNRs were PEG conjugated and complexed with a photosensitizer, Al(III) phthalocyanine chloride tetrasulfonic acid (GNRs-AlPcS4), for evaluating the fluorescence imaging and antitumor efficacy on mice bearing SSC7 squamous carcinoma tumor xenograft (Jang et al. 2011). The long circulating GNRs passively accumulated in tumors by EPR effect were used to induce the hyperthermia with the help of an externally applied NIR light to trigger cell death by PTT, while the GNR-associated photosensitizers were irradiated at suitable wavelengths to emit strong NIR fluorescence, enabling the visualizing of the tumor with a high signal-to-background ratio and to produce ROS as well for killing the cells by PDT. When the photosensitizers are located close to the surface of GNR, they are nonfluorescent as well as nontoxic. However, after accumulation into the tumor tissue, the photosensitizers are released and become highly fluorescent and phototoxic. Although PTT alone did not show a sufficient anticancer effect, a significant therapeutic efficacy was obtained with PDT with the highest improvement observed for the combination of PTT and PDT.

10.5.2.8 Combined MR/optical imaging and therapy

A multimodal imaging and drug delivery system was developed based on an oil-in-water nanoemulsion in which iron oxide crystals for MRI and the hydrophobic glucocorticoid prednisolone acetate valerate (PAV) as a potent anticancer drug were encapsulated inside the lipid core, while Cy7 dye and PEG were embedded on the surface for NIRF imaging and particle stabilizing, respectively (Gianella et al. 2011). The effectiveness of this theranostic nanodevice was evaluated on a colon cancer model. After intravenous injection, imaging by MR or NIRF demonstrated the preferential accumulation of the nanoemulsion in the tumors, whereas the therapeutic efficiency of nanoemulsion-loaded PAV was evident from a significant inhibition of the tumor growth.

10.5.2.9 Ultrasonograpy and chemotherapy

A theranostic approach using ultrasound as imaging modality was demonstrated by an oil-in-water emulsion system composed of a core of liquid perfluorooctyl bromide (PFOB), an effective ultrasound contrast agent, and an outer monolayer of phospholipids stably incorporating melittin, a non-specific cytolitic peptide capable of inducing cell lysis (Soman et al. 2009). After intravenous administration of the melittin-loaded nanoemulsion in an MDA-MB-435 xenograft model of breast cancer, the ultrasound imaging facilitated by a significant contrast enhancement by emulsion-loaded PFOB enabled monitoring the therapeutic efficacy, thus demonstrating a significant inhibition of the tumor growth.

10.5.3 Nanotheranostics in cardiovascular diseases

Current methods for early detection and therapy of cardiovascular diseases, particularly atherosclerosis, the prime cause of coronary heart disease, are limited and therefore development of nanotheranostics could pave the way to an effective solution.

10.5.3.1 MRI and antiangiogenic therapy

Neoangiogenesis is crucial for high plaque instability leading to myocardial infarction and stroke. Despite the development of various antiangiogenic drugs, such as fumagillin, the high dose required to produce the therapeutic effect is often associated with severe neurocognitive side effects. Since vascular epithelial cells widely express $\alpha v\beta 3$-integrin, $\alpha v\beta 3$-targeted Gd^{3+}-based nanoparticles with encapsulated fumagillin were designed as a potential theranostic system for better management of atherosclerosis and subsequently tested on hyperlipidemic New Zealand rabbits (Winter et al. 2006). Intravenous injection of the $\alpha v\beta 3$-targeted nanoparticles resulted in a significant contrast enhancement, allowing a clear visualization of the neovasculature even in the early-stage plaque induced by the high-cholesterol diet. A drastic reduction of MRI signal intensity after a second injection of the $\alpha v\beta 3$-targeted nanoparticles was attributed to the treatment outcome as also reflected in the reduction of angiogenesis and expression of $\alpha v\beta 3$ integrin.

10.5.3.2 Optical imaging and PDT

Macrophages promote degradation and rupture of the atherosclerotic plaque through the accumulation of oxidized lipoproteins and the release of inflammatory cytokines and extracellular proteinases. Since the cross-linked dextran-coated iron oxide (CLIO) nanoparticles possess high affinity toward the inflammatory macrophages, a theranostic nanodevice was developed by modifying CLIO with a NIR fluorophore for fluorescence imaging and a potent chlorin-based photosensitizer for PDT (McCarthy et al. 2010). Twenty-four hours after injection of the theranostic nanoparticles in a murine model of atherosclerosis, surgically exposed carotid arteries were imaged by intravital laser scanning fluorescence microscopy, revealing localization of the particles in macrophage-rich regions within the inflamed atherosclerotic lesions. In addition, subsequent irradiation of the mice with a 650-nm laser light in order to activate the photosensitizer resulted in extensive apoptosis of the targeted macrophages.

chapter eleven

Nanoparticles for therapeutic delivery in animal models of different cancers

11.1 Brain cancer

Treatment of brain tumors remains a great challenge with a critical correlation found between early detection and positive prognosis. Surgery, radiotherapy, and chemotherapy are the currently available treatment options. Most brain tumors are removed by surgery, while radiotherapy, which relies on high-powered rays, is often used to kill cancer cells or stop their growth in case the tumor tissue cannot be removed with surgery or will likely remain after surgery. Like radiotherapy, chemotherapy, which is based on intravenous or oral administration of anticancer drugs, can be used to kill cancerous or malignant tumors, such as glioblastoma multiforme (GBM), the most frequent primary central nervous system tumor. However, in addition to having an adverse effect on healthy dividing cells, many cancer drugs are unable to cross the blood–brain barrier (BBB). Nanoparticles have emerged as potential drug carriers in the treatment of brain tumors partly because of their potential ability to cross the BBB, in addition to offering favorable pharmacokinetics to an encapsulated drug (Nunes et al. 2012).

11.1.1 Taxane-loaded poly(ε-caprolactone) nanoparticles

Paclitaxel (PTX) as a powerful microtubule stabilizing agent has noticeable antitumor activity in glioma cell lines. However, because of the poor aqueous solubility and poor penetration capacity across the BBB, its clinical application to treat GBM is extremely limited. To overcome the shortcomings, PTX was encapsulated in methoxy polyethylene glycol (PEG)–poly(ε-caprolactone) (PCL) (MPEG–PCL) nanoparticles and the formulation was intravenously administered to intracranial C6 glioblastoma-bearing mice (Xin et al. 2010). The mean survival time of MPEG–PCL/PTX nanoparticles (28 days) was found to be much longer than that of Taxol alone (20 days) or PCL/PTX nanoparticles (23 days), which could be correlated to the higher accumulation of the latter in the tumor than two control groups. In an attempt to further improve the therapeutic efficacy, Angiopep-2, a specific ligand of low-density lipoprotein receptor–related protein that is overexpressed not only on the BBB but also on glioma cancerous cells, was coupled to PEG–PCL nanoparticles for PTX delivery, finally bringing about an enhanced anti-glioblastoma effect in vivo compared to Taxol and unmodified PEG–PCL nanoparticles (Xin et al. 2012). For targeted delivery of docetaxel (DTX), a taxane derivative similar to PTX, to glioblastoma, PEG–PCL nanoparticles were functionalized with GMT8, a DNA aptamer with affinity for U87 human glioma cells, which significantly prolonged the median survival time of glioblastoma-bearing mice compared to nontargeted nanoparticles or DTX (Gao et al. 2012).

11.1.2 Doxorubicin-loaded surfactant-coated nanoparticles

Biodegradable poly(butyl cyanoacrylate) (PBCA) nanoparticles, when overcoated with the nonionic surfactant polysorbate, could significantly enhance the enrichment of loaded doxorubicin (DOX) in the brain compared to the uncoated nanoparticles, after intravenous administration to rats (Gulyaev et al. 1999). When glioblastoma-bearing rats were subjected to the DOX-bound polysorbate-coated nanoparticles, a significant increase in rat survival time was observed compared to other control groups, with more than 20% of the animals in the former group showing a long-term remission (Steiniger et al. 2004). Surface functionalization of PBCA nanoparticles with other surfactants, such as poloxamer 188 (Pluronic F68), could also enhance the antitumor effect of DOX against intracranial glioblastoma (Ambruosi et al. 2006).

11.1.3 hTRAIL gene-loaded cationic albumin– conjugated PEGylated PLA nanoparticles

Proapoptotic Apo2 ligand/tumor necrosis factor–related apoptosis-inducing ligand (Apo2L/TRAIL) is a promising therapeutic gene for treating GBM, since most glioma cell lines express the agonist Apo2L/TRAIL receptors. TRAIL selectively induces apoptosis of the glioma cells by binding to its cognate DR4 and DR5 receptors, without affecting the normal cells that express DcR1 and DcR2 antagonistic receptors. On the other hand, cationic albumin–conjugated PEGylated PLA nanoparticles showed their accumulation in mouse brain cells upon intravenous administration. Therefore, hTRAIL plasmid DNA was incorporated into the nanoparticles to transport the bound plasmid across the BBB. Repeated intravenous administration of the resulting DNA-loaded construct to BALB/c mice bearing i.c. C6 gliomas induced apoptosis and significantly delayed tumor growth, with a moderate effect on median survival of tumor-bearing mice (Lu et al. 2006).

Chapter eleven: Nanoparticles for therapeutic delivery in cancers

11.1.4 Nanoliposomal irinotecan

Irinotecan (CPT-11) is a potent anticancer drug that inhibits the action of topoisomerase I. To reduce the toxicity and preserve the active form of the drug, irinotecan was encapsulated in liposome and administered intravenously in the murine intracranial U87MG brain tumor model (Noble et al. 2014). Nanoliposomal irinotecan led to an almost 10-fold increase in tumor drug accumulation compared with free irinotecan, with a total of 33% of the treated animals completely cured compared with no survivors in the other groups.

11.1.5 Bcl2L12-targeting spherical nucleic acids

Spherical nucleic acids (SNAs) consist of a 13-nm gold nanoparticle core decorated with a corona of thiolated small interfering RNA (siRNA) duplexes. Since the oncogene Bcl2-Like12 (Bcl2L12) is overexpressed prominently in GBM, while exhibiting low or undetectable levels in the surrounding glial cells of normal brain tissue, SNAs were designed to target the oncogene Bcl2-Like12 (Bcl2L12), an effector caspase and p53 inhibitor (Jensen et al. 2013). Upon systemic delivery, the SNAs reduced Bcl2L12 expression in intracerebral GBM, increasing intratumoral apoptosis and reducing tumor burden and progression in the xenografted mice.

11.2 Breast cancer

Breast cancer is the most common malignancy in women worldwide and is usually treated with chemotherapy often after surgical removal of the tumor to prevent disease recurrence. Among the drugs, anthracyclines (such as DOX) and taxanes (such as PTX A) are widely used as indispensable drug components in several combination regimens. Despite contributing to favorable patient outcomes, the current drug formulations possess significant shortcomings that predominantly stem from their lack of specificity or failure to discriminate between malignant and noncancerous cells. Cardiotoxicity after DOX administration and hypersensitivity reactions and neuropathic pain associated with PTX treatment are now well documented. Apart from increasing solubility of the hydrophobic drugs, the conventionally used excipient materials do not play any role in influencing the critical pharmacokinetic parameters required for drug accumulation and bioavailability within tumors (Blanco and Ferrari 2014). Human epidermal growth factor receptor 2 (HER2), which is overexpressed in approximately 15% of all breast cancers, has become one of the most important therapeutic targets, leading to the development of four licensed anti-HER2-based nanotherapeutics: trastuzumab, lapatinib, pertuzumab, and trastuzumab emtansine for increasing survival in both the metastatic and early-stage settings of the disease (Moya-Horno and Cortés 2015).

11.2.1 Anti-HER2 liposome formulations of DOX and topotecan

To enhance the targeting and internalization of DOX in HER2-overexpressing breast tumors, a recombinant Fab' derived from trastuzumab was conjugated to a DOX-loaded PEGylated liposome (Park et al. 2002). After single intravenous administration to rats, both the Fab' and DOX components of anti-HER2 immunoliposome-DOX showed concordant plasma PK values with terminal $t_{1/2}$ = 10 hours for both, whereas free DOX was undetectable after 5 minutes, and free rhuMAb HER2-Fab' had a terminal $t_{1/2}$ of 1–2 hours, thus clearly indicating that the immunoliposomes greatly prolonged the circulation of both Fab' and DOX. Furthermore, in contrast to free DOX that produced modest tumor growth inhibition, anti-HER2 immunoliposome-DOX showed highly potent anti-tumor efficacy with frequent cures (16%–50%) and markedly reduced host toxicity in four different HER2-overexpressing murine breast cancer xenograft models. In a separate study, anti-HER2 liposomes containing topotecan, a topoisomerase I inhibitor, were applied intravenously to the mice bearing BT474 breast tumors, resulting in significantly improved antitumor activity (without complete regression of tumors) compared to the nontargeted nanoliposomal topotecan (Drummond et al. 2010).

11.2.2 Rapamycin-loaded nanoparticles of elastin-like polypeptide diblock copolymers

Rapamycin (Rapa), a cyclic and hydrophobic macrolide antibiotic, has great potency in suppressing immune response by inhibiting proliferation of lymphocytes. Its antitumor potential in treating breast, prostate, and colon cancers is currently under clinical investigation. Through binding to its cognate receptor FKBP, Rapa inhibits the mTOR pathway that has essential roles in cell proliferation and growth and thus sequesters cancer cells in the G_1 phase. However, free Rapa treatment is associated with a number of drawbacks, such as severe cytotoxicity, low bioavailability, and rapid clearance. To address these concerns, diblock copolymers of two genetically engineered elastin-like polypeptides, with a hydrophobic to hydrophilic length ratio of 1:1, were used to form stable nanoparticles with the surface decorated with FKBP (Shi et al. 2013). Thus, the hydrophobic drug (Rapa) could be encapsulated into the nanoparticle core via hydrophobic interactions (similar to other micelle delivery systems), in addition to having specific drug binding at the corona. As a result, the terminal half-life of drug release from the construct was extended to approximately 60 hours. Intravenous injection of Rapa nanoparticles in

Chapter eleven: Nanoparticles for therapeutic delivery in cancers 175

an MDA-MB-468 breast cancer xenograft model led to an approximately fivefold increase in tumor growth inhibition and, thus, a significant prolongation of mouse survival.

11.2.3 Sorafenib-loaded nanoliposomal ceramide

Phosphatidylinositol 3-kinase/Akt and Ras/Raf/mitogen-activated protein kinase pathways that regulate apoptosis and proliferation, respectively, are frequently upregulated in breast cancer like other cancers, such as melanoma. Therefore, drugs that target the components of the two pathways are attractive candidates for cancer therapy. To explore the synergistic potential of targeting both the PI3K/Akt/mTOR pathway and the MAPK/ERK pathway in enhancing therapeutic activity against breast cancer, sorafenib, a Raf kinase inhibitor, was combined with nanoliposomal ceramide, an agent that causes dephosphorylation of Akt (Tran et al. 2008). Intraperitonieal injection of sorafenib on alternate days and daily intravenous administration of nanoliposomal ceramide in an MDA-MB-231 murine model of breast cancer resulted in an approximately 30% increase in tumor inhibition compared with sorafenib treatment alone and an approximately 58% reduction in tumor size compared with nanoliposomal ceramide monotherapy with negligible systemic toxicity.

11.2.4 Mdr-1 and surviving siRNA-loaded poly(β-amino esters)

P-glycoprotein (P-gp), one of the ATP-binding cassette transporters responsible for multidrug resistance (MDR), is overexpressed in the malignant tissues of almost 40%–50% of breast cancer patients. On the other hand, survivin, a member of the inhibitors of apoptosis family, with a role in both cell division and apoptosis control, is frequently upregulated in cancers. Therefore, combined delivery of siRNAs using a single vector for targeted knockdown of P-gp and survivin mRNAs is a promising strategy to synergistically overcome the MDR and sensitize the breast cancer cells to chemotherapy. A biodegradable polymer of poly(β-amino esters) (PAEs) containing amino segments that enable the PAEs to electrostatically complex with nucleic acid molecules and acid- or base-sensitive ester bonds that help the polymer release from endosomes to cytoplasm was employed to form complexes with P-gp short hairpin RNA (shRNA) plasmid and survivin shRNA plasmid (Yin et al. 2012). When administered via the tail vein in a mice xenograft model bearing MCF7/ADR tumors, DOX, which was given at 24 hours after administration of the PAEs–shRNA complexes, significantly suppressed the tumor growth, with the best tumor suppression efficiency achieved when DOX was combined with the PAEs–shRNA complex targeting both P-gp and survivin.

11.2.5 DOX-loaded folate-targeted pH-sensitive polymer micelle

The antimetastatic activity of DOX loaded in a pH-sensitive mixed polymeric micelle consisting of two block polymers, poly(L-lactide) (PLLA)-b-PEG-folate and poly(L-histidine) (PHis)-b-PEG was examined in an aggressive murine mammary carcinoma model (Gao et al. 2011). The targeted, pH-responsive micelles significantly inhibited the growth of 4T1 breast tumors in nude mice after intravenous injection, with 100% animal survival in the entire duration (43 days) of the experiment. The folic acid as a targeting moiety and the pH-sensitive core phase enabled the micelle to recognize the 4T1 cells and triggered the drug release at tumor sites, respectively, for the effective killing of the cancer cells. In addition, a significant decrease in tumor metastasis was observed for the folate-decorated, DOX-loaded micelle, unlike DOX carried by PBS, PLLA-b-PEG micelle, and PHis-b-PEG micelle, which showed significant metastasis to the lung and heart.

11.3 Colon cancer

Colorectal cancer (CRC) is one of the most common internal malignancies affecting millions of men and women in all age groups. Like many other cancers, conventional chemotherapies are ineffective in CRC because of the short half-lives of the anticancer drugs and their inability to reach the tumor site in effective concentrations. 5-Fluorouracil (5-FU), a fluorinated pyrimidine, is one of the most widely used therapeutic agents for the treatment of colon cancer, with folic acid, which stabilizes drug–enzyme binding applied as a common adjuvant in the treatment. When 5-FU is administered to cancer patients, 80% of the drug is metabolized in the liver and other extrahepatic tissues and 15% is excreted by the kidney while only 5% reaches the tumor (Prados et al. 2013). The emergence of nanotechnology would have a profound effect on enrichment of chemotherapy agents in cancer tissues, thereby enhancing the therapeutic efficacy and alleviating the side effects of the drugs.

11.3.1 Taxol-containing liposomes

To improve the efficacy and reduce the toxicity of Taxol-based anticancer therapy, Taxol was incorporated in a liposomal formulation and tested in a Taxol-resistant murine colon tumor model (Colon-26) (Sharma et al. 1993). While free Taxol given intravenously was ineffective at delaying tumor growth at the maximum tolerated dose (MTD), Taxol–liposomes were well tolerated at doses greater than or equal to the MTD of free Taxol and showed a significant delay of tumor progression at almost all dose levels.

Chapter eleven: Nanoparticles for therapeutic delivery in cancers 177

Taxol–liposome formulations demonstrated lower acute as well as chronic toxicity in tumor-bearing mice than the Taxol–Cremophor formulation.

11.3.2 Endostar-loaded PEG–PLGA nanoparticles

Endostar is a novel recombinant human endostatin that inhibits the growth of a variety of human tumors by inhibiting neovascularization. However, because of its short half-life and rapid metabolism, multiple injections are often required, causing poor patient compliance and, thus, limiting its clinical use. Encapsulation of Endostar into biodegradable PEG–PLGA nanoparticles was shown to enhance the accumulation of the protein and drastically reduce the vascular endothelial growth factor level in tumor tissue after intravenous injection in BALB/c mice (Hu and Zhang 2010). In addition, systemic administration of the Endostar-loaded nanoparticles significantly inhibited the tumor growth and reduced the tumor microvascular density in an HT-29 colon cancer model.

11.3.3 PEG-coated Bcl-2 siRNA lipoplex

Although 5-FU is traditionally used as the drug of choice for treating human CRC, overexpression of antiapoptotic proteins such as Bcl-2 in the cancer cells often causes 5-FU resistance, leading to treatment failure. To silence the expression of Bcl-2 and, thus, enhance the chemosensitivity of tumors, PEG-coated Bcl-2 siRNA lipoplex was formulated and intravenously administered every 2 days along with the daily oral administration of Tegafur (TF), a prodrug of 5-FU (to mimic 5-FU infusion) in a DLD-1 (a CRC) xenograft mouse model (Nakamura et al. 2011). The combined treatment resulted in superior tumor growth suppression with a tumor weight reduction of as much as 62% compared to each single treatment either with Bcl-2 siRNA lipoplex or S-1 accounting for 15% or 28% reduction in tumor weight, respectively. Moreover, the daily oral delivery of S-1 enhanced the accumulation of the PEGylated siRNA lipoplex in the tumor tissue.

11.3.4 Thymidylate synthase shRNA-expressing Ad

5-FU inhibits the enzyme thymidylate synthase (TS), leading to dTTP depletion and inhibiting RNA and DNA synthesis. Since a high level of intratumoral TS expression is also linked to resistance to 5-FU in CRC, another similar approach was undertaken by engineering Ad to express TS shRNA (Kadota et al. 2011). The combined treatment of TS shRNA-expressing Ad and S-1 demonstrated the strongest antitumor effect with significantly high apoptotic index against the DLD-1 xenografts in nude mice in comparison to each single treatment, that is, S-1 or Ad-TS shRNA.

11.3.5 Oxaliplatin-encapsulated chitosan micelle

Cancer stem-like cells (CSLCs) with self-renewal and pluripotent capabilities are a subpopulation of cells within a tumor, playing active roles in tumor initiation, growth, and metastasis. The non-CSLCs within the tumor can be transformed to CSLCs through epithelial–mesenchymal transition. The resistance of CSLCs to conventional chemotherapy drugs such as oxaliplatin (OXA) is partly accountable for the failure of cancer treatments. A polymeric micelle based on stearic acid-g-chitosan oligosaccharide (CSO-SA) was fabricated to encapsulate OXA for effectively killing both CSLCs and bulk cancer cells in CRC (Wang et al. 2011). Intravenous administration of CSO-SA/OXA micelles was found to effectively retard the tumor growth and reduce the CSLC (CD133+/CD24+) cells compared with free OXA treatment, which is responsible for CSLC enrichment in CRC xenografts.

11.4 Lung cancer

Lung cancer, which is the most prevalent type of cancer and is a major cause of death throughout the world, is caused by smoking and exposure to high levels of radiation, pollution, and asbestos. The various treatment modalities for lung cancer include chemotherapy, radiation, and surgery, with the surgical resection applicable mostly for early-stage and localized lung tumors (Chandolu and Dass 2013). However, usually by the time the tumor is diagnosed, it is found to have already metastasized, thus requiring radiation therapy or chemotherapy with or without involving an earlier surgical procedure to remove the tumor. Non–small cell lung cancer (NSCLC), which is the most common lung cancer, for instance, has the propensity to metastasize early and develop resistance to a wide range of anticancer drugs. Hence, innovative nanotechnology-based strategies are urgently required to effectively treat the lung cancer.

11.4.1 EGFR siRNA-containing anisamide–PEG–LPD nanoparticles

Epidermal growth factor receptor (EGFR) is overexpressed in a variety of tumors with diverse roles in increasing proliferation, decreasing apoptosis, enhancing metastasis, and inducing resistance to chemo- and radiation therapy. Silencing of EGFR expression, which was shown to induce cell cycle arrest, apoptosis, growth inhibition, and chemosensitization of tumor cells, is an alternative to the EGFR therapy, which is currently based on tyrosine kinase inhibitors and monoclonal antibody. To induce apoptosis by suppressing EGFR expressing in the NCI-H460 xenograft tumor of NCI-H460, a highly aggressive lung cancer cell line, siRNA targeting EGFR was loaded into liposome–polycation–DNA (LPD) nanoparticles composed of

Chapter eleven: Nanoparticles for therapeutic delivery in cancers   179

calf thymus DNA (a carrier DNA), protamine (a highly positively charged peptide), and cationic liposomes (consisting of DOTAP [1,2-bis(oleoyloxy)-3-(trimethylammonio)propane] and cholesterol at 1:1 molar ratio) and stabilized by using a PEGylated lipid (DSPE-PEG) tethered to anisamide, a targeting ligand with affinity for sigma receptor expressed in NCI-H460 cells (Li et al. 2008). Three daily intravenous injections of the formulation led to the silencing of EGFR in the xenograft tumor and induced apoptosis in ~15% of the tumor cells and 40% tumor growth inhibition when combined with cisplatin. The pharmacokinetic study showed that 4 h after intravenous injection of the formulation, 70%–80% of injected fluorescent siRNA accumulated in the tumor while ~10% and ~20% were detected in the liver and the lung, respectively. A significant improvement in tumor growth inhibition was observed with targeted delivery of EGFR siRNA after substituting the DOTAP in the LPD nanoparticles with DSGLA, a novel non–glycerol-based cationic lipid that more efficiently downregulated pERK in H460 cells than DOTAP (Chen et al. 2009).

11.4.2   Bcl-x SSO-loaded anisamide–PEG–LPD nanoparticles

Splice-switching oligonucleotides (SSOs) designed by altering the oligonucleotide sugar-phosphate backbone (e.g., 2′-O-methoxyethyl ribose modification) modulate alternative splicing of multiexon pre-mRNA transcripts by hybridizing to the pre-mRNA sequences and thereby blocking access by various splicing factors. Through alternative splicing, the Bcl-x gene yields two major protein isoforms with opposing functions: anti-apoptotic Bcl-x_L, which is upregulated in a number of cancers, enabling survival of cancer cells and conferring resistance to a broad range of chemotherapeutic drugs, and pro-apoptotic Bcl-x_S, which antagonizes the anti-apoptotic activity of Bcl-x_L and another anti-apoptotic protein, Bcl-2. Therefore, redirecting Bcl-x splicing from Bcl-x_L to Bcl-x_S would induce apoptosis and increase chemosensitivity in cancer cells. However, to harness the full therapeutic benefit, SSOs, which are readily degradable in blood, must be safely and efficiently delivered to the pharmacological site of action. Anisamide–PEG–LPD nanoparticles were employed to intravenously deliver Bcl-x SSOs to kill the B16F10 murine melanoma cells metastasized to the lungs (Bauman et al. 2010). Targeted delivery of the Bcl-x SSOs to the B16F10 cells that express the sigma receptor resulted in modification of Bcl-x pre-mRNA splicing in the lung metastases and reduced tumor load.

11.4.3   DOTAP/cholesterol–tumor suppressor plasmid complexes

Nitrogen permease regulator-like 2 (NPRL2) is a tumor suppressor gene that is inactivated in the early stages of lung cancer development, but upon reactivation, it can inhibit growth of tumor cells, stimulate their apoptosis, and downregulate the metastatic process. Cisplatin that can form platinum–DNA adducts and induce cytotoxicity by interfering with the DNA repair process is among the most effective cytotoxic agents for advanced cancer treatments including NSCLC. However, its therapeutic efficacy is severely limited by the intrinsic or acquired resistance developed by cancer cells. On the basis of the hypothesis that restoring normal NPRL2 function in cisplatin-resistant tumor cells might resensitize the cells to cisplatin, NPRL2 plasmid was electrostatically condensed with cationic DOTAP/cholesterol (DC) forming nanoparticles for subsequent assessment of their antitumor potential in an orthotopic mouse model of human NSCLC (H322 cell line) (Ueda et al. 2006). Intravenous administration of the NPRL2 nanoparticles together with intraperitoneal injection of cisplatin resulted in suppression of tumor growth with a reduction greater than 90% compared to other treatments, as a result of overcoming cisplatin-induced resistance. In another study, systemic coadministration of DC nanoparticles carrying tumor suppressor genes FUS1 and p53 via tail vein injection in human H322 orthotopic lung cancer xenograft mouse models was reported to synergistically suppress the development and growth of the tumors (Deng et al. 2007). DC nanoparticles were also applied to complex with the gene (in a plasmid) of 101F6, a tumor suppressor protein expressed in normal lung bronchial epithelial cells and fibroblasts but lost in most lung cancers (Ohtani et al. 2007). Intravenous injection of 101F6 nanoparticles and intraperitoneal administration of ascorbate synergistically inhibited both tumor formation and growth in human NSCLC H322 orthotopic lung cancer mouse models.

11.4.4 Poly-(γ-L-glutamylglutamine)–PTX nanoparticles

Conjugation of PTX to poly-(γ-L-glutamylglutamine) (PGG), which consists of poly(L-glutamic acid) to which a glutamate side chain is added to each of the monomers in the polymer, causes interactions of hydrophobic PTX molecules with each other resulting in spontaneous collapse of the polymer into nanoparticles of ~20 nm in aqueous environments. The pharmacokinetic studies in mice bearing NCI-H460 lung tumors showed that conjugation of PTX to the PGG polymer prolonged the plasma half-life and the period of accumulation in tumor, with reduced washout of the drug from the tumor. A single dose of intraperitoneally administered PGG–PTX resulted in more significant inhibition of tumor growth than Abraxane, a commercial nanoparticular formulation of PTX, at doses that produced equivalent degrees of acute weight loss (Feng et al. 2010).

Chapter eleven: Nanoparticles for therapeutic delivery in cancers 181

11.4.5 DOX-encapsulated PEG–PE micelle

Incorporation of DOX into micelles composed of PEG–phosphatidyletha-nolamine (PE) block copolymer improved penetration and antitumor effi-cacy in the Lewis lung carcinoma (LLC) mouse model, which could be attributed to the enhanced tumor accumulation through the enhanced permeability and retention (EPR) effect and subsequent efficient cellular uptake of the DOX-loaded PEG–PE micelles via endocytosis (Tang et al. 2007). Micelle-encapsulated DOX treatment via tail vein injection pro-longed survival in both mouse subcutaneous and pulmonary LLC models and reduced metastases in the pulmonary model, with demonstration of fewer signs of toxicity than free DOX.

11.5 Ovarian cancer

Ovarian cancer is the most lethal gynecologic malignancy in women with most of the patients diagnosed at an advanced stage of the disease, at which point the 5-year survival rate is below 30%. The treatment for the patients with advanced ovarian cancer includes maximal surgical cytoreduction followed by systemic platinum-based chemotherapy. However, use of cis-platin (CDDP) is greatly limited by severe side effects, poor half-life owing to rapid elimination, and development of acquired drug resistance. There has been substantial progress in development of nanotechnology-based approaches to increase the survival rate in ovarian cancer.

11.5.1 EphA2 siRNA-encapsulated neutral liposomes

The EphA2 gene encoding the epithelial cell receptor protein tyrosine kinase is overexpressed in multiple cancer types including ovarian can-cer, indicating the aggressive features of tumor growth and thus provid-ing a predictive marker for tumor recurrence and patient survival. In nude mice bearing intraperitoneal ovarian tumors, the EphA2 siRNAs encapsulated in DOPC (1,2-dioleoyl-*sn*-glycero-3-phosphatidylcholine)-based neutral liposomes effectively reduced EphA2 expression in the tumors 48 hours after a single dose administration. Furthermore, intra-venous delivery of DOPC-encapsulated EphA2-targeting siRNA reduced tumor growth in the same orthotopic mouse model of ovarian cancer and additionally, when combined with intraperitoneally administered PAX, led to a dramatic decline in the tumor growth (Landen et al. 2005).

11.5.2 Liposomal EphA2 siRNA-loaded silica particles

In order to facilitate tumor-specific delivery and sustained release of a therapeutic siRNA, positively charged porous silica particles were used to load the DOPC liposomes with encapsulated EphA2 siRNA (30–35 nm in diameter) into the nanopores (60 nm in diameter) (Shen et al. 2013). Treatment of the mice carrying metastatic SKOV3ip2 ovarian tumors with the resulting formulation delivered via the intravenous route caused a dose-dependent reduction of tumor weight and number of tumor nodules, whereas the tumor growth was completely inhibited by the treatment in combination with intraperitoneally delivered PAX. Combining the treatment in the same way with DTX was found to inhibit the growth of DTX-resistant HeyA8-MDR ovary tumors.

11.5.3 Multifunctional DOX-carrying lipid-based nanoformulations

In an effort to circumvent DOX resistance in ovarian cancer, multifunctional liposomes were fabricated by modifying the commercially available DOX-containing liposomes, Lipodox, with a cell-penetrating peptide, transactivator of transcription peptide, a pH-sensitive PEG–PE conjugate for shielding the peptide, and the antinucleosome monoclonal antibody 2C5 to target the nucleosomes overexpressed on the tumor cell surface, and finally the antitumor activity of the formulation on tumor xenografts developed subcutaneously in nude mice with MDR and drug-sensitive human ovarian cancer cells (SKOV-3) was evaluated (Apte et al. 2014). After systemic administration, these long-circulating liposomes were found to be efficiently taken up by both drug-resistant and drug-sensitive SKOV-3 tumor xenografts, resulting in significant enhancement of therapeutic efficacy compared to Lipodox in both resistant and sensitive models, as indicated by the reduction in the tumor volume and final tumor weight and the increase in apoptosis. In a separate study, phospholipid-based nanoparticle clusters coated with the glycosaminoglycan hyaluronan, the major ligand of CD44 that is upregulated in ovarian cancer cells like many other cancer cell types, showed a robust tumor accumulation and a superior therapeutic effect of encapsulated DOX over free DOX in a resistant human ovarian adenocarcinoma (NAR) mouse xenograft model, apparently by utilizing both the passive (EPR effect) and the active CD44 hyaluronan–mediated targeting (Cohen et al. 2014).

11.5.4 Hybrid micelles carrying cisplatin and PTX

Self-assembly of triblock copolymers containing the blocks of ethylene glycol, glutamic acid, and phenylalanine (PEG–PGlu–PPhe) was harnessed for fabrication of the hybrid micelles consisting of PPhe hydrophobic core, cross-linked ionic PGlu intermediate shell layer, and PEG corona (Desale et al. 2013). Co-incorporation of hydrophilic cisplatin (CDDP) and

Chapter eleven: Nanoparticles for therapeutic delivery in cancers 183

hydrophobic PTX, which are potent for combination chemotherapy owing to their having distinct mechanisms of action, into the multicompartment cross-linked micelles was shown to exert a superior antitumor activity after systemic administration of the (CDDP + PTX)/micelles, thereby significantly increasing overall survival of the animals compared to individual drug-loaded micelles or free cisplatin in A2780 human ovarian cancer xenograft-bearing female nude mice.

11.5.5 LHRH peptide- and PTX-conjugated dendrimer/CD44 siRNA

Since CD44 is responsible for tumor cell growth, MDR, and metastasis in ovarian and other cancers, suppression of the CD44 protein by employing siRNA targeted to CD44 mRNA in cancer cells could be a valuable therapeutic option. A multifunctional nanocarrier system containing a modified polypropylenimine dendrimer, the anticancer drug PTX as a cell death inducer, a synthetic analog of luteinizing hormone–releasing hormone peptide as a tumor targeting moiety, and siRNA targeted to CD44 mRNA was developed and administered intraperitoneally in a murine xenograft model of human ovarian carcinoma, resulting in dramatic enhancement in accumulation of the complexes in the tumor, downregulation of the expression of CD44, and finally almost complete shrinkage of the tumor (Shah et al. 2013).

11.6 Pancreatic cancer

Pancreatic cancer is one of the most lethal cancers worldwide with a 5-year relative survival rate recorded below 6% only (McCarroll et al. 2014). Patients diagnosed with distant metastases have extremely high mortality rates despite the combinations of therapies ranging from surgery, chemotherapy, and radiotherapy. High resistance to therapy imposed by the pancreatic tumor microenvironment is a major challenge in the treatment of pancreatic cancer. For example, the enriched tumor stromal component and disorganized vasculature of pancreatic cancer tissues prevent delivery of a sufficient amount of therapeutic agents into the cancer cells. Nanotherapeutics could be a viable option to treat the cancer by overcoming the barriers and effectively reaching the target site.

11.6.1 uPAR-targeted, gemcitabine-loaded iron oxide nanoparticles

The enriched tumor stroma significantly limits the delivery of anticancer drugs and also promotes proliferation, invasion, metastasis, and chemoresistance in pancreatic cancer cells. Since pancreatic cancer tissues have high levels of urokinase plasminogen activator (uPA) receptors (uPARs) in tumor cells, tumor endothelial cells, and tumor stromal fibroblasts and macrophages, magnetic iron oxide nanoparticles (IONPs) were conjugated to the amino-terminal fragment (ATF) peptide of the receptor-binding domain of uPA, a natural ligand of uPAR, and to the chemotherapy drug gemcitabine (Gem) via a lysosomally cleavable tetrapeptide linker, for targeted delivery and release of the cytotoxic drug into the tumor and stromal cells overcoming the physical barrier of the stroma (Lee et al. 2013). Intravenous administrations of ATF-IONP-Gem significantly inhibited the growth of orthotopic human pancreatic cancer xenografts in nude mice.

11.6.2 DACHPt-loaded polymeric micelle

To promote the tumor-directed delivery and enhance the therapeutic efficacy of DACHPt, the parent complex of OXA in a pancreatic tumor spontaneously developed in transgenic mice, DACHPt was encapsulated in polymeric micelles composed of PEG-*b*-poly(L-glutamic acid) via polymer–metal complexation between DACHPt and the carboxylic groups of glutamic acid residues, forming the core of the micelles (Cabral et al. 2013). Repeated systemic administration of the DACHPt-loaded micelle significantly inhibited the growth of pancreatic cancers as well as the incidence of metastasis and prolonged the overall survival time in the clinically relevant pancreatic cancer model, by enhancing the accumulation of drugs in the tumors.

11.6.3 PEGylated human recombinant hyaluronidase PH20

Desmoplastic reaction surrounding pancreatic tumor cells generates very high amounts of interstitial fluid pressure (IFP) leading to the induction of vascular collapse while presenting substantial barriers to perfusion, diffusion, and convection of therapeutic agents. Systemic administration of PEGylated human recombinant hyaluronidase PH20 (PEGPH20), an enzymatic agent that targets hyaluronic acid (HA), a critical component of the desmoplastic stroma, produced a marked decrease in tumor stroma and a consequential rapid and significant decrease in IFP along with an increase tumor blood vessel lumen diameter in a murine pancreatic cancer model (Provenzano et al. 2012). When the PEGPH20 treatment is combined with gemcitabine, a standard chemotherapeutic drug, there was a strong antitumor effect compared to the drug alone, owing to the higher concentrations of the drug that could reach the tumor after removal of the barrier through PEGPH20-mediated cleavage of HA.

Chapter eleven: Nanoparticles for therapeutic delivery in cancers 185

11.6.4 HER-2 siRNA-loaded immunoliposome

A nanoimmunoliposome composed of a pH-sensitive histidine–lysine peptide–incorporated cationic liposome with surface-decorated anti–transferrin receptor (TfR) single-chain antibody fragment (TfRscFv) was fabricated to electrostatically complex with modified hybrid (DNA–RNA) anti-HER-2 siRNA molecules to serve to target the complex to a pancreatic tumor and facilitate subsequent cytosolic accumulation of the associate siRNA (Pirollo et al. 2007). Systemic injection of the siRNA-loaded multifunctional liposome in a human pancreatic (PANC-1) subcutaneous xenograft mouse model significantly inhibited the tumor growth and sensitized the tumor to intraperitoneally delivered gemcitabine.

11.7 Skin cancer

Melanoma, a cancer of pigmented skin cells called melanocytes residing at the epidermal–dermal junction, is the most aggressive type of skin cancer accounting for the vast majority of skin cancer deaths. Although an early-stage melanoma can be surgically removed, with an overall survival rate of 99%, metastasized melanoma is difficult to cure, with 5-year survival rates reported below 20% (Chen et al. 2013). Metastasized melanoma is currently treated by chemotherapy, targeted therapy, immunotherapy, and radiotherapy, with poor therapeutic outcomes. Despite initially having a higher positive response rate to targeted therapy, most of the melanoma patients with a mutation in the oncogene v-Raf murine sarcoma viral oncogene homolog B1 (BRAF) ultimately develop drug resistance. Nanotechnology could help in melanoma treatment by decreasing the drug resistance, increasing the therapeutic efficacy, and reducing the off-target effects.

11.7.1 PTX/ETP-loaded lipid nanoemulsions

To decrease the toxicity and increase the therapeutic action of PTX and etoposide (ETP), a cytotoxic drug that inhibits DNA synthesis by forming a complex with topoisomerase II and DNA, cholesterol-rich nanoemulsions were employed as carriers to deliver the drugs. Intraperitoneal injection in B16F10 melanoma-bearing mice resulted in a greater reduction in the number of animals bearing metastases in the nanoemulsion-encapsulated PTX + ETP group (30%) compared to the PTX + ETP group (82%), with a reduction of cellular density and blood vessels and an increase of collagen fibers in the tumor tissues observed in the former group but not in the latter group (Kretzer et al. 2012). Earlier, the same research group reported that administration of the lipid nanoemulsion with an encapsulated etoposide derivative promoted tumor growth inhibition and survival of the melanoma-bearing mice, by drastically reducing the drug toxicity, as reflected by the MTD, which was approximately fivefold greater than the commercial etoposide, and causing longer drug retention in the blood-stream than the latter formulation, with consequential enhanced tumor accumulation of the drug (Lo Prete et al. 2006).

11.7.2 Nanoliposomal siRNA targeting B-Raf and Akt3

Deregulation of the B-Raf and Akt3 signaling cascades are important regulators in early melanoma development. Approximately 60% of early invasive cutaneous melanomas have mutated B-Raf ([V600E]B-Raf) with high enzymatic activity, leading to constitutive activation of the MAP kinase pathway, whereas approximately 70% of melanomas have elevated Akt3 signaling because of increased gene copy number and PTEN loss. Targeting [V600E]B-Raf and Akt3 therefore presents a highly promising strategy to prevent or treat the cutaneous melanomas (Tran et al. 2008). Cationic nanoliposomes were used to stably complex with the siRNA targeting [V600E]B-Raf or Akt3 in order to protect the siRNA from degradation and facilitate its entry into melanoma cells. Application of low-frequency ultrasound at the tumor site enabled penetration of the nanoliposomal–siRNA complex throughout epidermal and dermal layers of laboratory-generated skin, as well as cooperatively decreased melanocytic lesion development in the skin of nude mice for the combination treatment (i.e., siMutB-Raf–nanoliposomal complex along with siAkt3–nanoliposomal complex).

11.7.3 Antigenic peptide–encapsulated PLGA nanoparticles

In general, antigenic proteins or peptides that are encapsulated in poly-meric nanoparticles can be efficiently taken up via phagocytosis and presented by APC as peptide fragments in complex with MHC class I mol-ecules on cell surfaces to subsequently induce a strong and specific T-cell immunity. Aiming to induce potent cytotoxic T lymphocyte responses against melanoma-associated self-antigen, relatively less immunogenic melanoma antigen peptides, hgp100 and TRP2, and a toll-like receptor 4 (TLR4) agonist were encapsulated in PLGA nanoparticles (Zhang et al. 2011). Intradermal injection of the peptide-loaded PLGA nanoparticles into mice induced strong antigen-specific T-cell responses, while vaccina-tion with the nanoparticles carrying TRP2 and monophosphoryl lipid A, a TLR4 agonist, caused a delay in the growth of subcutaneously inoculated B16 melanoma cells, which was further augmented by combined treat-ment with interferon-γ.

Chapter eleven: Nanoparticles for therapeutic delivery in cancers 187

11.7.4 MART-1 mRNA-loaded mannosylated nanoparticles

Genetically modified DCs expressing one or more tumor antigens (from intracellularly delivered DNA or mRNA) can present the relevant epitopes via the MHC class I pathway, leading to the induction of tumor antigen–specific cytotoxic T lymphocytes. As an example, the mRNA encoding MART-1 melanoma antigen was loaded into mannosylated and histidylated lipopolyplexes by condensing it with PEGylated histidylated polylysine and forming the polyplexes, which was then added to the mannose-conjugated liposomes (Perche et al. 2011). After intravenous injection, the mannosylated construct was four times more efficient in transfecting the splenic DCs possessing mannose receptors than the non-mannosylated counterpart. Furthermore, a greater inhibition of B16F10 melanoma growth was observed when the tumor-bearing mice were immunized with the mannosylated polyplex carrying MRT-1 mRNA than the non-mannosylated one.

11.7.5 Anti-CD47 siRNA-encapsulated liposome–protamine–HA nanoparticles

Overexpression of CD47, a "self-marker" on the surface of cancer cells, enables them to escape immunosurveillance by inhibiting their phagocytosis by macrophages via interaction of CD47 with its receptor α(SIRPα) present in macrophage cell membrane. Since human melanoma significantly upregulates CD47, knocking down CD47 or blocking its function by a monoclonal antibody would subject the tumor cells to macrophage-mediated destruction. Although antibody-based therapy is highly specific and generally very effective, there may be some safety concerns because of the ubiquitous expression of CD47, particularly on hematopoietic cells. Hence, in a more promising strategy, CD47 mRNA in melanoma was targeted by intravenously delivering siRNA formulated in the liposome–protamine–HA (LPH) nanoparticles (Wang et al. 2013). Efficient silencing of CD47 in tumor tissues significantly inhibited the growth of B16F10 melanoma tumors, with the tumor volume finally reduced by ~93% compared to the untreated controls. Moreover, CD47 siRNA-loaded LPH nanoparticles efficiently inhibited lung metastasis of melanoma B16F10 cells to approximately 27% of the untreated control in a murine lung metastasis model.

chapter twelve

Nanoparticles for therapeutic delivery in animal models of other critical human diseases

12.1 Arthritis

Rheumatoid arthritis (RA) and osteoarthritis (OA), which belong to the group of chronic, noninfectious arthritis and for which the currently available therapeutic approaches relying on nonsteroidal anti-inflammatory drugs (NSAIDs) and disease-modifying antirheumatic drugs, such as methotrexate (MTX), are inefficient with severe side effects, could be effectively treated with more specific monoclonal antibody-based drugs as well as disease-modifying antirheumatic nanomedicines (DMARNs), by promoting their selective targeting to the inflamed joints characterized by leaky postcapillary venules and effector cells with overexpression of transmembrane receptors and thereby circumventing the damage caused by drug exposure to the healthy tissues (Rubinstein and Weinberg 2012). Three different monoclonal antibodies—infliximab, adalimumab, and etanercept—that are now clinically used for treating RA target tumor necrosis factor-alpha (TNF-α), a proinflammatory cytokine involved in the pathogenesis of RA by increasing proinflammatory cytokines and recruiting immune cells to stimulate cell proliferation, and mediate the destruction of bone and cartilage (Barnes and Moots 2007). There are a number of nanoformulations investigated recently in animal models of RA and OA.

12.1.1 Liposomally conjugated MTX (G-MLV)

A single intra-articular injection of MTX-conjugated liposome in a rat antigen-induced arthritis model was demonstrated to significantly reduce knee swelling and inhibit the histological progression of the arthritis in comparison with MTX alone, with a concomitant reduction in interleukin (IL)-1β messenger RNA (mRNA) expression in the synovial tissue (Williams et al. 2001). Liposomal conjugation of MTX also reduced the systemic toxicity observed with an equivalent dose of the free drug (Williams et al. 2000).

12.1.2 SOD enzymosomes

The therapeutic success of superoxide dismutase (SOD) with anti-inflammatory activity in treating arthritis has been hindered by its rapid clearance from blood by glomerular filtration. To improve its plasma half-life and delivery to the target site, SOD was attached as a water-soluble enzyme to the surface of liposomes (through a covalent linkage of SOD to the distal polyethylene glycol [PEG] ends at the surface of PEG liposomes) in order to allow faster action of SOD without a need of liposomal disruption to release the therapeutic enzyme. The therapeutic activity of the SOD enzymosomes in the rat adjuvant arthritis model showed a faster onset of therapeutic activity than SOD liposomes wherein SOD was encapsulated in the inner aqueous space of the liposome (Corvo et al. 2015).

12.1.3 VIP–SSM

Vasoactive intestinal peptide (VIP), a potent 28–amino acid mammalian immunomodulator, was passively loaded onto sterically stabilized phospholipid micelles (SSMs) to generate safe and long-acting immunomodulatory nanoparticles (VIP–SSM) for the treatment of RA (Sethi et al. 2002). Since the VIP receptors are expressed on the effector cells (e.g., activated T-lymphocytes, macrophages, and proliferating synoviocytes) in inflamed joints, but not on the capillary endothelial cells, interactions of the circulating VIP–SSM nanoconstructs within the capillaries of extra-articular tissues could be prevented. In addition, the improved stability of the VIP owing to its association with the micelle further enhanced the circulation time and bioavailability of the peptide. A single subcutaneous or intravenous injection of the VIP–SSM resulted in selective localization of the construct in the joints of mice with collagen-induced arthritis (CIA) after extravasating from the hyperpermeable postcapillary venules and thus ameliorated the arthritis by exerting the anti-inflammatory effects.

12.1.4 CPT–SSM–VIP

Self-association of camptothecin (CPT), a selective topoisomerase I inhibitor with SSM, increased the solubility and stability of the drug. Surface modification of the resultant CPT–SSM with covalently conjugated VIP enabled the final construct (CPT–SSM–VIP) to actively target the VIP receptors overexpressed on the effector cells located at the site of inflammation after a single, subcutaneous injection in the mice with CIA, thereby promoting the CPT efficacy in inhibiting synoviocyte proliferation, angiogenesis, and collagenase expression as well as the immunomodulatory effect of VIP in abrogating CIA without noticeable systemic toxicity. To circumvent the possible L-VIP–induced activation of intracellular signal

Chapter twelve: Nanoparticles for therapeutic delivery in human diseases   191

transduction pathways, D-VIP, the inactive enantiomer of naturally occurring L-VIP, was grafted through PEG molecule onto the CPT-loaded SSM to ensure the active targeting of CPT–SSM–VIP to the target effector cells (Koo et al. 2011). Thus, like VIP–SSM, CPT–SSM–VIP also represents a promising DMARN for RA.

12.1.5   Betamethasone-encapsulated PEGylated PLGA nanoparticles

Betamethasone, a synthetic glucocorticosteroid, was encapsulated in various biodegradable nanoparticles of poly(lactic-co-glycolic acid)/polylactide (PLGA/PLA) homopolymers and PEG-block-PLGA/PLA copolymers in order to investigate the anti-inflammatory roles of the nanoformulations in experimental arthritis models. The stealth nanosteroid composed of PLA (2.6 kDa) and PEG (5 kDa)–PLA (3 kDa), with a PEG content of 10%, exhibited the highest and prolonged anti-inflammatory activity after being intravenously administered to adjuvant arthritis rats and mice with anti–type II collagen antibody-induced arthritis (Ishihara et al. 2009). The observed therapeutic benefit could be attributed to prolonged circulation time, passive targeting to inflamed joints, and, finally, sustained release of the drug in the target tissues.

12.1.6   Gold nanoparticles

Intra-articular administration of gold nanoparticles measuring 13 and 50 nm to rats with established CIA was found to reduce the inflammation, joint swelling, and development of polyarthritis without demonstration of toxic effects on the internal organs. The anti-arthritic effect of the gold nanoparticles could be attributed to their antioxidant role as shown by the increased level of antioxidant enzyme catalase (Leonavičienė et al. 2012). In a similar study, intra-articular delivery of 13-nm nanogold resulted in attenuation of CIA in rats by exerting antiangiogenic activities through binding to vascular endothelial growth factors (VEGFs) and subsequently reducing macrophage infiltration and inflammation (Tsai et al. 2007).

12.1.7   PEGylated cyclodextrin–methylprednisolone conjugate

A linear cyclodextrin polymer composed of β-cyclodextrin and PEG was conjugated to a glycinate derivative of methylprednisolone, a synthetic glucocorticosteroid through an ester linker that is cleavable in the inflamed joint (Hwang et al. 2008). The resultant construct self-assembled into nanoparticles of 27 nm and showed a reduction in CIA in mice at significantly lower doses compared to the free drug, after intravenous injection.

12.2 Cardiovascular diseases

The development of atherosclerotic plaques, the major cause of morbidity and mortality in cardiovascular diseases, starts at the lesion-prone areas in arteries characterized by dysfunctional endothelium, causing increased permeation of macromolecules (e.g., lipoproteins), increased expression of chemotactic molecules and adhesion molecules, and enhanced recruitment of monocytes and resulting in plaque progression, cell apoptosis, and neovascularization over a period of several years or decades. Plaque neovascularization may either promote plaque progression causing the wall of the artery to be remodeled inward and leading to stenosis and tissue ischemia or contribute to plaque rupture resulting in thrombotic occlusions (Lobatto et al. 2011).

12.2.1 Nanotherapeutics in atherosclerosis

12.2.1.1 Modulating HDL levels

High-density lipoprotein (HDL), an endogenous lipidic nanoparticle, is known to protect against cardiovascular disease by carrying cholesterol from peripheral tissues including lipid-laden plaque macrophages to the liver. Direct infusions with recombinant HDL (rHDL) particles and augmenting the levels of apolipoprotein A-I (APOA1), the major protein component of HDL, were found promising in inducing regression of atherosclerosis. Intravenous administration of HDL on a weekly basis resulted in the inhibition of aortic fatty streaks and lipid deposition in the atherosclerotic wall or the regression of atherosclerotic plaques in a rabbit model, while intravascular infusion of rHDL containing a genetic variant of APOA1 led to a reduction in the lipid and macrophage content of plaques, reversing endothelial dysfunction and inhibiting the progression of atherosclerosis in a mice model (Lobatto et al. 2011).

12.2.1.2 Anti-inflammatory therapy

Considering the role of inflammation in the pathogenesis of atherosclerosis, several anti-inflammatory drugs are currently under investigation for treating atherosclerosis. Although corticosteroids have anti-inflammatory effects in atherosclerosis, they have a poor pharmacokinetic profile with demonstrable adverse side effects. Thus, to increase the anti-inflammatory action and decrease the adverse effects of glucocorticoids, such as prednisolone phosphate, a liposomal formulation of the drug was intravenously applied to a rabbit model of atherosclerosis, with the significant anti-inflammatory effects and reduction in atherosclerotic lesions observed as early as 2 days after the single dose administration. Another liposomal formulation with surface-decorated antibodies against vascular cell adhesion molecule-1 and loaded cyclopenterone prostaglandins (anti-inflammatory agents)

Chapter twelve: Nanoparticles for therapeutic delivery in human diseases **193**

was found to be directed toward the injured arterial wall cells of atherosclerotic mice and completely cured them from the vascular injuries (Homem de Bittencourt et al. 2007).

12.2.1.3 Anticoagulant therapy

Rupture of an atherosclerotic plaque exposes collagen and other plaque components to the bloodstream, leading to activation of thrombin and formation of a thrombus at the site of rupture. Elevated levels of the activated thrombin are also implicated in the progression of atherosclerosis. Since anticoagulants can reduce the formation of thrombus, micelles loaded with the anticoagulant drug bivalirudin (Hirulog) were targeted with the help of a conjugated clot-binding peptide, to the fibrin deposited on the luminal surface of atherosclerotic plaques. After intravenous administration to atherosclerotic mice, the micelles were mainly accumulated at the rupture-prone region of the plaque as shown, causing considerable antithrombin activity. In another study, streptokinase, a plasminogen activator, has been encapsulated in phospholipid vesicles in order to improve the half-life and pharmacokinetics of the therapeutic protein and thereby more effectively lyse the coronary thrombus and reestablish blood after intravenous infusion in an animal model of acute myocardial infarction. The results showed that the time required to restore vessel patency was significantly reduced (more than 50%) when compared with free streptokinase (Nguyen et al. 1990).

12.2.1.4 Antirestenotic therapy

Platelet-derived growth factor (PDGF) has been shown to play an important role in the pathogenesis of restenosis (renarrowing of an artery after treatment to clear a blockage) by acting as both a mitogen and a chemoattractant for smooth muscle cells (SMCs) and transforming SMCs from the contractile to the proliferative phenotype (Banai et al. 2005). PLA nanoparticles were employed to encapsulate tyrphostin AGL-2043, a potent inhibitor of PDGFβ receptor to selectively inhibit SMC proliferation. Intraluminal administration of the AGL-2043–encapsulated nanoparticles resulted in high drug levels in the rat arterial tissue 90 min after delivery and significantly decreased neointimal formation in the injured rat carotid arteries.

The antiangiogenic therapy for cardiovascular diseases utilizing nanoparticles have been discussed earlier in Chapter 10.

12.3 Diabetes

Diabetes mellitus (DM) is a chronic disease characterized by perturbation in intracellular glucose transport and consequential accumulation of glucose in blood (hyperglycemia) owing to defective insulin production by the pancreas (type 1 DM) or loss of insulin sensitivity or relative deficiency of insulin (type 2 DM). Long-term hyperglycemia can lead to multiorgan damage, accounting for significant morbidity and mortality in the current population (Mo et al. 2014). Insulin therapy, which is required to control the level of blood glucose, is essential for patients suffering from type 1 DM and type 2 DM especially in their late stage. Nowadays, insulin that is chemically identical to human insulin can be constructed using recombinant DNA technology. Current modes of delivering exogenous insulin include intravenous and subcutaneous injections with the subcutaneous insulin preparations, which include rapid-acting (insulin lispro, insulin aspart, and insulin glulisine), intermediate-acting (neutral protamine Hagedorn insulin and Lente insulin), and long-acting (Ultralente, insulin glargine, and insulin detemir) insulin being more commonly used. However, frequent subcutaneous injections are generally accompanied by pain, tenderness, necrosis, microbial contamination, and nerve damage in the injection sites. In view of the current limitations in insulin therapy, nanotechnologies are being developed to efficiently deliver functional insulin through oral, nasal, pulmonary, and transdermal routes (Ahmad et al. 2012, 2014; Mo et al. 2014).

12.3.1 Oral insulin delivery using nanotechnologies

12.3.1.1 Lipid-based nanocarriers

12.3.1.1.1 Solid and cationic liposomes The hypoglycemic effect of liposomal insulin depends on the lipid composition, surface charge, and physical state of the phospholipid bilayer, which, in turn, regulate the stability of insulin-loaded liposomes in the gastrointestinal (GI) tract. Thus, liposomes formed with high-melting dipalmitoyl phosphatidylcholine (DPPC) or negatively charged phosphatidylinositol, which could protect the bound insulin from enzymatic degradation, were found to cause a marked decrease in blood glucose concentration after oral administration. In a subsequent study, oral administration of insulin entrapped in the liposomes (solid at 37°C) consisting of both DPPC and dipalmitoyl phosphatidylethanol (1:1) to diabetic rats produced a rise in plasma insulin for 3.5 hours with the peak at 1.5 hours, followed by a significant fall in blood glucose levels almost by 50% (Mo et al. 2014).

12.3.1.1.2 Chitosan- and lectin-coated liposomes Chitosan-embedded liposomes consisting of DPPC and dicetyl phosphate showed high mucoadhesiveness with the degree of adhesion dependent on the amount of chitosan present on the surface of the liposomes. The blood glucose level was significantly reduced after oral administration of the insulin-containing liposomes to normal rats. To develop lectin-modified liposome, wheat germ agglutinin (WGA), a lectin having affinity to sialic acids present on both M cells and regular intestinal absorptive cells, was conjugated to

Chapter twelve: Nanoparticles for therapeutic delivery in human diseases **195**

phosphatidylethanolamine, and the formulated insulin-loaded liposomes after oral administration to diabetic mice or rats caused a decrease in the blood glucose level in proportion to the increase in serum insulin concentration (Zhang et al. 2005).

12.3.1.1.3 Liposomes containing bile salts Insulin-loaded liposomes containing bile salts, such as sodium glycocholate (SGC), sodium taurocholate, or sodium deoxycholate (SDC), that serve as a stabilizer and a permeation-enhancer elicited a mild and prolonged hypoglycemic effect in parallel with an increase in blood insulin levels for a period of more than 20 hours with the peak at approximately 8–12 hours after oral administration to nondiabetic and diabetic rats. The SGC-embedded liposomes showed a higher oral bioavailability as well as a greater hypoglycemic effect than the other liposomal formulations, which is attributable to the better protective effect of the former against enzymatic degradation of the encapsulated insulin (Mo et al. 2014).

12.3.1.1.4 Nanoemulsion Electrostatic interactions between cationic chitosan and anionic insulin lead to spontaneous formation of polyelectrolyte complexes (nanoparticles) that are easily dissociated in acidic fluid of the stomach, since both insulin and chitosan are soluble at acidic pH. Although water-in-oil (w/o) microemulsion–based formulation of insulin protects it from the enzymatic breakdown in the GI tract and increases its permeability through the intestinal wall, high doses of insulin is required to elicit a pharmacological response. To make a better system, a w/o nanoemulsion was prepared by dispersing the aqueous solution of polyelectrolyte complexes of chitosan–insulin in oleic acid/surfactant mixture. Insulin was protected from gastric enzymes by incorporation into the lipid-based formulation while retaining its biologically activity. A drastic reduction of blood glucose levels was observed in diabetic rats after oral administration of the formulation with a particle diameter of 108 nm in fasted and fed states, and the hypoglycemic effect was maintained for a longer time compared to subcutaneous injection. Other micro/nanoemulsion-based carriers with different oil components or surfactants also showed the enhanced oral bioavailability of insulin (Mo et al. 2014).

12.3.1.1.5 Solid lipid nanoparticles On the basis of the w/o/w double emulsion technique, insulin was encapsulated in SLNs composed of wax cetyl palmitate, an ester derived from palmitic acid and cetyl alcohol, and poloxamer 407, a stabilizing surfactant. The oral administration of the insulin-loaded SLN to diabetic rats led to a relative pharmacological bioavailability of 5.1% compared to subcutaneous injection of insulin with a 24-hours-lasting hypoglycemic effect. Coating the surface of SLNs with chitosan or WGA enhanced the pharmacological bioavailability compared with uncoated SLNs, apparently by promoting the intestinal binding and absorption of insulin (Mo et al. 2014).

12.3.1.2 Polymeric nanoparticles

12.3.1.2.1 Chitosan/poly(γ-glutamic acid) nanoparticles Blending of chitosan and poly(γ-glutamic acid) (PGA) with tripolyphosphate and $MgSO_4$ led to generation of multi-ion cross-linked nanoparticles that were highly compact and stable over a broader pH range from 2.0 to 7.2, compared to the chitosan/PGA complexes. After oral administration in diabetic rats, the nanoparticles showed a noticeable hypoglycemic effect for at least 10 hours with the relative bioavailability of insulin being approximately 15% compared to subcutaneous injection (Sonaje et al. 2009). Since some of the orally administered nanoparticles were retained in the stomach for a long duration, the insulin-encapsulated nanoparticles were freeze-dried and filled in an enteric-coated capsule (Sonaje et al. 2010). Upon oral administration, the enteric-coated capsule remained intact in the stomach and consequently traveled to the small intestine, thus enhancing the intestinal absorption of insulin and causing a prolonged reduction in blood glucose levels. A further modification of chitosan/PGA nanoparticles was made by conjugation of diethylene triamine pentaacetic acid (DTPA) known to disrupt intestinal tight junctions and inhibit intestinal proteases by chelating divalent metal ions to PGA. The oral intake of the enteric-coated capsule containing chitosan/PGA–DTPA nanoparticles caused prolonged hypoglycemia in diabetic rats, with the relative oral bioavailability of insulin being approximately 20% (Su et al. 2012). Taking into consideration the crucial role of calcium (Ca^{2+}) in maintaining the intestinal protease activity and in forming the apical junctional complex in the epithelium, ethylene glycol tetraacetic acid (EGTA), a Ca^{2+}-specific chelating agent, was conjugated to PGA, thereby forming nanoparticles with chitosan. Oral delivery of insulin with chitosan/PGA–EGTA nanoparticles produced a significant and prolonged hypoglycemic effect in diabetic rats (Chuang et al. 2013).

12.3.1.2.2 Thiolated trimethyl chitosan nanoparticles Trimethyl chitosan (TMC) is a partially quaternized derivative of chitosan with higher solubility and permeation-enhancing effect than the unmodified chitosan. However, TMC exhibits decreased intrinsic mucoadhesivity than chitosan. Thiolated polymers (thiomers), on the other hand, although insoluble in aqueous environment, can tightly adhere to the intestinal mucus layer for a prolonged time through covalent bonding with mucin glycoproteins, in addition to exerting their roles in permeation enhancing and inhibiting protein tyrosine phosphatase and intestinal P-glycoprotein. With a view to combining the mucoadhesion and permeation-enhancing effects of TMC and thiomers, TMC-cysteine/insulin nanoparticles (TMC-Cys NP)

Chapter twelve: Nanoparticles for therapeutic delivery in human diseases   197

were prepared through self-assembly (Yin et al. 2009). In vivo assessment of pharmacological efficacy showed good correlation with the improved mucoadhesion and permeation-enhancing effect of TMC-Cys nanoparticles by inducing a more potent and prolonged hypoglycemic effect that lasted until 8 and 7 hours than the corresponding TMC nanoparticles, after oral administration in normal rats.

    12.3.1.2.3 PLGA- and PLA-based nanoparticles   Hydrophilic insulin was combined with lipophilic soybean phosphatidylcholine, and the resultant insulin–phospholipid complex was loaded into PLGA nanoparticles with entrapment efficiency up to 90%. This formulation reduced fasting plasma glucose levels to 57.4% within the first 8 hours of intragastric administration, with 7.7% of oral bioavailability relative to subcutaneous injection (Cui et al. 2006). A similar approach was undertaken to enhance the liposolubility of insulin by complexing it with SDC, a bile salt to form an insulin–sodium deoxycholate complex (Ins–SD-Comp) that was encapsulated into PLGA nanoparticles with the maximal encapsulation efficiency being 93.6%. Oral administration of Ins–SD-Comp–PLGA nanoparticles produced significant plasma glucose reduction, with the relative bioavailability being approximately 11.7% (Sun et al. 2011). In order to improve the loading efficiency and alleviate the initial burst release of insulin, hydroxypropyl methylcellulose phthalate 55 (HPMCP55 or HP55) (commonly used to coat tablets and microspheres for enteric purposes) was introduced into the matrix of PLGA nanospheres. The final PLGA–HP55 nanoparticles enabled the encapsulated insulin to significantly avoid acid- and enzyme-mediated degradation in the GI tract and enhanced the relative oral bioavailability compared to PLGA nanoparticles in diabetic rats (Cui et al. 2007). By modifying insulin-loaded PLGA nanoparticles with folate (Feng et al. 2014), or decorating their surface with cell-penetrating peptide (CPP) oligoarginine (Kay 2011), the relative oral bioavailabilities and the hypoglycemic effects of insulin could also be improved. Insulin-encapsulated nanoparticles of poly(lactic acid)-*b*-pluronic-*b*-poly(lactic acid) and Fc fragment-embedded PLA–PEG targeting the neonatal Fc receptor in the intestinal epithelium were shown to produce prolonged hypoglycemic effects in diabetic mice and wild-type mice, respectively.

12.3.2 Nasal insulin delivery using nanotechnologies

12.3.2.1 Liposome-based formulation

DPPC liposomes modified with sterylglucoside (SG) demonstrated increased fluidity of the liposome bilayer compared with the liposomes modified with cholesterol (Ch) or soybean-derived sterol (SS) (Muramatsu et al. 1999). When administered nasally to rabbits, insulin-loaded DPPC/

SG liposomes significantly reduced blood glucose level for a prolonged period (8 h), whereas DPPC/SS or DPPC/Ch (7/4) liposomes caused a low reduction in blood glucose and insulin as a solution showed no hypoglycemic effect. Mucoadhesive liposomes prepared by coating with chitosan were subjected to intranasal delivery of insulin in diabetic rats, resulting in effective reduction of plasma glucose level up to 2 days.

12.3.2.2 Chitosan-based formulation

Intranasal instillation of insulin-loaded nanoparticles of chitosan and chitosan/alginate complex resulted in reduction of blood glucose level to 60% and 70% of the basal level, respectively, in rabbits, whereas thermosensitive hydrogels based on quaternized chitosan and PEG reduced blood glucose level to 40%–50% of the basal level in rats. Insulin-loaded, chitosan-reduced gold nanoparticles decreased blood glucose by 20.27% after nasal administration to diabetic rats. Polysaccharide nanoparticles consisting of chitosan and cyclodextrin derivatives were able to significantly decrease plasma glucose levels by more than 35% after nasal delivery of the insulin-associated hybrid nanoparticles (Teijeiro-Osorio et al. 2009).

12.3.3 Pulmonary insulin delivery

12.3.3.1 Liposome

Insulin-loaded liposomes consisting of DPPC and cholesterol (Chol) enhanced pulmonary uptake of insulin and hypoglycemic effect compared to free insulin, after intratracheal administration. Aerosol delivery of nebulizer-compatible liposomes composed of hydrogenated egg yolk PC (HPC)/Chol/PEG–DPPE (70:30:1) led to an effective homogeneous deposition with a high retention period of insulin in the alveolar lung, improving the therapeutic efficacy (Mo et al. 2014).

12.3.3.2 Solid lipid nanoparticles

The pharmacological bioavailability and relative bioavailability of insulin encapsulated in nebulizer-compatible SLNs were significantly increased compared to subcutaneous injection of insulin, after pulmonary administration to diabetic rats. On the other hand, a dry powder inhalation (DPI) system based on insulin-loaded SLNs after intratracheal instillation into diabetic rats demonstrated a relative pharmacological bioavailability of 44.40% with a prolonged hypoglycemic effect (Mo et al. 2014).

12.3.3.3 PLGA nanoparticles

Insulin-loaded PLGA nanospheres that were nebulized to discrete droplets by an ultrasonic nebulizer were shown to exert a hypoglycemic effect for a duration of 48 hours compared to 6 hours of nebulized aqueous solution of insulin, after administration into the trachea of guinea pigs (Mo et al. 2014).

Chapter twelve: Nanoparticles for therapeutic delivery in human diseases **199**

12.3.4 Transdermal insulin delivery

Treatment of insulin-loaded cationic liposomes consisting of 1,2-dioleoyl-3-trimethylammonium-propane (DOTAP), egg phosphatidylcholine (EPC), and Chol (2:2:1) coupled with iontophoresis (ITP; a tool to transiently permeabilize skin by applying electric current) on the skin of diabetic rats produced a gradual reduction of blood glucose level up to 20% over a duration of 24 hours. A combination of microneedles (a technology to create micrometer channels on the skin) and ITP was attempted for enhanced transdermal penetration of insulin encapsulated in a liposomal formulation. Finally, the transdermal delivery to diabetic rats resulted in reduction of blood glucose to 28.3% of the basal level at 6 hours after administration, which was comparable to that observed with subcutaneous injection of insulin (Mo et al. 2014).

12.3.5 Subcutaneous insulin delivery

Since even long-acting insulin, such as Lantus (insulin glargine) or Levemir (insulin detemir), is currently unable to completely avoid the risk of nocturnal/fasting hypoglycemia, a sustained release formulation is still being highly sought for subcutaneous administration of insulin. An enzymatically degradable thermosensitive gel consisting of poly(ethylene glycol)-block-poly(alanine-co-phenyl alanine) (PEG–PAF) was shown to undergo sol–gel transition as the temperature increased and degradation by proteases (cathepsin B, cathepsin C, and elastase) in the subcutaneous layer of rats, leading to a sustained release of insulin. A single subcutaneous injection of the insulin-loaded formulation demonstrated a hypoglycemic effect in rats over a period of 18 days. An injectable thermosensitive hydrogel containing nanoparticles of poly(3-hydroxybutyrate-co-3-hydroxyhexanoate) could also maintain a hypoglycemic effect for more than 5 days in diabetic rats (Mo et al. 2014).

12.4 Neurodegenerative diseases

Treatment strategies of neurological diseases, such as Alzheimer's disease (AD), Parkinson's diseases (PD), autism, traumatic brain injury, stroke, or schizophrenia have received limited successes so far because of the existence of blood–brain barrier (BBB), which is mainly formed by the polarized brain capillary endothelial cells sealed by complex tight junctions and supported by pericytes, astrocytes, and a basal membrane, thereby preventing the paracellular transport of drug molecules into the central nervous system. Large molecules, such as recombinant proteins, monoclonal antibodies, and nucleic acids, are unable to cross the BBB, whereas among the small molecules, only approximately 2% with high lipid solubility have the access to the BBB. Absorption of molecules into the brain parenchyma across the brain endothelium takes place via either passive transport, which predominantly includes transcellular diffusion of small lipophilic molecules (<400–600 Da), or active transport, which includes carrier-mediated transcytosis for relatively small molecules, absorptive-mediated transcytosis for positively charged peptides, and receptor-mediated transcytosis (RTM) for certain proteins. The best known efflux pump, P-glycoprotein in the BBB, is highly active in pumping out unwanted compounds, such as cytotoxic anticancer drugs and antibiotics. Nanoparticles have emerged as highly promising delivery systems in order to increase the translocation of small and macromolecular drugs through the BBB by exploiting the available transport routes (Nunes et al. 2012).

12.4.1 Nanoparticles as neuroprotective and therapeutic drugs for neurodegenerative diseases

Neurodegenerative disorders are characterized by progressive and persistent loss of neurons often caused by a decline in neurological functions with aging. Chronic inflammation is particularly associated with neurodegenerative AD and PD, leading to excessive production of proinflammatory molecules and ROS, thus causing neuronal cell death.

12.4.1.1 VP025 nanoparticles

A phosphatidylglycerol-based phospholipid nanoparticle, VP025, has shown to potentially induce a neuroprotective effect in the brain. Intramuscular administration of VP025 in a rat model of PD (6-hyroxydopamine [6-OHDA] model) revealed that the treatment protected against a decrease in concentrations of striatal dopamine and its metabolites and significantly inhibited the loss of dopaminergic neurons while preventing the activation of p38, an apoptotic-signaling molecule within those neurons (Nunes et al. 2012).

12.4.1.2 Poly(butyl cyanoacrylate) nanoparticles

The nerve growth factor (NGF), which is essential for the survival of peripheral ganglion cells and central cholinergic neurons in the basal forebrain, was adsorbed on poly(butyl cyanoacrylate) (PBCA) nanoparticles coated with polysorbate 80, and the resultant formulation was subjected to the pharmacological evaluation in the models of acute scopolamine-induced amnesia in rats and 1-methyl-4-phenyl-1,2,3,6-tetrahydropyridine (MPTP)-induced Parkinsonian syndrome. Intravenous administration of the NGF-bound nanoparticles led to the efficient transport of NGF across the BBB, reversing scopolamine-induced amnesia and improving recognition and memory. In addition, there was a significant reduction of the basic symptoms of Parkinsonism (Nunes et al. 2012).

Chapter twelve: Nanoparticles for therapeutic delivery in human diseases 201

12.4.1.3 Lactoferrin-conjugated PEG–PLGA nanoparticles

PEG–PLGA nanoparticles were conjugated to lactoferrin (Lf), a promising brain-targeting molecule, and used to load urocortin (UCN), a corticotropin-releasing hormone–related peptide. Intravenous administration demonstrated enhanced BBB penetration of the Lf-conjugated nanoparticles and significantly attenuated the striatum lesion caused by 6-hydroxydopamine in rats (Nunes et al. 2012).

12.4.1.4 Transferrin receptor–targeted PEG–chitosan nanospheres

Caspases act as mediators of cell death in acute and chronic neurological disorders, and peptide-based caspase inhibitors are attractive drug candidates for neuroprotection. Since the inhibitors cannot cross the BBB, PEG-coated chitosan nanospheres were conjugated to an antimouse transferrin receptor monoclonal antibody (TfRMAb) that selectively recognizes the TfR type 1 on the cerebral vasculature and loaded with N-benzyloxycarbonyl-Asp(OMe)-Glu(OMe)-Val-Asp(OMe)-fluoromethyl ketone (Z-DEVD-FMK), a relatively specific caspase-3 inhibitor. The nanospheres efficiently delivered Z-DEVD-FMK peptide into the brain after intravenous administration and decreased the infarct volume, neurological deficit, and ischemia-induced caspase-3 activity in the ischemic brain models (Nunes et al. 2012).

12.4.1.5 Dendrimers

High-mobility group box 1 (HMGB1) is a cytokine-like molecule that is critically involved in mediating local and systemic inflammation and activating various types of immune-related cells. Biodegradable arginine-PAMAM esters (e-PAM-R) were employed to electrostatically associate with HMGB1 small interfering RNA (siRNA), and the resultant complex was injected stereotaxically into the rat cortex, leading to the silencing of HMGB1 expression in more than 40% of neurons and astrocytes of the normal brain and reduction of infarct volume in the postischemic rat brain generated by occluding the middle cerebral artery (MCA) (Nunes et al. 2012).

12.4.1.6 Neurotensin polyplexes

The glial cell line–derived neurotrophic factor (GDNF) is a survival factor for midbrain dopamine neurons and thus a strong candidate for the treatment of PD. Neurotensin polyplexes, which consist of neurotensin–poly-L-lysine conjugate and electrostatically associated Vp1 nuclear localization signal of simian virus 40, were used to complex with human GDNF (hGDNF) gene-carrying plasmid DNA. A single dose of the formulation into the substantia nigra of hemiparkinsonian rats 1 week after a 6-OHDA injection into the ventrolateral part of the striatum resulted in expression of the hGDNF gene into the injection site and produced biochemical, anatomical, and functional recovery from hemiparkinsonism (Nunes et al. 2012).

12.4.1.7 Liposome

Liposome-encapsulated hemoglobin (LEH) was investigated in the treatment of stroke in a rat model of permanent MCA occlusion. Infusion of LEH in the brain significantly reduced edema formation in a wide area of the brain 24 hours after permanent occlusion of the MCA (Kawaguchi et al. 2009). LEH also notably reduced the area of infarction in the cortex after photochemically induced thrombosis of the MCA in the rat (Kawaguchi et al. 2007).

12.5 Degenerative retinal diseases

Retinal degenerative diseases, such as glaucoma, age-related macular degeneration (AMD), diabetic retinopathy (DR), or retinitis pigmentosa (RP), are the major causes of vision loss and blindness. The potential nanomedicine-based approaches to the treatment of such diseases are discussed in Sections 12.5.1 through 12.5.5.

12.5.1 Neurotrophic factor therapy

12.5.1.1 bFGF-loaded gelatin nanoparticles

Basic fibroblast growth factor (bFGF) is a protective neurotrophic factor activity with potent photoreceptor degeneration ability in rats. Despite its ability to delay photoreceptor degeneration with a single intravitreal or subretinal injection, long-term rescue of the photoreceptors is still challenging because of the short half-life of the trophic factor. Intravitreal injection of biodegradable gelatin nanoparticles with loaded bFGF increased the efficacy of bFGF in providing sustained retinal rescue by inhibiting apoptosis in the retina of Royal College of Surgeons (RCS) rats through improved targeting and sustained release. Some forms of RP (a group of inherited retinal disorders characterized by progressive loss of photoreceptors, leading to retina degeneration and atrophy) in humans arise from the same mutation that prevents proper outer-segment phagocytosis by retinal pigment epithelial (RPE) cells, resulting in progressive rod and cone photoreceptor degeneration in RCS rats (Zarbin et al. 2012).

12.5.1.2 GDNF gene-carrying AAV

A recombinant adeno-associated virus (AAV) vector was employed to transfer the gene of GDNF to the retinas in a transgenic rat model of RP, resulting in increased rod photoreceptor survival (McGee Sanftner et al. 2001). Intravitreal administration of an AAV variant carrying the GDNF

Chapter twelve: Nanoparticles for therapeutic delivery in human diseases 203

gene was also shown to generate high GDNF levels from the transducted glial cell in treated retinas, leading to sustained functional rescue by slowing down the progression of retinal degeneration in a rat model of RP (Dalkara et al. 2011).

12.5.2 Antioxidant therapy

12.5.2.1 Cerium oxide nanoparticles

Many retinal diseases including AMD, RP, DR, and inherited retinal degeneration are characterized partly by oxidative damage owing to the presence of excess reactive oxygen species (ROS). By mimicking the activities of SOD and catalase, cerium oxide nanoparticles, also known as nanoceria, can catalytically scavenge the ROS. A single intravitreal injection of nanoceria into the eye of a mouse that lacks the Vldlr gene, resulting in intraretinal and subretinal neovascular lesions and thus representing a form of AMD called retinal angiomatous proliferation, was shown to inhibit the rise in ROS in the Vldlr$^{-/-}$ retina, increases in VEGF in the photoreceptor layer, and the formation of intraretinal and subretinal neovascular lesions (Zhou et al. 2011).

12.5.3 Anti-inflammatory therapy

12.5.3.1 PLA and PLA–PEG nanosteroids

Inflammation and oxidative stress are common pathological features for many neurodegenerative diseases. Betamethasone phosphate (BP), a synthetic steroid with an anti-inflammatory activity, was encapsulated into PLA nanoparticles. Intravenous administration of the BP-loaded particles to Lewis rats with experimentally induced autoimmune uveoretinitis demonstrated ocular targeting and sustained release in the intraocular inflammation site, with the rapid and prolonged anti-inflammatory effects (equivalent to those of a five times higher dose of BP) apparently exerted by inhibiting the ocular infiltration of activated macrophages and T cells and activation of glial cells (Sakai et al. 2006). A relatively strong therapeutic effect of PLA–PEG nanosteroids subsequently observed over PLA nanosteroids could be attributed to the prolonged blood circulation time of the former along with sustained release in situ and targeting to the inflamed eyes (Sakai et al. 2011).

12.5.4 Inhibiting choroidal neovascularization

One of the major treatments for angiogenesis-related blindness is monthly, intravitreal injection of bevacizumab (Avastin), an antibody that targets the VEGF-A molecule involved in vascularization of the cornea, thus ultimately inhibiting corneal angiogenesis. Since antibody-based formulations are expensive with limited bioavailability, development of potent nanotherapeutics is currently in progress with the aim of effectively inhibiting the angiogenic factors that lead to blindness.

12.5.4.1 Dendrimer–anti-VEGF oligonucleotide conjugate

A lipophilic amino acid dendrimer was used to deliver an anti-VEGF oligonucleotide into rat eyes having laser-induced choroidal neovascularization (CNV). The dendrimer–oligonucleotide conjugate inhibited choroidal new vessel development by up to 95% for 4 to 6 months in the initial stages. The dendrimer–oligonucleotide conjugate was well tolerated in vivo with excellent biodistribution and no observable increase in inflammation-associated antigens (Marano et al. 2005).

12.5.4.2 Dual-functionalized, Flt23K plasmid-loaded PLGA nanoparticles

The anti-VEGF intraceptor (Flt23K) is a recombinant construct of VEGF-binding domains 2 and 3 of VEGFR-1/Flt-1 receptor coupled with an endoplasmic reticulum (ER) retention signaling sequence, thus binding VEGF intracellularly and sequestering it within the ER. Transferrin/ RGD peptide-coated PLGA nanoparticles were used for targeted delivery and expression of Flt23K plasmid into the neovascular eye on intravenous administration. The treatment increased the retinal delivery of the nanoparticles and subsequently gene expression of the intraceptor in retinal vascular endothelial cells, photoreceptor outer segments, and RPE cells, thereby inhibiting the progression of laser-induced CNV in a rodent model (Singh et al. 2009).

12.5.5 Retinal gene therapy

Retinal degeneration patients suffering from the diseases, such as RP, that arise from mutations in photoreceptor- or nonphotoreceptor-specific genes, become gradually blind as their photoreceptors die. One promising strategy involves *in vivo* expression of light-activated channels in surviving retinal neurons utilizing gene carriers. Channerhodopsin2 (ChR2) and melanopsin were successfully delivered to the retina with AAV, restoring light sensitivity to rodent models of blindness, such as rd1 mice and RCS rats. Vision restoration was shown by AAV-targeted expression of halorhodopsin in surviving non–light-sensitive photoreceptor inner segments, after loss of the light-sensing outer segments. Intravitreal delivery of the light-gated ionotropic glutamate receptor gene with AAV2 also restored light responsiveness to the retinal ganglion cells of a model of retinal degeneration (rd1 mouse) (Caporale et al. 2011).

12.6 Inflammatory bowel diseases

Inflammatory bowel diseases (IBDs), such as Crohn's disease and ulcerative colitis (UC), happen because of dysregulated immune responses in genetically predisposed individuals under various environmental conditions. Currently available medications for IBD involving 5-aminosalicylic acid drugs, corticosteroids, immunosuppressive agents, biologic therapies, and antibiotics are generally inefficient with many side effects and unable to offer complete remission (Pichai and Ferguson 2012). Since TNF-α is critically involved in IBD pathogenesis by acting as an important mediator of inflammation, many biological therapies comprising monoclonal antibodies or soluble receptors were tested in many clinical trials for reducing TNF-α activity in IBD. However, the consequential systemic depletion of TNF-α caused a number of adverse effects including immunosuppression, opportunistic infections, and decreased efficacy as a result of antibody formation against the biologics.

12.6.1 TNF-α siRNA-encapsulated polymeric nanoparticles

TNF-α siRNA was encapsulated in thioketal nanoparticles (TKNs) made from ROS-sensitive poly-(1,4-phenyleneacetone dimethylene thioketal) (PPADT). Orally delivered TNF-α TKNs were found stable in the GI tract, thereby protecting the siRNA and preventing its release to noninflamed tissues, but at sites of intestinal inflammation subjected to degradation in the presence of the elevated ROS levels, thus triggering the release of siRNA at the inflamed intestinal tissues in a murine model of UC, diminishing TNF-α mRNA levels in the colon and protecting the mice from UC (Wilson et al. 2010). In another study, the oral administration of TNF-α siRNA/PEI nanocomplexes specifically reduced the TNF-α expression/secretion in colonic tissue of a lipopolysaccharide (LPS)-treated UC mouse model (Laroui et al. 2011).

12.6.2 TNF-α–neutralizing nanobodies

Treatment of IBD patients with intravenously administered anti-TNF antibodies, such as infliximab, has become an established therapy for Crohn's disease and UC. The unwanted effects owing to systemic administration of infliximab could be resolved by local delivery at the site of inflammation. *Lactococcus lactis* strain MG1363 was engineered to secrete murine (m)TNF-neutralizing nanobodies as therapeutic proteins. Daily oral administration of nanobody-secreting *L. lactis* resulted in local and active delivery of anti-mTNF nanobodies at the mucosa of the colon without measurable levels in systemic circulation, significantly reducing inflammation in mice with dextran sulfate sodium (DSS)–induced chronic colitis, as well as improving established enterocolitis in IL-10 (–/–) mice (Vandenbroucke et al. 2010).

12.6.3 PHB gene–carrying Ad and PHB-entrapped PLA nanoparticles

Prohibitin 1 (PHB) protein sustains antioxidant expression and has anti-inflammatory properties in the intestinal epithelium. Proinflammatory cytokines, such as TNF-α, decrease expression of PHB. Sustained expression of PHB in intestinal epithelial cells was found to reduce TNF-α–stimulated NF-κB activation through inhibition of p65 nuclear translocation involving alteration of importin α3 levels, while overexpression of PHB reduced the severity of colitis by acting as a regulator of increased antioxidant response. To assess the therapeutic potential of restoring PHB, PHB gene–carrying Ad or PHB-encapsulated PLA nanoparticles were orally delivered, resulting in increased levels of PHB in the surface epithelial cells of the colon and, consequently, reduction of the severity of DSS-induced colitis in mouse models, through curtailment of NF-κB–mediated inflammatory reactions (Theiss et al. 2011).

12.6.4 NF-κB decoy ODN-loaded chitosan–PLGA nanoparticles

Chitosan-modified PLGA nanospheres were developed for oral delivery of NF-κB decoy oligodeoxyribonucleotide (ODN) in DSS-induced murine UC models. Oral administration of the ODN-loaded nanospheres improved UC, producing a significant anti-inflammation effect, suppressing diarrhea and rectal bleeding and preventing the DSS-induced colon shortening (Tahara et al. 2011).

12.6.5 Map4k4 siRNA-encapsulated glucan shells

Since macrophages promote pathogenic inflammatory responses in diseases, such as RA, atherosclerosis, IBD, and diabetes, they are attractive targets for RNA interference therapy. As an example, oral delivery of β1,3-D-glucan–encapsulated siRNA targeting the mitogen-activated protein kinase kinase kinase kinase 4 (Map4k4) gene in macrophages protected mice from LPS-induced lethality by inhibiting TNF-α and IL-1β production (Aouadi et al. 2009).

12.6.6 $β_7$ integrin-targeted, CyD1 siRNA-loaded liposomes

Cyclin D1 (CyD1) is a cell cycle regulatory protein that is aberrantly upregulated in both epithelial and immune cells in IBD. Liposome-based, $β_7$

Chapter twelve: Nanoparticles for therapeutic delivery in human diseases 207

integrin-targeted, stabilized nanoparticles (β_7I-tsNPs) were used to entrap protamine-condensed CyD1 siRNAs, before intravenous administration in murine models of DSS-induced colitis (Peer et al. 2008). β_7I-tsNP-delivered CyD1 siRNA potently reduced CyD1 mRNA in inflamed gut leukocytes, concomitantly suppressed mRNA expression of TNF-α and IL-12, reduced intestinal tissue damage, and inhibited leukocyte infiltration into the colon.

12.6.7 IL-10 gene-encapsulated NiMOS

Nanoparticles-in-microsphere oral system (NiMOS) is specifically designed for oral gene therapy, whereby, after oral delivery, the DNA-encapsulated gelatin nanoparticles are released in the intestine from the poly(ϵ-caprolactone) matrix because of lipase-induced degradation and subsequently endocytosed by the cells of the intestinal lumen resulting in efficient transgene expression. Since IL-10 is known to play a very important role in the immunological balance of the mucosal immune system, NiMOS was formulated with IL-10–expressing plasmid DNA and subjected to oral administration in a trinitrobenzenesulfonic acid (TNBS)–induced acute colitis mouse model (Bhavsar and Amiji 2008). The locally expressed IL-10 caused a reduction of the proinflammatory cytokines and chemokines required for recruitment and proliferation of macrophages and monocytes at the site of inflammation, leading to an increase in body weight, clinical activity score, restoration in colon length and weight, and suppression of inflammatory response.

12.6.8 Mesalamine (5-ASA)-loaded nanoparticles

Increased adhesion of nanoparticle drug carriers to the inflamed colonic tissue along with an intensified particle uptake owing to enhanced permeability and availability of a highly increased number of immune-related cells inside the tissue would allow for specific drug targeting in IBD. Hence, conventional IBD small drugs orally delivered with nanoparticles showed greater therapeutic impact as compared to the free drugs in animal models. For example, covalently linked to the poly(caprolactone) nanoparticles, the anti-inflammatory IBD drug mesalamine (5-ASA) was found to be 60 times more efficient than the free solution of 5-ASA in treating TNBS-induced colitis in murine models (Pertuit et al. 2007). Similarly, 5-ASA loaded in silicon nanoparticles was shown to have a sixfold increased ability to adhere to inflamed tissues when compared to tissues in healthy controls, inducing a positive impact on clinical activity score at reduced drug doses, as compared to conventional delivery in TNBS-induced murine colitis models (Moulari et al. 2008).

12.7 Obstructive respiratory diseases

Obstructive respiratory diseases generally impede the flow rate into and out of the lungs, with the most common obstructive diseases being asthma, chronic obstructive pulmonary disease, respiratory allergies, occupational lung diseases, and pulmonary hypertension (Swai et al. 2009). Nanoparticle-based drugs that are currently being developed for respiratory applications are expected to overcome the limitations of conventional therapy by enhancing systemic bioavailability and improving long-term therapeutic effects of small and macromolecular drugs.

12.7.1 Nanomedicine for allergic asthma

Asthma is a chronic disease characterized by allergen-driven airway inflammation leading to the infiltration of inflammatory cells, hypersecretion of mucus, and development of airway hyperresponsiveness (AHR) to a variety of environmental stimuli. The consequential abnormal low airflow rates in the airways are usually restored by bronchodilator and anti-inflammatory medications, such as NSAIDs, and immunosuppressive agents, such as MTX, cyclosporine, and azathioprine generally prescribed for more advanced stages of the disease. Allergic asthma is associated with airway inflammation and hyperresponsiveness caused by the dysregulated production of cytokines secreted by allergen-specific helper T type 2 (Th2) cells. Allergic subjects produce relatively low amounts of interferon gamma (IFN-γ), a pleiotropic Th1 cytokine that downregulates Th2-associated responses (Swai et al. 2009).

12.7.1.1 IFN-γ gene therapy

Dysregulated production of the cytokines secreted by allergen-specific helper T type 2 (Th2) cells is responsible for allergic asthma-associated airway inflammation and hyperresponsiveness. IFN-γ that is produced in relatively low amounts by allergic subjects downregulates the Th2-associated responses. Thus, recombinant IFN-γ has emerged as an excellent candidate for asthma therapy. However, its short half-life as a cytokine-based drug requires repeated and frequent dosing, resulting in antidrug antibodies that block the therapeutic effectiveness. To overcome the limitation, intranasal IFN-γ gene transfer was made by a recombinant replication-deficient Ad in a BALB/c mouse model of established allergic asthma. The treatment effectively attenuated ovalbumin (OVA)-induced airway inflammation, exhibiting significantly lower levels of Th2 cytokines IL-4 and IL-5, OVA-specific serum IgE, lung eosinophilia, and AHR compared with control mice (Behera et al. 2002). Similar findings were obtained with intranasally delivered chitosan-IFN-γ pDNA nanoparticles in effectively

Chapter twelve: Nanoparticles for therapeutic delivery in human diseases 209

reducing the OVA-induced airway inflammation and AHR in mouse asthma models (Kumar et al. 2003).

12.7.1.2 Theophylline-adsorbed thiolated chitosan nanoparticles

Theophylline is a widely prescribed antiasthmatic agent that reduces the inflammatory effects of allergic asthma. However, it also exerts side effects, such as nausea, headache, and cardiac arrhythmias, at the dose necessary to achieve bronchodilation in asthma. Thiolated chitosan nanoparticles (TCNs), because of their high mucoadhesiveness and permeability properties, were employed to load theophylline in order to enhance its absorption through bronchial mucosa (Lee et al. 2006). Intranasal delivery of the theophylline-adsorbed TCNs augmented the anti-inflammatory effects of the drug compared to the free drug in a mouse model of OVA-induced allergic asthma.

12.7.2 Nanomedicine for respiratory syncytial virus

Respiratory syncytial virus (RSV), an enveloped RNA virus, is the leading cause of severe bronchiolitis and pneumonia in infants and also responsible for lower respiratory tract infections in immunodeficient and elderly adults. Repeated RSV infections are common owing to the incomplete immunity produced by natural infection. Nonstructural proteins NS1 and NS2 have an important role in RSV replication, with deletion of either NS1 or NS2 severely attenuating the viral infection.

12.7.2.1 siNS1/chitosan nanoparticles

Since RSV NS1 is believed to antagonize the type-1 IFN-mediated antiviral response, the NS1 gene was subjected to knockdown with the help of RNA interference to attenuate RSV replication and thus to provide an effective antiviral and immune enhancement therapy. Mice treated intranasally with chitosan nanocomplex of a plasmid-borne siRNA targeting the NS1 gene (siNS1) before or after infection with RSV showed substantially decreased virus titers in the lung and decreased inflammation and AHR, suggesting that siNS1/chitosan nanoparticles present a promising prophylactic and therapeutic agent against RSV infection in humans (Zhang et al. 2005).

12.8 Hepatic fibrosis and infections

Hepatic fibrosis, microbial infections, and hepatocellular carcinoma (HCC) are the main causes of hepatic diseases. The existing medications have the major drawback of delivering insufficient amount of therapeutics to the affected liver tissue and contributing to undesirable side effects.

12.8.1 Hepatic fibrosis

Hepatic fibrosis originates from the chronic liver injury owing to alcohol consumption, hepatitis B virus (HBV) or hepatitis C virus (HCV) infection, or genetic abnormalities, leading to the activation of Kupffer cells, endothelial cells, and platelets and the influx of leucocytes, thereby inducing the generation of lipid peroxides, ROS/free radicals, nitric oxide (NO), and diverse cytokines. As a result, hepatic stellate (HS) cells undergo activation, transforming into myofibroblasts that produce extracellular matrix proteins including fibrillar collagen and inducing the release of diverse matrix metalloproteinases (MMPs) that degrade the normal matrix, with the overall consequence of fibrillar collagen deposition or fibrosis, which may end up in cirrhosis and even HCC. To prevent the nonspecific drug disposition of traditional antifibrotic drugs, ligand-anchored nanoparticles can be fabricated to deliver the encapsulated drugs in targeted fashion to the HS cells of the fibrotic region expressing or overexpressing the ligand-specific receptors (Reddy and Couvreur 2011).

12.8.1.1 RGD-coupled sterically stabilized liposome with entrapped IFN-α1b

IFN-α, which can inhibit HCV replication and improve liver injury, is the most common and effective drug for treating viral hepatitis C. It was also shown to inhibit the progression of hepatic fibrosis in patients with hepatitis C infection. However, IFN-α has many adverse effects that many patients cannot tolerate, resulting in discontinuation of treatment. Therefore, targeted delivery of IFN-α is required to simultaneously improve its therapeutic effects and reduce the adverse effects. Aiming to achieve the goal, a cyclic peptide containing Arg–Gly–Asp (cRGD), which is specific to collagen type VI receptor expressed on hematopoietic stem cells (HSCs), was coupled to PEGylated liposomes (SSLs). After intravenous delivery, IFN-α1b–loaded cRGD-SSLs were found to accumulate in the HSCs of BDL rats 10-fold more efficiently than the unlabeled SSLs and significantly reduce the extent of liver fibrosis compared with the BDL rats untreated or treated with IFN-α1b entrapped in the SSLs (Du et al. 2007).

12.8.1.2 Vitamin A–conjugated liposomes with loaded gp46 siRNA

The collagen-specific chaperone heat shock protein 47 (HSP47) is involved in proper triple-helix formation of procollagen in the ER as well as translational regulation of procollagen synthesis in HS-derived myofibroblasts, favoring the process of fibrillar collagen-mediated fibrosis. Targeted knockdown of HSP47 mRNA in the HS cells is therefore a potentially viable approach to the treatment of fibrosis. Since uptake of vitamin A by HS cells is mediated by the receptors for retinol binding protein, vitamin

Chapter twelve: Nanoparticles for therapeutic delivery in human diseases   211

A–coupled liposomes were employed to encapsulate HSP47 siRNA and administered intravenously, with a considerable level of suppression of collagen secretion and reduction of fibrosis in three animal models of liver cirrhosis induced by dimethylnitrosamine (DMN), CCl_4, or bile duct ligation (Sato et al. 2008).

12.8.2   Hepatic infections

Chronic hepatitis B and C infections are the largest risk factors for development of HCC. Treatment of hepatitis infections includes administration of both PEGylated IFN-α and ribavirin, a nucleoside analog. Although PEGylated IFN-α has shown prolonged half-life with enhanced therapeutic index compared to IFN-α, the former is often accompanied by dose-limiting side effects, while nucleoside analogs are associated with resistance. The majority of the intravenously administered nanoparticles without surface modification are rapidly taken by liver macrophages and other reticuloendothelial organs, but not by hepatocytes, the major target for anti-HBV drugs/genes. Targeted delivery of biofunctional nanoparticles could be beneficial in potentially delivering the anti-HBV agents at the disease site, thus reducing the side effects and resistance and improving the treatment outcome.

12.8.2.1   HBV RNA-specific siRNA loaded into cationic PEGylated liposomes

Replication of HBV is susceptible to RNAi-mediated inhibition, unlike HIV-1 or HCV. This is because HBV has a very compact genome that limits its sequence plasticity through mutation. Accordingly, HBV genome-specific siRNAs were loaded into cationic PEGylated liposomes (80–100 nm size) with acidic pH–labile oxime linkages to facilitate the siRNA release from hepatic endosomes and intravenously injected into an HBV transgenic murine model, resulting in threefold higher suppression of HBV replication markers, such as circulating viral particle and intrahepatic HBV mRNA, depending on the siRNA target sequences (Carmona et al. 2009).

12.8.2.2   SNALP with loaded HBV RNA-specific siRNA

The HBV RNA-targeted siRNAs chemically stabilized for nuclease resistance by substituting all 2′-OH residues on the RNA with 2′-F, 2′-O-Me, or 2′-H residues were incorporated into a specialized liposome forming a stable nucleic acid–lipid particle (SNALP). The lipid bilayer containing a mixture of cationic and fusogenic lipids assisted in cellular uptake of the SNALP and subsequent endosomal escape of the siRNA payload, while the surface-coated diffusible PEG-lipid component provided a hydrophilic exterior for favorable pharmacokinetics of the particle. Three daily intravenous administration of the SNALP to the mice carrying replicating

HBV caused a specific, dose-dependent reduction in the serum HBV DNA, lasting for up to 7 days after dosing. The improved efficacy of the siRNA formulation correlated well with its longer half-life in plasma and liver and reduced toxic and immunostimulatory effects (Morrissey et al. 2005).

12.8.2.3 Locked nucleic acid–modified oligonucleotide complementary to miR-122

MicroRNA-122 (miR-122), a liver-expressed microRNA that binds to two closely spaced target sites in the 5' noncoding region of the HCV genome, is essential for HCV RNA accumulation in liver cells. Intravenous delivery of a locked nucleic acid–modified oligonucleotide complementary to miR-122 to the chimpanzees with chronic HCV infection resulted in long-lasting suppression of HCV in blood without demonstration of viral resistance or side effects. In addition, the treatment showed derepression of the target mRNAs, downregulation of IFN-regulated genes, and improvement of HCV-induced liver pathology (Lanford et al. 2010).

12.9 Malaria

Malaria, a major public health problem in tropical and subtropical countries, is caused by protozoan parasites belonging to the genus *Plasmodium* and transmitted from host to host by the bite of an infected *Anopheles* mosquito. Treatment of malaria is currently threatened by an increasing number of drug-resistant parasites. Nanotechnology can potentially help improve the solubility and poor bioavailability of the existing antimalarial drugs (i.e., quinine and artemisinin derivatives), lower the drug resistance, and enhance the therapeutic efficacy through targeted delivery of the drugs (Aditya et al. 2013).

12.9.1 Curcumin nanoparticles

Curcumin, a natural polyphenolic compound, has prophylactic and therapeutic potential in anti-inflammatory, anticarcinogenic, and anti-infectious activities. The compound displayed inhibitory activity against a *Plasmodium berghei*–infected murine model of malaria, as well as a synergistic antimalarial activity with artemisinin (Nandakumar et al. 2006). To improve the pharmacokinetic and safety profile of curcumin, nanotized curcumin of 20–50 nm was formulated, and subsequently, its oral bioavailability and antimalarial role were tested (Ghosh et al. 2014). After oral administration to mice, the nanotized curcumin demonstrated a significantly higher curcumin in blood than the native curcumin. In addition, oral treatment of *P. berghei*–infected mice with the nanotized curcumin prolonged the survival of the animals with complete clearance of parasites in comparison to the animals treated with the native form of curcumin.

Chapter twelve: Nanoparticles for therapeutic delivery in human diseases   213

12.9.2   Curcumin-entrapped chitosan nanoparticles

Chitosan nanoparticles were used to encapsulate curcumin in order to improve its bioavailability and chemical stability. When fed to mouse orally, curcumin-entrapped chitosan nanoparticles were able to pass through the mucosal barrier intact to reach the bloodstream and enhanced the antimalarial activity in the mice infected with a lethal strain of *Plasmodium yoelii* (N-67) (Akhtar et al. 2012).

12.9.3   Artemisinin- and curcumin-loaded liposomes

Artemisinin and its derivatives are potent and rapid-acting antimalarial drugs. However, like curcumin, it has a poor bioavailability and a short half-life. Thus, in order to prolong blood circulating time and enhance half-life, artemisinin alone or in combination with curcumin was encapsulated in conventional and PEGylated liposomes (Isacchi et al. 2012). Intraperitoneal treatments of *P. berghei* NK-65–infected mice with the different liposomal formulations of the individual or combined drugs showed an immediate antimalarial effect compared to artemisinin alone, which began to decrease parasitemia levels only 7 days after the start of the treatment. Moreover, the liposomal formulations of artemisinin and artemisinin plus curcumin were found to efficiently cure all malaria-infected mice within the same postinoculation period.

12.9.4   β-Artemether–loaded liposomes

β-Artemether (ARM), a potent antimalarial drug of the artemisinin derivatives, was encapsulated into the liposomes composed of EPC and cholesterol. The resultant formulation was used to intraperitoneally treat the mice infected with the virulent rodent malaria parasite *Plasmodium chabaudi*, which effectively cured all the mice by clearing the recrudescent parasitemia (Chimanuka et al. 2012).

12.9.5   Chloroquine-encapsulated immunoliposomes

Chloroquine was the most widely used drug for malaria treatment until the development of resistance against it by the malarial parasites. New strategies are required for effectively treating drug-resistant malaria with existing drugs such as chloroquine. In an effort to achieve the goal, chloroquine was encapsulated in the liposomes made of EPC, Chol, and gangliosides and surface coated with F(ab′)2 fragments of a mouse monoclonal antibody raised against the cell membranes of erythrocytes (isolated from the *P. berghei*–infected mouse). Upon loading with chloroquine, the antibody-bearing liposomes were subjected to intravenous delivery, finally eliminating the chloroquine-resistant *P. berghei* infections (75% to 90%) in vivo. The immunoliposomes showed better antimalarial activity at lower dose and specific binding to the infected erythrocytes (Owais et al. 1995). In another study, chloroquine-loaded liposomes (lipCQ) with surface-anchored antimouse red blood cell (anti-mRBC) Fab′ were found significantly more effective than lipCQ or free chloroquine in delaying or preventing a potent infection caused by intravenous injection of parasitized mouse red blood cells in rats (Peeters et al. 1989).

12.9.6 Liposome-coupled TNF-α

TNF-α plays important roles in antitumor activities, immunological responses, and inflammatory reactions. In malaria, much of the pathology in severe and cerebral malaria is mediated by high levels of TNF-α, whereas low amounts of TNF-α mediate protection against malaria. However, recombinant human TNF-α (rhTNF-α) with a short half-life is rapidly cleared from the blood circulation. To enhance the protective efficacy of rhTNF-α against *P. berghei*–induced experimental cerebral malaria (ECM) in mice, rhTNF-α was coupled to the outer surface of liposomes or encapsulated into liposomes. After intravenous administration, rhTNF-α coupled to either conventional or sterically stabilized liposomes exhibited an enhanced protective effect against ECM compared with free rhTNF-α, seemingly by stabilizing the bioactive form of rhTNF-α and thus prolonging its blood residence time (Postma et al. 1999). However, the intravenously injected rhTNF-α encapsulated into the liposomes did not improve the protective efficacy against ECM.

12.9.7 ARM-loaded lipid nanoparticles

Artemisinin and its derivatives are among the most potent antimalarial drugs with the ability to act on both chloroquine-sensitive and -resistant *Plasmodium falciparum* strains. However, the drug has a short half-life with poor aqueous solubility, contributing to low oral bioavailability. Intramuscular formulations currently available in the market are associated with low patient compliance and inconsistent assimilation and unsuited to treat cerebral malaria. To develop a better system, ARM, a derivative of artemisinin, was loaded into lipid nanoparticles (LNPs) consisting of a solid matrix prepared by blending solid lipid (glyceryl trimyristate) with increasing ratio of liquid lipid (soybean oil) (Aditya et al. 2010). Compared to plain drug solution and marketed formulation, intraperitoneally injected ARM–LNPs showed better antimalarial efficiency, significantly prolonging the life span of the *P. berghei*–infected mice.

12.9.8 Primaquine-loaded nanoemulsions

Primaquine (PQ) is one of the most widely used antimalarial I drug, with higher doses of PQ limited by severe hematological and GI-related side effects. Oral delivery of PQ-loaded lipid nanoemulsion showed effective antimalarial activity against *P. berghei* infection in mice at a 25% lower dose compared to conventional oral dose. Moreover, the lipid nanoemulsion formulation exhibited improved oral bioavailability of PQ with preferential uptake by the liver (Singh and Vingkar 2008).

12.10 Regeneration of tissues

12.10.1 Nanotechnology in wound healing

Wound healing is a complex cascade of events leading to the anatomical reconstitution of the biological barrier damaged as a result of trauma, compression, burns, or vascular diseases (Cortivo et al. 2010). The normal acute wound healing process involves three overlapping steps: inflammation, proliferation, and remodeling. However, depending on the type of wound, a pathological deviation from the physiological process of tissue repair can lead to an excessive or insufficient wound reparation. Nanotechnology can play diverse roles in accelerating the wound healing process while ensuring minimal scarring and maximal function.

12.10.1.1 Wound healing process

Inflammation is the first event that begins immediately after injury and lasts for 1–4 days. It is characterized by formation of fibrin clot providing 3D scaffold through which inflammatory cells, such as neutrophils, macrophages, and lymphocytes, migrate to the wound, cleansing the wound and secreting a host of chemoattractants and growth factors. The proliferation phase, which normally takes place between 4 and 21 days after wounding, is driven by the growth factors secreted in the inflammation phase. The growth factors FGF and PDGF stimulate fibroblasts to invade the wound site and produce extracellular matrix components, such as collagen, elastin, and glycosaminoglycans, thus generating granulation tissue. Fibroblasts are also induced to secrete FGF, which together with VEGF secreted by platelets and neutrophils stimulate proliferation and migration of endothelial cells, thus promoting vascularization at the healing site. The last phase is remodeling of the extracellular matrix by a coordinated process of collagen production by fibroblasts and its breakdown attributed primarily to MMP and tissue inhibitor of MMP, causing the wound to be gradually less vascularized and strengthening the granulation tissue. The presence of bacteria, such as *Staphylococcus aureus*, exacerbates the tissue-damaging processes and also accounts for the non-healing wounds in the mucosa infected with *Helicobacter pylori*, giving rise to GI ulcers (Cortivo et al. 2010).

12.10.1.2 Thrombin-conjugated iron oxide nanoparticles

Thrombin, which is traditionally used for topical hemostasis and wound management in clinical setting, is essentially required to convert fibrinogen to fibrin, promote aggregation of blood platelets, activate other hemostatic factors, and increase vascular permeability for allowing cells and fluid to enter the wounded tissue. In order to increase the half-life of thrombin in plasma, which is shorter than 15 seconds, it was conjugated to BSA-coated iron oxide nanoparticles for the treatment of incisional wounds on rat skin (Ziv-Polat et al. 2010). In the course of 28 days of healing, the highest values of skin tensile strength were observed with the nanoparticle-bound thrombin after the treatment, while the significantly lower values were recorded for the free thrombin.

12.10.1.3 Antibiotics-carrying nanoparticles

Nanoparticles can be employed to carry antibiotics for controlled release or depot delivery, thereby decreasing the number of drug doses required to achieve a clinical effect. The most effective treatments for *H. pylori* infection are based on combinations of two antibiotics (e.g., clarithromycin, tetracycline, or metronidazole) and a proton pump inhibitor (Ramteke et al. 2009). The (L)-fucose bound chitosan–glutamate nanoparticles were used to deliver three antibiotics, amoxicillin, clarithromycin, and omeprazole in combination (triple therapy), targeting the lectin receptors on *H. pylori* grown in the mouse stomach. After oral administration, the mean bacterial counts for the ligand-conjugated nanoparticle-mediated triple therapy were found to be lower than the ligand-free nanoparticle-aided triple therapy. Moreover, the targeted triple therapy demonstrated the complete eradication of *H. pylori* from the gut with the higher dose, which was further supported by the negative results of Gram staining and urease test. This could be attributed to the targeting potential of the (L)-fucose–containing nanoparticle formulation, enabling the local release of the antibiotics on *H. pylori* surface.

12.10.1.4 Silver nanoparticles

Silver sulfadiazine has become the standard treatment for burns. However, pure silver nanoparticles (SNPs) could be more beneficial in healing the burn wounds by markedly increasing the rate of silver ion release. In an animal model, it was shown that SNPs promoted rapid healing and reduced scar appearance in a dose-dependent manner through their antimicrobial properties, reduction in wound inflammation, and modulation of fibrogenic cytokines (Tian et al. 2007).

Chapter twelve: Nanoparticles for therapeutic delivery in human diseases 217

12.10.1.5 Nitric oxide–delivering nanoparticles

NO exerts antibacterial effects and plays key roles in the natural wound healing process by regulating cell proliferation, collagen formation, and wound contraction. However, its clinical application is restricted by its gaseous nature and short half-life. Despite current development of NO-releasing nanoparticles, controlled release of NO from those nano-formulations over an extended period has yet to be achieved. Recently, NO-releasing PLGA–PEI nanoparticles (NO/PPNPs) were fabricated by reacting the secondary amine groups of PEI with NO and incorporating the resultant PEI/diazeniumdiolate (PEI/NONOate) in the matrix of PLGA nanoparticles. In a mouse model of an *S. aureus*–infected wound, NO/PPNPs exhibited a significantly reduced wound area without any scab, apparently attributed to the bactericidal effect as well as wound healing activity of the released NO (Nurhasni et al. 2015).

12.10.1.6 PLGA–curcumin nanoparticles

Lactate generated from biodegradable PLGA can accelerate angiogenesis and wound healing processes by stimulating collagen synthesis in fibroblasts and VEGF transcription in endothelial cells. On the other hand, curcumin aids in wound healing by increasing granulation tissue formation and enhancing the biosynthesis of transforming growth factor-β1 and proteins in the extracellular matrix and also scavenges the free radicals and inhibits the oxidative damage in keratinocytes and fibroblasts by downregulating the PI3K/AKT/NFκB pathway. Thus, PLGA–curcumin nanoparticles were synthesized and applied in a full-thickness excisional wound healing mouse model, resulting in a twofold higher wound healing activity compared to that of PLGA or curcumin and exhibiting higher reepithelialization, granulation tissue formation, and anti-inflammatory potential (Chereddy et al. 2013).

12.10.2 Bone regeneration

Both osteoblasts, the bone-forming cells, and osteoclasts, the bone-resorbing cells, act in concert to regulate bone homeostasis. Regeneration of bone tissues could be promoted by inducing growth of osteoblasts through nanoparticle-based growth factor delivery and inhibiting osteoclast resorption through nanoparticle-mediated local delivery and release of specific inhibitors (Tautzenberger et al. 2012).

12.10.2.1 Plekho1 siRNA-loaded liposomes

In order to treat bone defects in metabolic skeletal disorders, one of the approaches is to silence bone-formation–inhibitory genes by siRNAs in osteogenic-lineage cells of the bone-formation surfaces. Accordingly, a cationic liposome was attached to six repetitive sequences of an oligopeptide consisting of aspartate, serine, and serine (AspSerSer) for specific delivery to bone-formation surfaces and complexed with an osteogenic siRNA targeting casein kinase-2 interacting protein-1 gene (Plekho1) (Zhang et al. 2012). Systemic delivery of the Plekho1 siRNA-loaded liposomal formulation resulted in selective uptake by the osteogenic cells, inhibition of Plekho1 expression, and promotion of bone formation in both healthy and osteoporotic rats.

12.10.2.2 RGD-modified alginate hydrogels/calcium phosphate–BMP-2 plasmid DNA

Another approach to repair bone defects is delivery of bone morphogenetic proteins (BMPs) among which BMP-2 was particularly shown to have a strong osteoinductive effect by promoting the differentiation of mesenchymal stem cells into osteoblasts. However, to elicit an effect, a large amount of the protein needs to be delivered because of its degradation and poor retention at the site of injury. An alternative way could be localized gene delivery allowing sustained expression of BMP-2 by the host or transplanted cells near the defect site. Thus, an injectable system was developed on the basis of RGD-modified alginate hydrogels containing preosteoblastic cells and nanoparticles of calcium phosphate–BMP-2 plasmid DNA and injected subcutaneously in the backs of mice, resulting in the formation of bony tissue in two and a half weeks (Krebs et al. 2010). The RGD amino acid sequence allowed the cells to adhere to the biodegradable hydrogel and promotes their proliferation, while the nanoparticles facilitated expression of BMP-2 through transfection and thus promoted bone formation.

12.10.2.3 AAV carrying RANKL and VEGF genes

Gene expression profiling studies showed that allografts are deficient in VEGF and receptor activator of nuclear factor κB ligand (RANKL) that dominantly regulate angiogenesis and osteoclastic bone resorption, respectively, during skeletal repair. With this in mind, recombinant AAVs carrying RANKL and VEGF genes were immobilized onto the cortical surface of the allografts, which, after being transplanted into mice, showed marked remodeling and vascularization, leading to a new bone collar around the graft (Ito et al. 2005).

chapter thirteen

Nanomedicine in clinical trials

Since the approval of the first nanomedince, Adagen (Sigma-Tau Pharmaceuticals, Inc., Maryland, USA), a PEGylated bovine adenosine deaminase for the treatment of severe combined immunodeficiency disease in 1990, the field has exploded with substantial government and industrial funding promoting huge fundamental research for drug target identification and validation and extensive preclinical/clinical testing of potential products generated via chemical synthesis/engineering, recombinant DNA technology, and hybridoma (or phage display) technology. Clinical trials use human subjects so as to see whether a drug is effective and what sort of adverse effects it may cause. According to federal regulations, an Investigational New Drug application must be submitted to the Food and Drug Administration (FDA) for reviewing before human clinical trials for a drug are conducted, regardless of whether it is nanomedical or conventional. During the review process, the FDA may request any additional supporting information, such as particle size data if deemed relevant in order to ensure the adequacy of the study design or protect the safety of the clinical trial participants. Considering the huge cost incurred, industries are frequently engaged to sponsor the clinical trials by actively recruiting patients.

13.1   Different phases of clinical trials

There are three main phases of clinical trials (from phase 1 to phase 3) that are carried out sequentially, although some trials may include phase 4 after a drug is licensed or involve two phases together (phase 1/3 or phase 2/3). Phase 1 trials are done by slowly recruiting a few patents with the aim of finding out the safe dose range of a potential drug, its possible side effects, the way a body copes with and gets rid of the drug, and the therapeutic response. A very small dose of the drug is initially given to the first few patients, and depending on the progress, the next batch of patients receives a slightly higher dose. Thus, the dose is gradually increased with each group in a dose escalation study until the best dose is obtained. On the basis of the outcome in phase 1, the prospective drug can be moved to the next stage, phase 2, which is usually larger than phase 1, in order to further explore the side effects and the ways to manage them and justify whether the new treatment should be tested in larger phase 3 trials. The new treatment is sometimes compared in this phase with an existing treatment, or with a dummy drug (placebo), and depending on the positive outcomes, that is,

219

if the new medication is assumed to be as good as the existing one or at least better, it may be brought to phase 3. The phase 3 trials, which usually involve a larger number of patients than phase 2, compare the new treatment with the best currently available treatment and establish a standard treatment for the new drug. Some trials may stop recruiting into the respective groups in case the treatment is not working or if it is causing more side effects. To speed up the development of a promising new drug, phase 0 studies are sometimes carried out by giving very small doses of the drug to a small number of people. Although the dose of the drug is too small to treat a disease, the study is an effective approach to obtaining useful information relatively quickly. Phase 4, which is conducted after approval of a drug, aims to gather information on the drug's effects in various populations in addition to the side effects associated with its long-term use.

Approximately 20%–30% of phase 1 drugs successfully pass through phase 2 to phase 3, while approximately 40%–50% of drugs in phase 3 enter the market after approval by the relevant regulatory authority. The failure of most of the nanoparticulate drugs to reach phase 3 is mainly attributed to their toxicity observed in clinical settings or insignificant therapeutic potential relative to the existing non-nanoparticulate drug(s). Most of the nanoparticulate drugs (70%–80%) in clinical phases 1, 2, and 3 are classified as soft nanoparticles represented mainly by lipid- and polymer-based formulations. Among the diseases, cancer is the main therapeutic target of the nanoparticulate drug candidates being tested in clinical trials involving phases 1, 2, and 3 (Schütz et al. 2013).

13.2 Nanoparticulate drug delivery systems in clinical trials

Table 13.1 shows various nanoparticulate drug delivery systems in clinical trials.

13.3 Monoclonal antibodies as therapeutics in clinical trials (selected)

Table 13.2 shows monoclonal antibodies used as therapeutics in selected clinical trials.

Apart from the above clinical trials, there are numerous clinical trials currently being conducted by combining multiple drugs, one of which can be at least an approved nanomedicine or a nanoparticulate drug under development, while the other drug(s) can belong to the small-molecule therapeutics already approved or in the process of development. Some clinical trials involve both an approved nanomedicine and a nanoparticulate drug candidate.

Table 13.1 Nanoparticulate drug delivery systems under clinical trials, with their therapeutic components, sponsors, and disease targets

Drug name	Nanomaterial/ nanocarrier	Therapeutic	Sponsor	Disease target	Current status
ABI-007	Albumin	Paclitaxel	Celgene Corporation	Taxol-resistant patients with metastatic breast cancer	Phase 2
ABI-009 (Nab-rapamycin)	Albumin	Rapamycin (mTOR inhibitor)	Celgene Corporation	Malignant perivascular epithelioid cell tumor (PEComa)	Phase 2
			Aadi, LLC	Non-muscle-invasive bladder cancer	Phase 1/2
Actinium-225-labeled anti-CD33 antibody	Humanized anti-CD33 monoclonal antibody, HuM195	Actinium-225 radioisotope	Memorial Sloan Kettering Cancer Center	Advanced myeloid malignancies	Phase 1 (completed)
ADI-PEG 20	PEG–protein conjugate	Arginine deiminase	Polaris Group	Advanced hepatocellular carcinoma previously unsuccessfully treated with systemic therapy	Phase 3
				Acute myeloid leukemia	Phase 2
				Non-Hodgkin's lymphoma previously unsuccessfully treated with systemic therapy	Phase 2

(Continued)

Table 13.1 (Continued) Nanoparticulate drug delivery systems under clinical trials, with their therapeutic components, sponsors, and disease targets

Drug name	Nanomaterial/ nanocarrier	Therapeutic	Sponsor	Disease target	Current status
			Ludwig Institute for Cancer Research	Small-cell lung cancer	Phase 2
			Barts & The London NHS Trust	Malignant pleural mesothelioma	Phase 2
			University of Miami Sylvester Comprehensive Cancer Center	Metastatic melanoma that cannot be removed by surgery	Phase 2 completed
			FDA Office of Orphan Products Development	Hepatocellular carcinoma	Phase 2 completed
ALN-CC5	GalNAc–siRNA conjugate	siRNA targeting complement component C5	Alnylam Pharmaceuticals	Paroxysmal nocturnal hemoglobinuria (PNH)	Phase 1/2
ALN-PCS02	Stable nucleic acid lipid nanoparticles (SNALP)	siRNA targeting gene proprotein convertase subtilisin/kexin type 9 (PCSK9)	Alnylam Pharmaceuticals	Elevated LDL-cholesterol (LDL-C) (Hypercholesterolemia)	Phase 1 completed

(Continued)

Table 13.1 (*Continued*) Nanoparticulate drug delivery systems under clinical trials, with their therapeutic components, sponsors, and disease targets

Drug name	Nanomaterial/ nanocarrier	Therapeutic	Sponsor	Disease target	Current status
ALN-AS1	GalNAc–siRNA conjugate	siRNA targeting aminolevulinic acid synthase-1 (ALAS-1)	Alnylam Pharmaceuticals	Acute intermittent porphyria (AIP)	Phase 1
ALN-AT3sc	GalNAc–siRNA conjugate	siRNA targeting antithrombin	Alnylam Pharmaceuticals	Hemophilia A, hemophilia B	Phase 1
SLIT cisplatin	Liposome	Cisplatin	Insmed Incorporated	Osteosarcoma metastatic to the lung	Phase 1/2 completed
ALN-RSV01	Unmodified siRNA sequence	siRNA targeting the nucleocapsid "N" gene of the respiratory syncytial virus genome	Alnylam Pharmaceuticals	Lung transplant patients infected with respiratory syncytial virus (RSV)	Phase 2 completed
ALN-TTR01	SNALP	siRNA targeting transthyretin gene	Alnylam Pharmaceuticals	Transthyretin-mediated amyloidosis (ATTR)	Phase 1 completed
ALN-TTRSC (Revusiran)	siRNA–GalNAc conjugate	RNAi targeting transthyretin (TTR)	Alnylam Pharmaceuticals	Transthyretin (TTR)-mediated familial amyloidotic cardiomyopathy (FAC)	Phase 3
ALN-VSP (ALN-VSP02)	SNALP	siRNA targeting VEGF and kinesin spindle protein (KSP)	Alnylam Pharmaceuticals	Solid tumors	Phase 1 completed

(*Continued*)

Table 13.1 (Continued) Nanoparticulate drug delivery systems under clinical trials, with their therapeutic components, sponsors, and disease targets

Drug name	Nanomaterial/ nanocarrier	Therapeutic	Sponsor	Disease target	Current status
AMG 223	Amine-functional polymer	Phosphate binder	Amgen	Chronic kidney disease on hemodialysis with hyperphosphatemia	Phase 2 completed
Arikace	Liposome	Amikacin	Insmed Incorporated	Bronchiectasis	Phase 1/2 completed
				Cystic fibrosis	Phase 1/2 completed
ANX-514	Nanoemulsion	Docetaxel	Mast Therapeutics, Inc.	Advanced cancer	Phase 1 completed
ANX-530	Nanoemulsion	Vinorelbine	Mast Therapeutics, Inc.	Breast cancer, non–small-cell lung cancer, non-Hodgkin's lymphoma	Phase 1 completed
ARC1779	PEGylated aptamer	A blocker of von Willebrand factor (vWF)–dependent platelet function	Archemix Corp.	vWF-related platelet function disorders	Phase 2 completed
Aroplatin (L-NDDP)	Liposome	Bis-neodecanoate diamino-cyclohexane platinum (NDDP)	Agenus, Inc.	Advanced solid malignancies or B-cell lymphoma	Phase 1 completed

(Continued)

Table 13.1 (Continued) Nanoparticulate drug delivery systems under clinical trials, with their therapeutic components, sponsors, and disease targets

Drug name	Nanomaterial/ nanocarrier	Therapeutic	Sponsor	Disease target	Current status
			Aronex Pharmaceuticals	Advanced colorectal cancer resistant to standard therapies	Phase 2
			New York University School of Medicine	Malignant pleural mesothelioma	Phase 2 completed
ASG-22ME	Antibody–drug conjugate	Monomethyl auristatin E (MMAE) conjugated to human anti–Nectin-4 monoclonal antibody	Agensys, Inc.	Metastatic urothelial cancer and other malignant solid tumors	Phase 1
ASG-5ME	Antibody–drug conjugate	MMAE conjugated to SLC44A4-directed human monoclonal antibody	Astellas Pharma Inc.	Prostate neoplasms	Phase 1 completed
			Seattle Genetics, Inc. Agensys, Inc.	Gastric neoplasms, pancreatic neoplasms	Phase 1 completed
ATI-1123	Liposome	Docetaxel	Azaya Therapeutics, Inc.	Solid tumors	Phase 1 completed

(Continued)

Table 13.1 (*Continued*) Nanoparticulate drug delivery systems under clinical trials, with their therapeutic components, sponsors, and disease targets

Drug name	Nanomaterial/ nanocarrier	Therapeutic	Sponsor	Disease target	Current status
Atu027	Liposome	siRNA targeting PKN3	Silence Therapeutics GmbH	Advanced solid tumors	Phase 1 completed
AurImmune (CYT-6091)	Gold	Recombinant human TNF-α	National Institutes of Health Clinical Center (CC)	Advanced solid tumors	Phase 1 completed
				Patients undergoing surgery for primary cancer or metastatic cancer	Phase 0 completed
AuroShell (AuroLase Therapy)	Gold-silica nanoshell	Laser therapy	Nanospectra Biosciences, Inc.	Head and neck cancer	Phase 1 completed
ABI-008 (Nab-docetaxel)	Albumin	Docetaxel	Celgene Corporation	Hormone-refractory prostate cancer	Phase 1/2
2B3-101	Glutathione PEGylated liposome	Doxorubicin	BBB Therapeutics B.V.	Solid tumors and brain metastases or recurrent malignant glioma	Phase 1/2 completed
BC-819 (DTA-H19)/ PEI	Polyethylenimine (PEI)	BC-819 (plasmid carrying diphtheria toxin A subunit gene)	BioCancell Ltd.	Patients with intermediate-risk superficial bladder cancer	Phase 2

(*Continued*)

Table 13.1 *(Continued)* Nanoparticulate drug delivery systems under clinical trials, with their therapeutic components, sponsors, and disease targets

Drug name	Nanomaterial/ nanocarrier	Therapeutic	Sponsor	Disease target	Current status
BIND-014	Polymer	Docetaxel	Bind Therapeutics	Urothelial carcinoma, cholangiocarcinoma, cervical cancer, squamous cell carcinoma of head and neck	Phase 2
				KRAS-positive patients with non–small-cell lung cancer, squamous cell non–small-cell lung cancer	Phase 2
Brakiva	Liposome	Topotecan	Gynecologic Oncology Group	Locally advanced cervical cancer	Phase 1 completed
				Stage III or stage IV ovarian epithelial cancer or primary peritoneal cancer	Phase 3 completed
CDP 791	PEGylated antibody fragment	Anti-VEGFR2 Fab fragment	UCB Pharma	Non-squamous non–small-cell lung cancer	Phase 2 completed
CPX-1	Liposome	Irinotecan: floxuridine	Celator Pharmaceuticals	Advanced colorectal cancer	Phase 2 completed

(Continued)

Table 13.1 (Continued) Nanoparticulate drug delivery systems under clinical trials, with their therapeutic components, sponsors, and disease targets

Drug name	Nanomaterial/ nanocarrier	Therapeutic	Sponsor	Disease target	Current status
CPX-351	Liposome	Cytarabine: daunorubicin	M.D. Anderson Cancer Center	Acute myeloid leukemia (AML)	Phase 2
			Bruno C. Medeiros	Relapsed or refractory acute myeloid leukemia or myelodysplastic syndrome	Phase 2
			Children's Hospital Medical Center, Cincinnati	Relapsed leukemia or lymphoma	Phase 1
			Celator Pharmaceuticals	Newly diagnosed AML in elderly patients	Phase 2 completed
			Celator Pharmaceuticals	Acute leukemia and myelodysplastic syndromes (MDS) with moderate hepatic impairment	Phase 2
			Celator Pharmaceuticals	Advanced hematologic cancer	Phase 1
CRLX101	A cyclodextrin-based polymer	Camptothecin	Cerulean Pharma Inc.	Advanced non–small-cell lung cancer	Phase 2 completed

(Continued)

Table 13.1 (Continued) Nanoparticulate drug delivery systems under clinical trials, with their therapeutic components, sponsors, and disease targets

Drug name	Nanomaterial/nanocarrier	Therapeutic	Sponsor	Disease target	Current status
CT-2106	Poly-L-glutamate	Camptothecin	CTI BioPharma	Advanced ovarian cancer previously unsuccessfully treated with platinum- and taxane-based regimen	Phase 2 completed
Fovista (E10030)	PEGylated aptamer	Anti-PDGF aptamer	Ophthotech Corporation	Neovascular age-related macular degeneration	Phase 1 completed
Genexol-PM	Polymeric micelle	Paclitaxel	Asan Medical Center	Advanced urothelial cancer previously treated with gemcitabine and platinum	Phase 2 completed
Glembatumumab vedotin (CDX-011)	Antibody–drug conjugate	MMAE conjugated to glycoprotein NMB (GPNMB)-targeting human monoclonal antibody	National Cancer Institute (NCI)	Recurrent or refractory osteosarcoma	Phase 2
				Metastatic or locally recurrent uveal melanoma	Phase 2
			Celldex Therapeutics	Advanced melanoma	Phase 2
				Metastatic, gpNMB overexpressing, triple negative breast cancer (METRIC)	Phase 2
				Advanced gpNMB-expressing breast cancer (EMERGE)	Phase 2 completed

(Continued)

Table 13.1 (Continued) Nanoparticulate drug delivery systems under clinical trials, with their therapeutic components, sponsors, and disease targets

Drug name	Nanomaterial/ nanocarrier	Therapeutic	Sponsor	Disease target	Current status
Hemospan	PEG–protein conjugate	Hemoglobin	Sangart	Chronic critical limb ischemia	Phase 2 completed
				Orthopedic surgery patients	Phase 1/2 completed
				Prostatectomy patients	Phase 2 completed
				Hypotension in hip arthroplasty	Phase 3 completed
IHL-305	Liposome	Irinotecan	Yakult Honsha Co., Ltd.	Advanced solid tumors	Phase 1
ILY101	Polymer	Phosphorus-binding agent	Ilypsa	Dialysis patients with hyperphosphatemia	Phase 2
INGN 241	A nonreplicating adenovector	Melanoma differentiation-associated gene-7 (mda-7/IL24)	Introgen Therapeutics	Malignant melanoma, neoplasm metastasis	Phase 2
Iodine I 131–labeled 81C6 antibody	Anti-tenascin antibody	Iodine 131 radioisotope	Duke University	Malignant primary brain tumors	Phase 1/2 (completed)
Iodine I 131–labeled F19 antibody	Monoclonal antibody, F19 (BIBH-1)	Iodine 131 radioisotope	Memorial Sloan Kettering Cancer Center	Colorectal cancer	Phase 1 (completed)

(Continued)

Table 13.1 (Continued) Nanoparticulate drug delivery systems under clinical trials, with their therapeutic components, sponsors, and disease targets

Drug name	Nanomaterial/ nanocarrier	Therapeutic	Sponsor	Disease target	Current status
131-I–Labeled antibody G250	Chimeric monoclonal antibody, G250 that recognizes carbonic anhydrase IX (CAIX)	Iodine 131 radioisotope	Memorial Sloan Kettering Cancer Center	Advanced kidney cancer	Phase 1/2 (completed)
Iodine 131–labeled tositumomab	Tositumomab (TST), a murine monoclonal antibody against the CD20	Iodine 131 radioisotope	GlaxoSmithKline	B-cell lymphomas	Phase 1 (completed)
LE-DT	Liposome	Docetaxel	INSYS Therapeutics Inc.	Solid tumors	Phase 1 completed
				Locally advanced or metastatic pancreatic cancer	Phase 2 completed
LEP-ETU	Liposome	Paclitaxel	INSYS Therapeutics Inc.	Metastatic breast cancer	Phase 2 completed
				Neoplasm	Phase 1 completed

(Continued)

Table 13.1 (*Continued*) Nanoparticulate drug delivery systems under clinical trials, with their therapeutic components, sponsors, and disease targets

Drug name	Nanomaterial/nanocarrier	Therapeutic	Sponsor	Disease target	Current status
LErafAON-ETU	Liposome	c-Raf antisense oligodeoxynucleotides	INSYS Therapeutics Inc.	Advanced cancer	Phase 1 completed
LiPlaCis	Liposome	Cisplatin	LiPlasome Pharma	Advanced or refractory tumors	Phase 1
Liposomal annamycin	Liposome	Annamycin	Callisto Pharmaceuticals	Acute lymphocytic leukemia	Phase 1/2
L-9NC (9NC-LP)	Liposome	9-Nitrocamptothecin	University of New Mexico	Metastatic endometrial cancer Non-small-cell lung cancer	Phase 2 completed Phase 2 completed
L-NDDP	Liposome	Cisplatin	New York University School of Medicine	Malignant pleural mesothelioma	Phase 2 completed
Lutetium (Lu)-177-anti-J591 antibody	Anti-J591 monoclonal antibody	Lu radioisotope	Memorial Sloan Kettering Cancer Center	Metastatic androgen-independent prostate cancer	Phase 2 (completed)
MAG-CPT (PNU 166148)	Polymer–drug conjugate	Camptothecin	University of Glasgow	Metastatic solid tumors	Phase 1

(*Continued*)

Table 13.1 (Continued) Nanoparticulate drug delivery systems under clinical trials, with their therapeutic components, sponsors, and disease targets

Drug name	Nanomaterial/ nanocarrier	Therapeutic	Sponsor	Disease target	Current status
MBP-426	Transferrin-conjugated liposome	Oxaliplatin (L-OHP)	Mebiopharm Co., Ltd.	Advanced or metastatic solid tumors	Phase 1 completed
				Gastric, gastroesophageal, or esophageal adenocarcinoma	Phase 1/2
Nanocrystalline silver cream (NPI)	Silver nanoparticles	Antimicrobial and anti-inflammatory agent	Nucryst Pharmaceuticals	Atopic dermatitis (eczema)	Phase 2 completed
Nanoxel	Polymeric micelle	Paclitaxel	Fresenius Kabi Oncology Ltd.	Advanced breast cancer	Phase 1
NBTXR3	Hafnium oxide–containing nanoparticles	Radioenhancement	Nanobiotix	Adult soft tissue sarcoma	Phase 1 completed
				Head and neck cancer	Phase 1
				Adult soft tissue sarcoma	Phase 2/3
NC 4016	Polymeric micelle	Oxaliplatin	M.D. Anderson Cancer Center	Advanced cancers, lymphoma	Phase 1
NK012	Polymeric micelle	SN-38 (irinotecan active metabolite)	Nippon Kayaku Co., Ltd.	Relapsed small-cell lung cancer	Phase 2 completed
				Advanced, metastatic triple negative breast cancer	Phase 2 completed
				Refractory solid tumors	Phase 1 completed

(Continued)

Table 13.1 (Continued) Nanoparticulate drug delivery systems under clinical trials, with their therapeutic components, sponsors, and disease targets

Drug name	Nanomaterial/ nanocarrier	Therapeutic	Sponsor	Disease target	Current status
NK105	Polymeric micelle	Paclitaxel	Nippon Kayaku Co., Ltd.	Metastatic or recurrent breast cancer	Phase 3
NKTR-102	PEG–drug conjugate	Irinotecan	Nektar Therapeutics	Advanced or metastatic solid tumors in patients with hepatic impairment	Phase 1
				Locally recurrent breast cancer, metastatic breast cancer	Phase 3
				Malignant solid tumor	Phase 2
				Colorectal cancer	Phase 2 completed
				Metastatic or locally advanced ovarian cancer	Phase 2 completed
				Metastatic or locally advanced breast cancer	Phase 2 completed
				Bevacizumab resistant high-grade glioma	Phase 2 completed
			Lawrence Recht	Recurrent high-grade gliomas	Phase 1 completed

(Continued)

Table 13.1 (Continued) Nanoparticulate drug delivery systems under clinical trials, with their therapeutic components, sponsors, and disease targets

Drug name	Nanomaterial/ nanocarrier	Therapeutic	Sponsor	Disease target	Current status
NKTR-118	PEG–drug conjugate	Naloxone	AstraZeneca	Renal impairment	Phase 1 completed
				Impaired hepatic function	Phase 1 completed
				Opioid-induced constipation (OIC)	Phase 3 completed
NL CPT-11	Liposome	Irinotecan	University of California, San Francisco	Recurrent small-cell lung cancer	Phase 2 completed
OSI-211	Liposome	Lurtotecan	Astellas Pharma Inc.	Relapsed epithelial ovarian cancer	Phase 2 completed
				Advanced solid tumors	Phase 1 completed
			NCIC Clinical Trials Group	Advanced or recurrent ovarian epithelial cancer	Phase 2 completed
				Metastatic or locally recurrent head and neck cancer	Phase 2 completed
			European Organisation for Research and Treatment of Cancer (EORTC)	Gastric or gastroesophageal (GEJ) cancer	Phase 2 completed

(Continued)

Table 13.1 (Continued) Nanoparticulate drug delivery systems under clinical trials, with their therapeutic components, sponsors, and disease targets

Drug name	Nanomaterial/ nanocarrier	Therapeutic	Sponsor	Disease target	Current status
OSI-7904L	Liposome	Thymidylate synthase inhibitor	OSI Pharmaceuticals	Biliary tract cancer	Phase 2 completed
				Head and neck cancer treated unsuccessfully earlier with first-line therapy	Phase 2 completed
				Rheumatoid arthritis	Phase 2 completed
Paxceed	Polymeric micelle	Paclitaxel	Angiotech Pharmaceuticals	Severe psoriasis	Phase 2 completed
			National Cancer Institute (NCI)	Advanced arginine auxotrophic solid tumors	Phase 1
PEG-BCT-100	PEG–protein conjugate	Recombinant human arginase	Bio-Cancer Treatment International Limited	Advanced hepatocellular carcinoma	Phase 2
				Locally advanced or metastatic cancer of the esophagus	Phase 2 completed
Pegylated interferon alfa-2a	PEG–protein conjugate	IFN alfa-2a	Merck Sharp & Dohme Corp.	Melanoma	Phase 2
			Hoffmann-La Roche	Chronic hepatitis C	Phase 3 completed

(Continued)

Table 13.1 (Continued) Nanoparticulate drug delivery systems under clinical trials, with their therapeutic components, sponsors, and disease targets

Drug name	Nanomaterial/ nanocarrier	Therapeutic	Sponsor	Disease target	Current status
			French National Agency for Research on AIDS and Viral Hepatitis	Hepatitis C	Phase 3 completed
			University Hospital Tuebingen	Malignant melanoma stage IIA-IIIB	Phase 3
			National Institute on Drug Abuse (NIDA)	Acute hepatitis C virus infection in injection drug users	Phase 4 completed
			PV-Nord	Polycythemia vera	Phase 2
			National Institute of Allergy and Infectious Diseases (NIAID)	HIV infection	Phase 2 completed
			National Institute of Diabetes and Digestive and Kidney Diseases (NIDDK)	Chronic hepatitis D	Phase 2 completed

(*Continued*)

Table 13.1 (Continued) Nanoparticulate drug delivery systems under clinical trials, with their therapeutic components, sponsors, and disease targets

Drug name	Nanomaterial/ nanocarrier	Therapeutic	Sponsor	Disease target	Current status
Pegylated interferon alfa 2b (PEG-Intron)	PEG–protein conjugate	IFN alfa 2b	European Organisation for Research and Treatment of Cancer (EORTC)	Stage III melanoma	Phase 3
			Memorial Sloan Kettering Cancer Center	Advanced kidney cancer	Phase 1/2 completed
			M.D. Anderson Cancer Center	Myeloproliferative disorders	Phase 2
			National Cancer Institute (NCI)	Diffuse intrinsic pontine glioma	Phase 2 completed
			Emory University	Juvenile pilocytic astrocytomas, optic pathway gliomas	Phase 2
			Weill Medical College of Cornell University	Early primary myelofibrosis	Phase 2
			Memorial Sloan Kettering Cancer Center	Advanced low-grade non-Hodgkin's lymphoma	Phase 2 completed

(Continued)

Chapter thirteen: Nanomedicine in clinical trials   239

Table 13.1 (Continued) Nanoparticulate drug delivery systems under clinical trials, with their therapeutic components, sponsors, and disease targets

Drug name	Nanomaterial/ nanocarrier	Therapeutic	Sponsor	Disease target	Current status
			The Wistar Institute	HIV-1 infection	Phase 2
			AOP Orphan Pharmaceuticals AG	Polycythemia vera	Phase 3
Pegylated interferon beta-1a	PEG–protein conjugate	IFN beta-1a	Biogen	Relapsing multiple sclerosis	Phase 3 completed
PNU-93914	Liposome	Paclitaxel	Memorial Sloan Kettering Cancer Center	Metastatic or locally advanced unresectable transitional cell carcinoma of the urothelium	Phase 2 completed
PPX (CT2103, OPAXIO)	Polyglutamate– drug conjugate	Paclitaxel	Brown University	Glioblastoma multiforme	Phase 2 completed
			University of Southern California	Metastatic colorectal cancer	Phase 1 completed
			Bennett, James P.	Amyotrophic lateral sclerosis	Phase 1/2 completed

(Continued)

Table 13.1 (Continued) Nanoparticulate drug delivery systems under clinical trials, with their therapeutic components, sponsors, and disease targets

Drug name	Nanomaterial/ nanocarrier	Therapeutic	Sponsor	Disease target	Current status
S-8184	Emulsion	Paclitaxel	OncoGenex Pharmaceuticals	Stage III or IV colorectal adenocarcinoma	Phase 2 completed
				Stage III or IV ovarian cancer	Phase 2 completed
				Relapsed stage IIIB or IV non–small-cell lung cancer	Phase 2 completed
				Advanced malignancies	Phase 1 completed
S-CKD602	PEGylated liposome	CKD-602, a camptothecin analog	University of Pittsburgh	Crohn's disease	Phase 2 completed
Semapimod (CNI-1493, CPSI-2364)	Nanocrystal drug	A TNF-α inhibitor and MAPK blocker	Ferring Pharmaceuticals	CD70-positive non-Hodgkin's lymphoma or renal cell carcinoma	Phase 1 completed
SGN-75	Antibody–drug conjugate	Humanized anti-CD70 monoclonal antibody conjugated to monomethyl auristatin F (MMAF)	Seattle Genetics, Inc.	CD33-positive acute mycloid leukemia (AML)	Phase 1

(Continued)

Chapter thirteen: Nanomedicine in clinical trials

Table 13.1 (Continued) Nanoparticulate drug delivery systems under clinical trials, with their therapeutic components, sponsors, and disease targets

Drug name	Nanomaterial/nanocarrier	Therapeutic	Sponsor	Disease target	Current status
SGN-CD33A	Antibody–drug conjugate	Humanized anti-CD33 monoclonal antibody conjugated to pyrrolobenzodiazepine (PBD) dimer	Seattle Genetics, Inc.	Recurrent glioblastoma	Phase 2
SGT-53	TfRscFv-liposome	p53 gene	SynerGene Therapeutics, Inc.	Refractory or recurrent solid tumors	Phase 1
				Recurrent ovarian cancer	Phase 2 completed
SPI-77	Cisplatin	PEGylated liposome	New York University School of Medicine	Neuroendocrine tumors (NET), adrenocortical carcinoma (ACC)	Phase 1/2
Thermodox	Liposome	Doxorubicin	Celsion	Recurrent regional breast cancer	Phase 1/2
				Hepatocellular carcinoma (HCC)	Phase 3

(Continued)

Table 13.1 (Continued) Nanoparticulate drug delivery systems under clinical trials, with their therapeutic components, sponsors, and disease targets

Drug name	Nanomaterial/ nanocarrier	Therapeutic	Sponsor	Disease target	Current status
TKM-080301	SNALP	siRNA directed against polo-like kinase 1 (PLK1)	Tekmira Pharmaceuticals Corporation	Advanced hepatocellular carcinoma	Phase 1/2
				Primary or secondary liver cancer	Phase 1 completed
			National Cancer Institute (NCI)	Gout, hyperuricemia	Phase 1
Trastuzumab emtansine (T-DM1)	Antibody–drug conjugate	Mertansine (DM1) conjugated to anti-HER2/Neu monoclonal antibody	West German Study Group	Breast cancer	Phase 2
			Genentech, Inc.	Metastatic breast cancer	Phase 2 completed
			Hoffmann-La Roche	Advanced gastric cancer	Phase 3
				Locally advanced or metastatic non–small-cell lung cancer	Phase 2
Uricase-PEG 20	Polymer–PEG conjugate	Uricase (urate oxidase)	EnzymeRx	Small-cell lung cancer, non–small-cell lung cancer	Phase 1
XMT-1001	Polymer–drug conjugate	Camptothecin analog	Mersana Therapeutics	Advanced solid tumors	Phase 1

(*Continued*)

Table 13.1 (Continued) Nanoparticulate drug delivery systems under clinical trials, with their therapeutic components, sponsors, and disease targets

Drug name	Nanomaterial/ nanocarrier	Therapeutic	Sponsor	Disease target	Current status
XMT-1107	Polymer–drug conjugate	Fumagillin analog	Mersana Therapeutics	Metastatic breast cancer	Phase 2
EZN-2208 (PEG-SN38)	PEG–drug conjugate	SN-38	Enzon Pharmaceuticals, Inc.	Metastatic colorectal carcinoma	Phase 2
				Pediatric patients with solid tumors	Phase 1/2
				Advanced solid tumors or lymphoma	Phase 1 completed
VivaGel (SPL7013 Gel)	A poly(L-lysine) dendrimer	SPL7013, a dendrimer with HIV and HSV antiviral activity	Starpharma Pty Ltd.	Bacterial vaginosis	Phase 3 completed
^{90}Yttrium-labeled anti-CEA antibody	MN-14, a humanized anti-CEA monoclonal antibody	^{90}Yttrium radioisotope	Garden State Cancer Center at the Center for Molecular Medicine and Immunology	Relapsed or refractory small-cell lung cancer	Phase 1
^{90}Yttrium-labeled anti-HMFG1 antibody	Anti-HMFG1 antibody	^{90}Yttrium radioisotope	Jonsson Comprehensive Cancer Center	Ovarian cancer and primary peritoneal cavity cancer	Phase 3

Table 13.2 Monoclonal antibodies under clinical trials, with their target antigens, sponsors, and disease targets

Monoclonal antibody	Sponsor	Disease target	Current status
ABT-874, an anti IL-12 antibody	AbbVie	Chronic plaque psoriasis	Phase 2 (completed)
105AD7, a human antibody (anti-idiotypic vaccine) that mimics the complement regulatory protein, CD55	Onyvax Limited at St. George's Hospital Medical School	Metastatic adenocarcinoma of the colon or rectum	Phase 1/2
Adalimumab (D2E7), a human anti-TNF antibody	Abbott	Psoriatic arthritis	Phase 2 (completed)
Alemtuzumab (Campath), an antibody that binds to CD52	National Institute of Neurological Disorders and Stroke (NINDS)	Sporadic inclusion body myositis	Phase 2 (completed)
AME-133v (LY2469298), an anti-CD20 antibody	Applied Molecular Evolution	Non-Hodgkin's lymphoma	Phase 1/2 (completed)
Antibody Me1-14 F(ab')2 fragment	Duke University	Brain metastases	Phase 1/2 (completed)
Anti-CD3 antibody	InSpira Medical AB	Chronic hepatitis C	Phase 2
4B5, anti-idiotypic antibody vaccine mimicking the GD2 antigen	University of Alabama at Birmingham	Stage III or stage IV melanoma	Phase 1/2 (completed)
Anti-tumor necrosis factor (TNF) alpha antibody	Assistance Publique - Hôpitaux de Paris	Uveitis in juvenile idiopathic arthritis	Phase 2/3
Apolizumab (Hu1D10), a humanized anti-human leukocyte antigen-DR beta-chain antibody	National Cancer Institute (NCI)	Recurrent non-Hodgkin's lymphoma	Phase 1 (completed)
APX005M, a CD40 agonistic antibody	Apexigen, Inc.	Solid tumors	Phase 1
AV-203, an ERBB3 inhibitory antibody	AVEO Pharmaceuticals, Inc.	Advanced solid tumors	Phase 1 (completed)
AVE1642, an anti-IGF-1R antibody	Sanofi	Advanced multiple myeloma	Phase 1 (completed)

(Continued)

Table 13.2 *(Continued)* Monoclonal antibodies under clinical trials, with their target antigens, sponsors, and disease targets

Monoclonal antibody	Sponsor	Disease target	Current status
Bavituximab, a chimeric anti-phosphatidylserine antibody	Peregrine Pharmaceuticals	Chronic hepatitis C	Phase 1 (completed)
BI 505, a human anti–intercellular adhesion molecule 1 antibody	BioInvent International AB	Relapsed/refractory multiple myeloma	Phase 1 (completed)
Bevacizumab, a humanized anti-VEGF antibody	National Cancer Institute (NCI)	Metastatic kidney cancer	Phase 2
cA2, an anti–tumor necrosis factor chimeric antibody	FDA Office of Orphan Products Development	Crohn's disease	Phase 3 (completed)
CAL, a humanized antibody to the parathyroid hormone-related protein (PTHrP)	Chugai Pharmaceutical	Breast cancer and bone metastases	Phase 1 (completed)
CAT-192, a human anti–TGF-beta1 antibody	Genzyme (a Sanofi Company)	Early-stage diffuse systemic sclerosis	Phase 1/2 (completed)
Chimeric anti-CD20 antibody	Sinocelltech Ltd.	B-cell non-Hodgkin's lymphoma	Phase 1 (completed)
11-1F4, a chimeric fibril-reactive antibody	Columbia University	Amyloid light chain (AL) amyloidosis	Phase 1
Cixutumumab (IMC-A12), an antibody to IGF-1R	National Cancer Institute (NCI)	Recurrent Ewing sarcoma/peripheral primitive neuroectodermal tumor	Phase 1 (completed)
ChAgly CD3, a humanized nonmitogenic CD3 antibody	AZ-VUB	Type 1 diabetes	Phase 2 (completed)

(Continued)

Table 13.2 (*Continued*) Monoclonal antibodies under clinical trials, with their target antigens, sponsors, and disease targets

Monoclonal antibody	Sponsor	Disease target	Current status
Chi Lob 7/4, an anti-CD40 antibody	Cancer Research UK	Advanced malignancies refractory to conventional anticancer treatment	Phase 1 (completed)
CNTO 328, an antibody against interleukin-6 (IL-6)	Southwest Oncology Group	Metastatic prostate cancer not responding to hormone therapy	Phase 2 (completed)
CP-675,206 (tremelimumab), an anti–CTLA-4 human antibody	Clinica Universidad de Navarra, Universidad de Navarra	Advanced hepatocellular carcinoma	Phase 2 (completed)
CT-011, a humanized anti-PD-1 antibody	CureTech Ltd.	Metastatic melanoma	Phase 2 (completed)
Daclizumab, a humanized anti-CD25 antibody	Facet Biotech	Ulcerative colitis	Phase 2 (completed)
GSK1070806, a humanized antibody directed against the soluble cytokine interleukin-18 (IL-18)	GlaxoSmithKline	Inflammatory bowel disease	Phase 1 (completed)
Edrecolomab (Panorex), an antibody directed against the 17-1A antigen	National Cancer Institute (NCI)	Stage II colon cancer	Phase 3 (completed)
Etaracizumab (Abegrin), a humanized antibody against alphavbeta3 integrin receptor	University of Wisconsin, Madison	Refractory advanced solid tumors or lymphoma	Phase 1 (completed)
3F8, an anti-ganglioside GD2-specific mouse antibody	Memorial Sloan Kettering Cancer Center	Metastatic neuroblastoma in second remission	Phase 2 (completed)

(*Continued*)

Table 13.2 (Continued)   Monoclonal antibodies under clinical trials, with their target antigens, sponsors, and disease targets

Monoclonal antibody	Sponsor	Disease target	Current status
F19, an antibody against fibroblast activation protein-alpha (FAP-α)	Memorial Sloan Kettering Cancer Center	FAP-α–positive advanced or metastatic cancer	Phase 1 (completed)
F105, a human antibody reacting with the CD4 binding region of gp120	National Institute of Allergy and Infectious Diseases (NIAID)	HIV infection	Phase 1 (completed)
GC1008, a human anti–transforming growth factor-beta (TGFβ) antibody	Genzyme	Renal cell carcinoma or malignant melanoma	Phase 1 (completed)
HCD122, an anti-CD40 antibody	Novartis Pharmaceuticals	Multiple myeloma	Phase 2 (completed)
HeFi-1, an anti-CD30 antibody	Beth Israel Deaconess Medical Center	Refractory anaplastic large cell lymphoma or Hodgkin's lymphoma	Phase 1 (completed)
huHMFG1, a humanized human milk fat globule-1 that binds to the extracellular MUC1 peptide sequence	Jonsson Comprehensive Cancer Center	Locally advanced or metastatic breast cancer	Phase 1 (completed)
HuM195 (lintuzumab), a humanized anti-CD33 antibody	Jonsson Comprehensive Cancer Center	Acute myelogenous leukemia	Phase 2
HuMV833 (bevacizumab), a humanized anti-VEGF antibody	European Organisation for Research and Treatment of Cancer	Relapsed or refractory solid tumors	Phase 1 (completed)
HuZAF, a humanized anti–interferon-γ antibody	Facet Biotech	Crohn's disease	Phase 2 (completed)

(Continued)

Table 13.2 (Continued) Monoclonal antibodies under clinical trials, with their target antigens, sponsors, and disease targets

Monoclonal antibody	Sponsor	Disease target	Current status
IDEC-152 (lumiliximab), an anti-CD23 antibody	Memorial Sloan Kettering Cancer Center	Relapsed or refractory chronic lymphocytic leukemia or small lymphocytic lymphoma	Phase 1 (completed)
Infliximab (cA2), an anti-TNF chimeric antibody	Centocor, Inc.	Fistulizing Crohn's disease	Phase 3 (completed)
ING-1(heMAb), a human-engineered antibody (MAb) that specifically targets the epithelial cell adhesion molecule (Ep-CAM)	XOMA (US) LLC	Advanced adenocarcinomas	Phase 1 (completed)
IPH2101, an anti-KIR (1-7F9) human antibody	Innate Pharma	Acute myeloid leukaemia	Phase 1 (completed)
Iratumumab (MDX-060), a human antibody against CD 30	Memorial Sloan Kettering Cancer Center	Refractory or relapsed lymphoma	Phase 1/2 (completed)
J591, a humanized antibody	Memorial Sloan Kettering Cancer Center	Prostate cancer	Phase 2 (completed)
J695, a human antibody to interleukin-12	National Institute of Allergy and Infectious Diseases (NIAID)	Crohn's disease	Phase 1 (completed)
KB003, an anti-GM-CSF antibody	KaloBios Pharmaceuticals	Asthma inadequately controlled by corticosteroids	Phase 2 (completed)
KRN330, a fully human antibody directed against A33	Kyowa Hakko Kirin Pharma, Inc.	Colorectal cancer	Phase 1 (completed)

(Continued)

Table 13.2 (Continued) Monoclonal antibodies under clinical trials, with their target antigens, sponsors, and disease targets

Monoclonal antibody	Sponsor	Disease target	Current status
KW-0761, an anti-CCR4 antibody	Kyowa Hakko Kirin Pharma, Inc.	Peripheral T-cell lymphoma	Phase 1/2 (completed)
L19–IL2, an antibody–cytokine fusion protein directed against the ED-B domain of fibronectin (L19)	Philogen S.p.A.	Advanced solid tumors	Phase 1/2 (completed)
MBL-HCV1, a human antibody against hepatitis C	MassBiologics	Hepatitis C virus (HCV) infected patients undergoing liver transplantation	Phase 2 (completed)
MDX-1100, an anti-CXCL10 human antibody	Bristol–Myers Squibb	Ulcerative colitis	Phase 1 (completed)
MEDI-522 (etaracizumab), a humanized antibody directed against the human alpha V beta 3 integrin	Memorial Sloan Kettering Cancer Center	Irinotecan-refractory advanced colorectal cancer	Phase 1/2 (completed)
MGAH22, an Fc-optimized chimeric anti-HER2 antibody	MacroGenics	Relapsed or refractory HER2-positive advanced breast cancer	Phase 2
Mik-Beta 1 antibody directed to the IL-2R beta subunit	National Cancer Institute (NCI)	Chronic lymphocytic leukemia	Phase 1 (completed)
Milatuzumab, an anti-CD74 humanized antibody	Kaplan Medical Center	Chronic lymphocytic leukemia	Phase 1/2
Mono-dgA-RFB4, a deglycosylated ricin A chain (dgA) conjugated anti-CD22 (RFB4) monoclonal antibody	National Cancer Institute (NCI)	Relapsed and refractory CD22+ B-cell lymphoma	Phase 1 (completed)

(Continued)

Table 13.2 (Continued) Monoclonal antibodies under clinical trials, with their target antigens, sponsors, and disease targets

Monoclonal antibody	Sponsor	Disease target	Current status
MORAb-004, a humanized antibody recognizing endosialin (TEM-1)	Morphotek	Solid tumors	Phase 1 (completed)
MRA, a humanized anti–IL-6 receptor antibody	National Institute of Arthritis and Musculoskeletal and Skin Diseases (NIAMS)	Systemic lupus erythematosus	Phase 1 (completed)
MSL-109, a human anti-CMV antibody	Johns Hopkins Bloomberg School of Public Health	HIV infections and cytomegalovirus retinitis	Phase 2/3 (completed)
muJ591 antibody that specifically binds to a prostate-related surface antigen	Weill Medical College of Cornell University	Hormone-independent prostate cancer	Phase 1 (completed)
Nimotuzumab, a humanized antibody against EGFR	National Institute of Cancerología	Recurrent or metastatic cervical cancer	Phase 1/2 (completed)
NPC-1C (NEO-101; Ensituximab), a chimeric antibody whose target appears to be a variant of MUC5AC	Precision Biologics, Inc.	Pancreatic and colorectal cancer	Phase 1/2
OKT3, an anti-CD3 antibody	Hadassah Medical Organization	Metabolic syndrome	Phase 2 (completed)
Omalizumab, an antibody targeting the high-affinity receptor binding site on human immunoglobulin (Ig)E	Shanghai Zhangjiang Biotechnology Limited Company	Allergic asthma	Phase 2/3
Oregovomab, a murine monoclonal antibody that binds to the tumor-associated antigen CA125	AltaRex	Ovarian, fallopian tube, or peritoneal cancer	Phase 2

(Continued)

Table 13.2 (Continued) Monoclonal antibodies under clinical trials, with their target antigens, sponsors, and disease targets

Monoclonal antibody	Sponsor	Disease target	Current status
Panitumumab, a fully human antibody specific to EGF receptor	Jonsson Comprehensive Cancer Center	Renal, prostate, pancreatic, non-small-cell lung, colon, rectal, esophageal, or gastroesophageal junction cancer	Phase 1 (completed)
Pascolizumab, an anti–IL-4 antibody	Facet Biotech	Symptomatic steroid-naive asthma	Phase 2 (completed)
PF-03446962, a fully human antibody against TGFβ receptor ALK1	Fondazione IRCCS Istituto Nazionale dei Tumori, Milano	Relapsed or refractory urothelial cancer	Phase 2
Ramucirumab, a human antibody against VEGF receptor-2 or anti–PDGFR alpha monoclonal antibody IMC-3G3	Sidney Kimmel Comprehensive Cancer Center	Recurrent glioblastoma multiforme	Phase 2 (completed)
RENCAREX (Girentuximab), an antibody that binds specifically to the protein carbonic anhydrase IX (CA IX, MN, or G250 antigen	Wilex	Kidney cancer	Phase 3 (completed)
Rituximab, a chimeric antibody against CD20	Anders Svenningsson	Progressive multiple sclerosis	Phase 2
3S193, a humanized anti–Lewis-Y antibody	Memorial Sloan Kettering Cancer Center	Advanced colorectal carcinoma	Phase 1 (completed)
	Ludwig Institute for Cancer Research	Progressive small-cell lung cancer	Phase 1 (completed)

(Continued)

Table 13.2 (Continued)   Monoclonal antibodies under clinical trials, with their target antigens, sponsors, and disease targets

Monoclonal antibody	Sponsor	Disease target	Current status
SCT200, a full human anti-EGFR antibody	Sinocelltech Ltd.	Metastatic colorectal cancer	Phase 1
Sevacizumab: a humanized anti-VEGF antibody	Jiangsu Simcere Pharmaceutical Co., Ltd.	Advanced or metastatic solid tumors	Phase 1
Sevirumab (MSL 109; Protovir), a human anti-CMV antibody	National Institute of Allergy and Infectious Diseases (NIAID)	Cytomegalovirus retinitis and HIV infection	Phase 2 (completed)
Siltuximab, an anti–IL-6 antibody	Janssen Research & Development, LLC	High-risk smoldering multiple myeloma	Phase 2
TB-403, an antibody directed against placental growth factor (PlGF)	BioInvent International AB	Solid tumors	Phase 1 (completed)
TRC093, a humanized antibody that specifically binds cleaved collagen	Tracon Pharmaceuticals Inc.	Locally advanced or metastatic solid tumors	Phase 1 (completed)
Tremelimumab, a fully human anti–CTLA-4 antibody	Azienda Ospedaliera Universitaria Senese	Malignant mesothelioma	Phase 2
TRM-1, a fully human antibody to the TRAIL-R1	Human Genome Sciences Inc.	Relapsed or refractory non-Hodgkin's lymphoma	Phase 2 (completed)
Visilizumab (HuM291), a humanized anti-CD3 antibody	Stanford University	Advanced or recurrent lymphoma	Phase 1 (completed)
VRC-HIVMAB060-00-AB (VRC01) antibody with broad HIV-1 neutralizing activity	National Institute of Allergy and Infectious Diseases	HIV-1 infection	Phase 1

chapter fourteen

Approved and commercialized nanomedicine

The Food and Drug Administration (FDA) assesses the clinical trial data of a potential nanoparticulate drug like any other small-molecule drug candidate before it decides as to whether the drug should be approved. The number of such nanomedicine products being submitted for approval by the FDA is rapidly growing. Commercialization of the approved nano-drugs for treating critical human diseases has also made a significant progress, with many nanotherapeutics currently seen in clinics. Cancer nanotherapeutics represent the biggest group, constituting more than 20% of the therapeutic nanoparticulate drugs in the clinics and predominantly include liposomal and micellar forms of poorly bioavailable anticancer drugs, such as paclitaxel, or anticancer agents with serious side effects, such as doxorubicin or daunorubicin. Indeed, the nanoparticle-based drug delivery systems have enhanced the therapeutic activities of these drugs by dramatically improving their pharmacokinetics, bioavailability, and tumor-directed delivery through either passive or active targeting, while simultaneously reducing their side effects. The clinically approved nanotherapeutics predominantly include liposomal or albumin-based formulations of anticancer drugs; antibody–drug conjugates; therapeutic monoclonal antibodies (e.g., Herceptin); therapeutic proteins that are gly-cosylated (e.g., Aranesp), PEGylated, amino acid substituted, or Fc-fusion based (e.g., Enbrel); and recombinant vaccines (Table 14.1).

As depicted in Figure 14.1, protein-based therapeutics and drug carri-ers constitute ~90% and liposomal, polymeric, and inorganic carriers rep-resent ~9% of the total FDA-approved drugs. Among them, monoclonal antibodies that are directed against target antigens or serve as drug carri-ers account for 32% of the total drugs. On the other hand, Figure 14.2 shows the number of drugs approved for each of the target human diseases, with cancer of different forms collectively being at the top of the list, followed by genetic diseases. With respect to the individual disease type, arthritis and leukemia, for which the highest number of drugs have been approved so far (8 for each), are followed in the list by diabetes (7), lymphoma (5), multiple sclerosis (5), age-related macular degeneration (4), melanoma (4), anemia (4), growth failure (4), and infertility (4) (Figure 14.2).

Table 14.1 Nanotherapeutics in clinical trials

Drug name	Nanomaterial/ nanocarrier	Therapeutic activity	Company	Disease target	Year of approval
Abraxane	Albumin-stabilized nanoparticle formulation of paclitaxel	Paclitaxel binds to and prevents depolymerization of microtubule, inhibiting cellular motility, mitosis, and replication	Celgene	Non–small-cell lung cancer	2012
Abelcet	Amphotericin B complexed with two phospholipids	Killing a fungus by penetrating its cell wall	Sigma-Tau Pharmaceuticals	Invasive fungal infections	1995
Abthrax (raxibacumab)	A monoclonal antibody	Neutralizing toxins produced by *Bacillus anthracis*	GlaxoSmithKline	Anthrax	2012
Accretropin	A recombinant human growth hormone (r-hGH)	Stimulating growth	Cangene	Growth failure in pediatrics	2008
Actemra (tocilizumab)	A humanized monoclonal antibody	Blocking IL-6 activity	Genentech	Rheumatoid arthritis	2010
				Systemic juvenile idiopathic arthritis	2011
				Polyarticular juvenile idiopathic arthritis	2013

(Continued)

Table 14.1 (*Continued*) Nanotherapeutics in clinical trials

Drug name	Nanomaterial/nanocarrier	Therapeutic activity	Company	Disease target	Year of approval
Adagen (pegademase)	Monomethoxy PEG–protein conjugate	Bovine adenosine deaminase (ADA)	Sigma-Tau Pharmaceuticals/ Enzon	Severe combined immunodeficiency disease (SCID) patients with ADA deficiency	1990
Adcetris (brentuximab vedotin)	An antibody–drug conjugate directed against the tumor necrosis factor (TNF) receptor CD30	Upon internalization by CD30-positive tumor cells, enzymatically released monomethyl auristatin E (MMAE) inhibits tubulin polymerization and thus arrests cell cycle	Seattle Genetics	Hodgkin lymphoma and anaplastic large cell lymphoma	2011
Aldurazyme (laronidase)	A recombinant form of human lysosomal alpha-L-iduronidase enzyme	Catalyzing hydrolysis of terminal α-L-iduronic acid residues of dermatan sulfate and heparan sulfate	Genzyme	Mucopolysac-charidosis I	2003
AlphaNine SD	A preparation of factor IX derived from human plasma	Prevention and control of bleeding	Alpha Therapeutic Corporation	Hemophilia B	1996
Alprolix	A recombinant, Fc-fusion coagulation factor IX	Ensuring normal blood clotting	Biogen Idec	Hemophilia B	2014

(*Continued*)

Table 14.1 (Continued) Nanotherapeutics in clinical trials

Drug name	Nanomaterial/ nanocarrier	Therapeutic activity	Company	Disease target	Year of approval
Ambisom	A liposomal formulation of amphotericin B	Antifungal effect	Astellas Pharma USA/Gilead Ltd.	Invasive fungal infections	1997
Amevive (alefacept)	An immunosuppressive dimeric fusion protein consisting of the extracellular CD2-binding portion of human leukocyte function antigen-3 (LFA-3) linked to the Fc portion of human IgG1	Interfering with lymphocyte activation by binding to the lymphocyte antigen, CD2, and inhibiting LFA-3/CD2 interaction	Biogen Idec	Moderate to severe chronic plaque psoriasis	2003
Arzerra (ofatumumab)	A fully human monoclonal antibody directed against the B-cell CD20 cell surface antigen	Complement-dependent cell lysis (CDCL) and antibody-dependent cell-mediated cytotoxicity (ADCC) of CD20-overexpressing B cells	GlaxoSmithKline	Chronic lymphocytic leukemia	2009
ATryn	A recombinant antithrombin	Neutralizing activity of thrombin and factor Xa by forming a complex and thereby inhibiting coagulation process	GTC BioTherapeutics	Perioperative and peripartum thromboembolic events	2009

(Continued)

Chapter fourteen: Approved and commercialized nanomedicine 257

Table 14.1 (Continued) Nanotherapeutics in clinical trials

Drug name	Nanomaterial/ nanocarrier	Therapeutic activity	Company	Disease target	Year of approval
Avastin (bevacizumab)	A humanized monoclonal antibody directed against the VEGF	Inhibition of VEGF receptor activation prevents growth and maintenance of tumor blood vessels	Genentech	Renal cell carcinoma	2009
Avonex	A recombinant human beta-1a interferon	Antiviral, immunomodulatory, and antiproliferative effects	Biogen Idec	Multiple sclerosis	1995
Benlysta (belimumab)	A human monoclonal antibody: inhibiting human B lymphocyte stimulator protein (BlyS)	Inhibiting human B lymphocyte stimulator protein (BlyS)	Human Genome Sciences	Systemic lupus erythematosus	2011
Berinet	A human plasma-derived C1 esterase inhibitor protein	Control of inflammation	CSL Behring	Acute abdominal or facial attacks of hereditary angioedema (HAE)	2009

(Continued)

Table 14.1 (*Continued*) Nanotherapeutics in clinical trials

Drug name	Nanomaterial/ nanocarrier	Therapeutic activity	Company	Disease target	Year of approval
Bexxar (tositumomab, iodine I 131 tositumomab)	A murine monoclonal antibody, either unradiolabeled or iodine I labeled, directed against the human B-cell–specific CD20	A predose of unradiolabeled antibody is administered first to promote complement-dependent cytotoxicity (CDC) and antibody-dependent cellular cytotoxicity (ADCC), followed by administration of iodine-labeled 131 antibody to kill cancer cells with radiation	Corixa	CD20-positive, follicular low-grade non-Hodgkin's lymphoma (NHL)	2003
Blincyto (blinatumomab)	A recombinant, single-chain, bispecific monoclonal antibody	Anti-CD19/anti-CD3 activities leading to CTL- and HTL-mediated cell death of CD19-expressing B-lymphocytes	Amgen	Philadelphia chromosome–negative (Ph–) relapsed/refractory B-cell precursor acute lymphoblastic leukemia	2014

(*Continued*)

Table 14.1 (*Continued*) Nanotherapeutics in clinical trials

Drug name	Nanomaterial/nanocarrier	Therapeutic activity	Company	Disease target	Year of approval
Botox (onabotulinumtoxinA)	A highly purified botulinum toxin protein refined from the bacterium *Clostridium botulinum*, serotype A	Blocking neuromuscular transmission by binding to acceptor sites on motor or sympathetic nerve terminals and inhibiting acetylcholine release	Allergan	Upper limb spasticity	2010
				Chronic migraine	2010
Byetta (exenatide)	A functional analog of glucagon-like peptide-1 (GLP-1)	Enhancing insulin secretion in response to elevated plasma glucose levels	Amylin/Eli Lilly	Type 2 diabetes	2005
Campath (alemtuzumab)	A humanized monoclonal antibody directed against CD52	Binding to CD52 triggers a host immune response resulting in lysis of CD52-expressing cells	Berlex Laboratories	B-cell chronic lymphocytic leukemia	2001
Cervarix	A recombinant, bivalent, human papillomavirus (HPV) vaccine	Humoral and cellular immunity against HPV-16 and HPV-18 antigens, preventing cervical infection against HPV types 16 and 18	GlaxoSmithKline	Cervical cancer and cervical intraepithelial neoplasia	2009

(*Continued*)

Table 14.1 (Continued) Nanotherapeutics in clinical trials

Drug name	Nanomaterial/ nanocarrier	Therapeutic activity	Company	Disease target	Year of approval
Cimzia (certolizumab pegol)	A PEGylated Fab' fragment of a humanized tumor necrosis factor alpha (TNF-α) inhibitor monoclonal antibody	Suppressing immune responses by blocking TNF-α activity	UCB	Crohn's disease	2008
				Rheumatoid arthritis	2009
Cinryze	A C1 esterase inhibitor, one of the serine proteinase inhibitors (serpins), derived from human plasma	Regulating activation of complement and intrinsic coagulation (contact system) pathway and fibrinolytic system	Lev Pharmaceuticals	Angioedema attacks in adolescent and adult patients with hereditary angioedema (HAE)	2008
Cosentyx (secukinumab)	A human monoclonal antibody that selectively binds to the interleukin-17A (IL-17A) cytokine and inhibits its interaction with the IL-17 receptor	Inhibiting release of proinflammatory cytokines and chemokines	Novartis	Plaque psoriasis	2015
Cyramza (ramucirumab)	A fully human monoclonal antibody	Anti–VEGFR-2 activity (antiangiogenesis role)	Eli Lilly	Gastric cancer	2014
DaunoXome	A liposomal formulation of daunorubicin	Antineoplastic activity	Galen Ltd.	Advanced HIV-associated Kaposi's sarcoma	1996

(Continued)

Table 14.1 (Continued) Nanotherapeutics in clinical trials

Drug name	Nanomaterial/ nanocarrier	Therapeutic activity	Company	Disease target	Year of approval
DepoCyt	A liposomal formulation of cytarabine	Antineoplastic activity	Pacira Pharma/ Sigma-Tau Pharmaceuticals	Lymphomatous meningitis	1999
DepoDur	A liposomal formulation of morphine sulfate	Acting as an opiate agonist in pain management	EKR Therapeutics	Pain after major surgery	2004
Doxil	A liposome-encapsulated form of doxorubicin	Doxorubicin intercalates between DNA base pairs, preventing DNA replication, inhibits topoisomerase II, and forms oxygen free radicals	Alza	Ovarian cancer refractory to other first-line therapies	1999
Egrifta (tesamorelin)	An analog of human growth hormone–releasing factor (GRF)	Stimulating human GRF receptors	Theratechnologies	HIV-associated lipodystrophy	2010
Elaprase (idursulfase)	A recombinant form of human lysosomal enzyme, iduronate 2-sulfataseA	Increasing catabolism of certain glycosaminoglycans (GAG) and thus replacing the natural enzyme	Shire Pharmaceuticals	Mucopoly-saccharidosis II	2006

(Continued)

Table 14.1 (*Continued*) Nanotherapeutics in clinical trials

Drug name	Nanomaterial/nanocarrier	Therapeutic activity	Company	Disease target	Year of approval
Elelyso (taliglucerase alfa)	A recombinant lysosomal glucocerebrosidase	Hydrolysis of the beta-glucosidic linkage of glucocerebroside in lysosome	Pfizer	Gaucher disease	2012
Eligard (leuprolide acetate)	A synthetic nonapeptide analog of gonadotropin-releasing hormone (GnRH)	Continuous, prolonged administration results in pituitary GnRH receptor desensitization, leading to a significant decline in testosterone production	Atrix Laboratories	Advanced prostate cancer	2002
Elitek (rasburicase)	A recombinant urate oxidase	Catalysis of oxidation of uric acid to excretable metabolite allantoin	sanofi-aventis	Hyperuricemia in adults with malignancies	2009
Enbrel (etanercept)	A fusion protein consisting of human IgG1 Fc region fused to a soluble portion of TNF receptor p75A	A TNF antagonist with anti-inflammatory effects	Immunex	Psoriatic arthritis	2002
Entyvio (vedolizumab)	A humanized monoclonal antibody	By targeting alpha-4-beta-7, an integrin on inflammatory cells, adhesion of those cells to the gastrointestinal tract is prevented	Millennium Pharmaceuticals	Ulcerative colitis and Crohn's disease	2014

(*Continued*)

Chapter fourteen: Approved and commercialized nanomedicine 263

Table 14.1 (Continued) Nanotherapeutics in clinical trials

Drug name	Nanomaterial/ nanocarrier	Therapeutic activity	Company	Disease target	Year of approval
Erbitux (cetuximab)	A chimeric monoclonal antibody directed against EGFR	Preventing activation and subsequent dimerization of EGFR	BMS	EGFR-expressing, metastatic colorectal cancer	2004
Erwinaze (asparaginase *Erwinia chrysanthemi*)	Asparaginase, an enzyme critical to protein synthesis in leukemic cells	Hydrolyzing L-asparagine in leukemic cells and thus, blocking protein synthesis and cell proliferation	Eusa Pharma	Acute lymphoblastic leukemia	2011
Exparel	A long-acting, sustained-release liposomal formulation	Bupivacaine, a local anesthetic	Pacira Pharmaceuticals	Postsurgical analgesia	2011
Extavia (interferon beta-1b)	A recombinant interferon beta-1b	Limiting activation of immune cells, suppressing production of inflammatory cytokines and stimulating production of anti-inflammatory cytokines	Novartis	Relapsing multiple sclerosis	2009

(Continued)

Table 14.1 (Continued) Nanotherapeutics in clinical trials

Drug name	Nanomaterial/ nanocarrier	Therapeutic activity	Company	Disease target	Year of approval
Eylea (aflibercept)	A recombinant fusion protein consisting of portions of human VEGF receptors 1 and 2 extracellular domains fused to the Fc portion of human IgG1	Binding to circulating VEGFs and thus suppressing choroidal neovascularization	Regeneron Pharmaceuticals	Neovascular (wet) age-related macular degeneration	2011
Fabrazyme (agalsidase beta)	A recombinant form of human alpha-galactosidase A	Catalyzing hydrolysis of globotriaosylceramide (GL-3) and other α-galactyl-terminated neutral glycosphingolipids	Genzyme	Fabry disease	2003
Feraheme	A superparamagnetic iron oxide nanoparticle coated with a low-molecular-weight semisynthetic carbohydrate	Restoring iron to the body	AMAG	Iron deficiency anemia associated with chronic kidney disease	2009
Ferrlecit	A stable macromolecular iron complex	Restoring iron to the body	R&D Laboratories	Anemia	1999

(Continued)

Chapter fourteen: Approved and commercialized nanomedicine

Table 14.1 (Continued) Nanotherapeutics in clinical trials

Drug name	Nanomaterial/ nanocarrier	Therapeutic activity	Company	Disease target	Year of approval
Fertinex	A preparation of highly purified human follicle-stimulating hormone	Stimulating ovarian follicular growth	Serono Laboratories	Infertility	1996
Flublok (seasonal influenza vaccine)	A recombinant vaccine containing recombinant HA proteins of the three strains of influenza virus	Inducing a humoral immune response	Protein Sciences	Influenza virus subtypes A and type B	2013
Follistim (TM)	A recombinant human follicle-stimulating hormone	Stimulating ovaries to produce egg(s)	Organon	Infertility	1997
Gardasil	A recombinant vaccine prepared from the highly purified virus-like particles (VLPs) of the major capsid (L1) protein of HPV types 6, 11, 16, and 18	Immunoprophylactic activity via development of humoral immune responses	Merck	Cervical cancer associated with human papillomavirus	2006

(Continued)

Table 14.1 (Continued) Nanotherapeutics in clinical trials

Drug name	Nanomaterial/ nanocarrier	Therapeutic activity	Company	Disease target	Year of approval
Gazyva (obinutuzumab)	A glycoengineered fully humanized monoclonal antibody	Binding affinity to the CD20 antigen on malignant human B cells, causing ADCC and caspase-independent apoptosis	Genentech	Previously untreated chronic lymphocytic leukemia	2013
Genotropin (somatropin)	A recombinant human growth hormone	Growth-promoting role	Pharmacia & Upjohn	Growth failure in children who were born small for gestational age (SGA)	2001
Glucagon	A polypeptide hormone identical to human glucagon	Increasing blood glucose level by converting liver glycogen into glucose	Eli Lilly	Severe hypoglycemia associated with diabetes	1998
Gonal-F	A recombinant human follicle-stimulating hormone (FSH)	Contributing to egg development in ovaries	Serono Laboratories	Infertility	1997
Herceptin (trastuzumab)	A humanized monoclonal antibody directed against HER2	Antibody-dependent cell-mediated cytotoxicity against HER2-overexpressing tumor cells	Genentech	Gastric cancer	2010

(Continued)

Table 14.1 (Continued) Nanotherapeutics in clinical trials

Drug name	Nanomaterial/ nanocarrier	Therapeutic activity	Company	Disease target	Year of approval
Hiberix	A polysaccharide conjugate vaccine consisting of capsular polysaccharide (polyribosyl–ribitol–phosphate, PRP), covalently bound to tetanus toxoid	Preventing *Haemophilus influenzae* type b (Hib) infection in children	GlaxoSmithKline	*Haemophilus influenzae* type b (Hib) infection	2009
Humalog (insulin lispro)	A rapid-acting recombinant human insulin analog	Regulation of glucose metabolism	Eli Lilly	Diabetes mellitus	1996
Humira (adalimumab)	A recombinant human monoclonal antibody	Binding specifically to TNF-α and blocking its interaction with p55 and p75 cell surface TNF receptors	Abbott Laboratories	Rheumatoid arthritis	2002
Ilaris (canakinumab)	A human monoclonal anti-human IL-1β antibody	Blocking IL-1β interaction with IL-1 receptors	Novartis	Cryopyrin-associated periodic syndromes	2009
				Systemic juvenile idiopathic arthritis	2013
Increlex (mecasermin)	A recombinant human insulin-like growth factor-1 (rhIGF-1)	Promoting normalized statural growth	Tercica	Growth failure in pediatric patients	2005

(Continued)

Table 14.1 (Continued) Nanotherapeutics in clinical trials

Drug name	Nanomaterial/ nanocarrier	Therapeutic activity	Company	Disease target	Year of approval
Intron A (recombinant interferon alfa-2b)	A nonglycosylated recombinant interferon	Mediating antiviral, antiproliferative, and immune-modulating effects	Schering-Plough	Non-Hodgkin's lymphoma	1997
Jetrea (ocriplasmin)	A recombinant truncated form of human plasmin	Dissolving fibrin blood clots	Thrombogenics	Symptomatic vitreomacular adhesion	2012
Kadcyla (ado-trastuzumab emtansine)	An antibody–drug conjugate consisting of HER2 monoclonal antibody trastuzumab conjugated to maytansinoid DM1	HER2-targeted intracellular delivery of DM1, which upon release binds to tubulin, disrupting microtubule assembly/disassembly dynamics	Genentech	HER2-positive metastatic breast cancer	2013
Keytruda (pembro-lizumab)	A humanized monoclonal antibody	Interacting with PD-1 (programmed death-1) receptor, with potential immune checkpoint inhibitory and antineoplastic activities	Merck	Unresectable or metastatic melanoma	2014

(Continued)

Table 14.1 (Continued) Nanotherapeutics in clinical trials

Drug name	Nanomaterial/ nanocarrier	Therapeutic activity	Company	Disease target	Year of approval
Kineret (anakinra)	A recombinant, nonglycosylated form of human interleukin-1 receptor antagonist (IL-1Ra)	Blocking the biologic activity of IL-1 alpha and beta by competitively inhibiting IL-1 binding to the interleukin-1 type I receptor (IL-1RI)	Swedish Orphan Biovitrum	Cryopyrin-associated periodic syndromes	2013
Kogenate FS	A formulation of recombinant factor VIII	Temporarily replacing the missing clotting factor VIII needed for effective hemostasis	Bayer	Hemophilia A	2000
Krystexxa (pegloticase)	A PEG-modified recombinant uricase	Catalyzing oxidation of uric acid to allantoin, thus lowering serum uric acid	Savient Pharma	Chronic gout (hyperuricemia)	2010
Lantus (insulin glargine)	The first FDA-approved long-acting (basal) recombinant human insulin analog	Regulation of glucose metabolism	sanofi-aventis	Diabetes mellitus	2000
Lemtrada (alemtuzumab)	A monoclonal antibody targeting CD52 on T and B cells	Depleting circulating T and B lymphocytes	Genzyme	Multiple sclerosis	2014

(Continued)

Table 14.1 (Continued) Nanotherapeutics in clinical trials

Drug name	Nanomaterial/ nanocarrier	Therapeutic activity	Company	Disease target	Year of approval
Leukine (sargramostim)	A recombinant granulocyte macrophage colony-stimulating factor (GM-CSF)	Stimulating white blood cell production	Immunex	Leukopenia or neutropenia	1996
Livatag (Doxorubicin Transdrug)	Polymer	Doxorubicin	BioAlliance Pharma S.A.	Primary liver cancer after treatment with Nexavar (sorafenib)	2014
Lucentis (ranibizumab)	Humanized monoclonal antibody fragment	Preventing growth and maintenance of tumor blood vessels by binding to VEGF alpha	Genentech, Inc.	Wet age-related macular degeneration	2006
				Macular edema after retinal vein occlusion	2010
				Diabetic macular edema (DME)	2012
				Diabetic retinopathy	2015
Macugen (pegaptanib)	A PEGylated modified oligonucleotide that binds with high specificity and affinity to extracellular VEGF	Suppressing neovascularisation as a selective VEGF antagonist	Pfizer/Eyetech Pharmaceuticals	Neovascular (wet) age-related macular degeneration	2004

(Continued)

Table 14.1 (Continued) Nanotherapeutics in clinical trials

Drug name	Nanomaterial/ nanocarrier	Therapeutic activity	Company	Disease target	Year of approval
Marqibo	A liposomal formulation of vincristine sulfate	Vincristine irreversibly binds to and stabilizes tubulin, preventing formation of mitotic spindle and arresting cell cycle in metaphase	Talon Therapeutics	Ph–acute lymphoblastic leukemia	2012
Mircera (methoxy polyethylene glycol-epoetin beta)	An erythropoiesis-stimulating agent (ESA) conjugated to methoxy PEG	An erythropoietin receptor activator for erythroid development	Roche	Anemia associated with chronic renal failure	2007
Myalept (metreleptin)	A recombinant human leptin analog	By activating leptin receptor, energy homeostasis is maintained	Bristol-Myers Squibb	Lipodystrophy	2014
Mylotarg (gemtuzumab ozogamicin)	A humanized anti-CD33 monoclonal antibody conjugated to cytotoxic antitumor antibiotic, calicheamicin	CD33 antigen-targeted delivery of calicheamicin which after intracellular release binds to the minor groove of DNA, causing double-strand DNA breaks and resulting in inhibition of DNA synthesis	Wyeth	CD33-positive acute myeloid leukemia (AML)	2000

(Continued)

Table 14.1 (*Continued*) Nanotherapeutics in clinical trials

Drug name	Nanomaterial/ nanocarrier	Therapeutic activity	Company	Disease target	Year of approval
Myozyme (alglucosidase alfa)	A recombinant formulation of human lysosomal glycogen-specific enzyme, acid alfa-glucosidase (GAA)	Degrading lysosomal glycogen by catalyzing hydrolysis of glycosidic linkages	Genzyme	Pompe disease	2006
Naglazyme (galsulfase)	A recombinant form of human enzyme N-acetylgalactosamine 4-sulfatase	Increasing catabolism of glycosaminoglycans (GAG)	BioMarin Pharmaceuticals	Mucopoly-saccharidosis VI	2005
Natrecor (nesiritide)	A recombinant form of human B-type natriuretic peptide (hBNP)	Binding to guanylate cyclase receptor leads to increased intracellular concentrations of cGMP and smooth muscle cell relaxation	Scios	Acutely decompensated congestive heart failure (CHF)	2001
Neulasta	Recombinant methionyl human granulocyte colony-stimulating factor (G-CSF) (filgrastim)–monomethoxy PEG conjugate	Stimulating neutrophil progenitor proliferation and differentiation and selected neutrophil functions	Amgen	Infection by febrile neutropenia in patients receiving chemotherapy	2002

(*Continued*)

Table 14.1 (Continued) Nanotherapeutics in clinical trials

Drug name	Nanomaterial/ nanocarrier	Therapeutic activity	Company	Disease target	Year of approval
Neumega (oprelvekin)	A recombinant interleukin 11 (IL-11)	Upon binding to its receptor, IL-11 promotes immune responses, modulates antigen-specific antibody reactions, and prevents apoptotic cell death	Genetics Institute	Thrombocytopenia	1997
Neutroval (tbo-filgrastim)	A recombinant, nonglycosylated human G-CSF	Activating G-CSF receptors and thus controlling production, differentiation, and function of neutrophilic granulocyte progenitors	Teva Pharmaceutical	Severe chemotherapy-induced neutropenia	2012
Norditropin	A recombinant polypeptide hormone	Stimulating normal skeletal, connective tissue, muscle and organ growth	Novo Nordisk	Growth failure in children	1997
NovoLog (insulin aspart)	A rapid-acting recombinant human insulin analog	Regulation of glucose metabolism	Novo Nordisk	Diabetes mellitus	2000

(Continued)

Table 14.1 (Continued) Nanotherapeutics in clinical trials

Drug name	Nanomaterial/ nanocarrier	Therapeutic activity	Company	Disease target	Year of approval
Nplate (romiplostim)	An Fc-peptide fusion protein (peptibody)	Activating intracellular transcriptional pathways and thus causing increased platelet production	Amgen	Thrombocytopenia in patients with chronic immune (idiopathic) thrombocytopenic purpura	2008
Omontys (peginesatide)	A PEGylated peptide-based erythropoiesis stimulating agent (ESA)	Activating human erythropoietin receptor and thus, stimulating erythropoiesis in human red cell precursors	Affymax	Anemia attributed to chronic kidney disease	2012
Oncaspar	L-Asparaginase conjugated to monomethoxyPEG	Selective killing of leukemic cells owing to depletion of plasma asparagine	Enzon	Acute lymphoblastic leukemia (ALL)	1994
Opdivo (nivolumab)	A fully human monoclonal antibody	A human programmed death receptor-1 (PD-1) blocking antibody	Bristol-Myers Squibb (BMS)	Unresectable or metastatic melanoma	2014
				Metastatic squamous non–small-cell lung cancer	2015

(Continued)

Table 14.1 (Continued) Nanotherapeutics in clinical trials

Drug name	Nanomaterial/nanocarrier	Therapeutic activity	Company	Disease target	Year of approval
Orencia (abatacept)	A soluble fusion protein consisting of the extracellular domain of human cytotoxic T-lymphocyte–associated antigen 4 (CTLA-4) linked to the modified Fc portion of IgG1	A selective costimulation modulator with inhibitory activity on T lymphocytes	Bristol-Myers Squibb	Rheumatoid arthritis Juvenile idiopathic arthritis	2005 2008
Ovidrel	A recombinant chorionic gonadotropin (r-hCG)	Follicular maturation, induction of ovulation	Serono Laboratories	Infertility	2000
Pegasys	A PEGylated interferon alfa-2a	Antiviral effects	Roche	Hepatitis C	2002
Peg-Intron (peginterferon alfa-2b)	A PEGylated recombinant alpha interferon	Antiviral and immunomodulatory effects	Schering-Plough	Chronic hepatitis C	2001
Perjeta (pertuzumab)	A humanized monoclonal antibody targeting extracellular dimerization domain of HER-2	Preventing activation of HER signaling pathways	Genentech	HER2+ metastatic breast cancer	2012

(Continued)

Table 14.1 *(Continued)* Nanotherapeutics in clinical trials

Drug name	Nanomaterial/ nanocarrier	Therapeutic activity	Company	Disease target	Year of approval
Praluent (alirocumab)	A human monoclonal antibody that binds to proprotein convertase subtilisin kexin type 9 (PCSK9)	By inhibiting the binding of PCSK9 to low-density lipoprotein (LDL) receptors (LDLR), the drug increases the number of LDLRs, thus clearing LDL and lowering LDL-C levels	Sanofi-Aventis	Heterozygous familial hypercholes-terolemia or atherosclerotic cardiovascular disease	2015
Prolia (denosumab)	A fully human monoclonal antibody	Specifically binding to and inhibiting the receptor activator of NF-kappaB ligand (RANK ligand), the primary mediator of bone resorption	Amgen	Postmenopausal women with osteoporosis at high risk for fracture	2010
Pulmozyme (dornase alfa)	A recombinant human deoxyribonuclease I (rhDNase)	Selectively cleaving DNA	Genentech	Cystic fibrosis	1998
Soliris (eculizumab)	A recombinant humanized monoclonal IgG2/4delta antibody	Binding to complement protein C5	Alexion	Atypical hemolytic uremic syndrome	2011

(Continued)

Table 14.1 *(Continued)* Nanotherapeutics in clinical trials

Drug name	Nanomaterial/ nanocarrier	Therapeutic activity	Company	Disease target	Year of approval
Synagis (palivizumab)	The first monoclonal antibody approved for treating any infectious disease	Inhibiting the actions of respiratory syncytial virus (RSV) and preventing the disease	MedImmune	Pediatric patients at high risk of RSV disease	1998
Rebif (interferon beta-1a)	A recombinant human interferon-beta 1a (IFN-β)	Upon binding to its receptors, IFN-β initiates a complex cascade of intracellular events, leading to the expression of numerous interferon-induced gene products	Serono Laboratories	Multiple sclerosis	2002
Remicade (infliximab)	A chimeric monoclonal antibody targeting tumor necrosis factor alpha (TNF-α)	By inactivating TNF-α, the inflammatory process is significantly reduced	Centocor Ortho Biotech	Rheumatoid arthritis	2002
Rituxan (rituximab)	A chimeric murine/ human antibody directed against CD20	Triggering a host cytotoxic immune response against CD20-positive cells	Biogen Idec, Genentech	Non-Hodgkin's lymphoma	1997
Rixubis	A recombinant coagulation factor IX (rFIX) protein	Preventing or reducing frequency of bleeding episodes	Baxter International	Hemophilia B	2013

(Continued)

Table 14.1 (Continued) Nanotherapeutics in clinical trials

Drug name	Nanomaterial/ nanocarrier	Therapeutic activity	Company	Disease target	Year of approval
Ruconest	A recombinant C1 esterase inhibitor belonging to the class of serine-protease inhibitors or serpins	Regulating several inflammatory pathways by inhibiting certain proteases that are part of the human immune system	Pharming Group	Hereditary angioedema	2014
Simponi (golimumab)	A human monoclonal antibody that targets TNF-α	Preventing binding of TNF-α to its receptors and thereby inhibiting its biological activity	Centocor Ortho Biotech	Rheumatoid arthritis, psoriatic arthritis, and ankylosing spondylitis	2009
Soliris (eculizumab)	A recombinant humanized monoclonal IgG2/4delta antibody	Binding to complement protein C5	Alexion	Paroxysmal nocturnal hemoglobinuria	2007
Somavert (pegvisomant)	A PEGylated recombinant analogue of human growth hormone (GH)	Blocking growth hormone binding and thus interfering with intracellular growth hormone signal transduction	Pharmacia & Upjohn	Acromegaly	2003

(Continued)

Table 14.1 (Continued) Nanotherapeutics in clinical trials

Drug name	Nanomaterial/ nanocarrier	Therapeutic activity	Company	Disease target	Year of approval
Stelara (ustekinumab)	A human monoclonal antibody directed against p40 protein subunit used by both interleukin (IL)-12 and IL-23 cytokines	Interfering with normal inflammatory responses through suppression of IL-12 and IL-23 functions	Centocor Ortho Biotech	Plaque psoriasis	2009
Sylatron (peginterferon alfa-2b)	PEG–recombinant interferon alpha conjugate	Interferon alfa induces expression of proteins that mediate antiviral, antiproliferative, and immune-modulating effects	Merck	Melanoma	2011
Sylvant (siltuximab)	A chimeric monoclonal antibody	Anti-interleukin-6 (IL-6) activity	Janssen Biotech	Multicentric Castleman's disease	2014
Tretten	A recombinant coagulation factor XIII A-subunit	Active role in formation of blood clots	Novo Nordisk	Congenital factor XIII (FXIII) A-subunit deficiency	2013

(Continued)

Table 14.1 (Continued) Nanotherapeutics in clinical trials

Drug name	Nanomaterial/ nanocarrier	Therapeutic activity	Company	Disease target	Year of approval
Tysabri (natalizumab)	A humanized monoclonal antibody that binds to the cell surface receptors known as alpha-4-beta-1 (VLA-4) and alpha-4-beta-7 expressed on activated vascular endothelium	Blocking immune cell adhesion to blood vessel walls	Elan Pharmaceuticals/ Biogen Idec	Multiple sclerosis	2005
			Biogen Idec	Crohn's disease	2008
Unituxin (dinutuximab)	A disialoganglioside, GD2-binding chimeric monoclonal antibody	Antibody-dependent cell-mediated and complement-dependent cytotoxicity against GD2-expressing tumor cells	United Therapeutics	High-risk neuroblastoma	2015
Vectibix (panitumumab)	A human monoclonal antibody against transmembrane EGF receptor	Inhibition of autocrine EGF stimulation in EGF receptor-expressing tumor cells	Amgen	Colorectal cancer	2006
				Metastatic carcinoma of the colon or rectum	2004

(Continued)

Chapter fourteen: Approved and commercialized nanomedicine

Table 14.1 (Continued) Nanotherapeutics in clinical trials

Drug name	Nanomaterial/ nanocarrier	Therapeutic activity	Company	Disease target	Year of approval
Victoza (liraglutide)	A recombinant polypeptide analog of human glucagon-like peptide-1 (GLP-1)	Acting as a GLP-1 receptor agonist, the drug increases intracellular cyclic AMP (cAMP), leading to insulin release in response to elevated glucose levels	Novo Nordisk	Type 2 diabetes mellitus	2010
Visudyne	Liposomal formulation of verteporfin, a hydrophobic photosensitizer	Acting as a photosensitizer in photodynamic therapy	Novartis	Wet age-related macular degeneration (AMD)	2000
Voraxaze (glucarpidase)	A recombinant enzyme that rapidly lowers blood levels of methotrexate	Catalyzing methotrexate to its inactive metabolites, DAMPA and glutamate, which are subsequently metabolized by the liver	BTG International	Toxic plasma methotrexate concentrations in patients with impaired renal function	2012
Vpriv (velaglucerase alfa)	A human recombinant hydrolytic lysosomal glucocerebrosidase enzyme	Catalyzing hydrolysis of glucocerebroside	Shire Pharmaceuticals	Gaucher disease	2010

(Continued)

Table 14.1 (Continued) Nanotherapeutics in clinical trials

Drug name	Nanomaterial/ nanocarrier	Therapeutic activity	Company	Disease target	Year of approval
Welchol (Colesevelam)	A nonabsorbed, polymeric, lipid-lowering and glucose-lowering agent	Having a high capacity for bile acid binding. The mechanism of glycemic control is unknown	Daiichi Sankyo Co. Ltd.	Glycemic control in type 2 diabetes mellitus	2000
Wilate	A human plasma-derived von Willebrand factor (VWF)/coagulation factor VIII complex	Providing primary hemostasis by the shortening of the bleeding time	Octapharma	von Willebrand disease	2009
Xiaflex	A formulation of purified collagenase	Degrading collagen within the connective tissue	Auxilium	Peyronie's disease	2013
Xeomin (incobotulin-umtoxinA)	A botulinum toxin (BT) type A drug without complexing proteins (Cps)	Preventing nerves from releasing acetylcholine essential for the nerves to communicate with muscle cells	Merz Pharmaceutical	Cervical dystonia and blepharospasm	2010
Xgeva (denosumab)	A humanized monoclonal antibody directed against the receptor activator of nuclear factor kappa beta ligand (RANKL)	Blocking interaction of RANKL with RANK and thus causing inhibition of osteoclast activity and a decrease in bone resorption	Amgen	Giant cell tumor of bone	2013
				Skeletal-related events in patients with bone metastases	2010

(Continued)

Table 14.1 (Continued) Nanotherapeutics in clinical trials

Drug name	Nanomaterial/ nanocarrier	Therapeutic activity	Company	Disease target	Year of approval
Xiaflex	A formulation of purified collagenase	Degrading collagen within the connective tissue	Auxilium Pharmaceuticals	Dupuytren's contracture	2010
Xolair (omalizumab)	A recombinant humanized monoclonal antibody that selectively binds to human IgE	Limiting release of mediators of allergic response	Genentech	Asthma	2003
Yervoy (ipilimumab)	A human monoclonal antibody against human T-cell receptor cytotoxic T-lymphocyte–associated antigen 4 (CTLA4)	Inhibition of CTLA4-mediated downregulation of T-cell activation leads to a cytotoxic T-lymphocyte (CTL)–mediated immune response against cancer cells	Bristol-Myers Squibb	Metastatic melanoma	2011
Zaltrap (ziv-aflibercept)	A recombinant protein composed of segments of the extracellular domains of human VEGFR1 and VEGFR2 fused to Fc of human IgG1	Disrupting binding of VEGFs to their cell receptors, thereby inhibiting tumor angiogenesis	Sanofi-aventis	Metastatic colorectal cancer	2012

(Continued)

Table 14.1 (*Continued*) Nanotherapeutics in clinical trials

Drug name	Nanomaterial/ nanocarrier	Therapeutic activity	Company	Disease target	Year of approval
Zemaira (alpha1-proteinase inhibitor)	A highly purified human alpha1-proteinase inhibitor (A1-PI), also known as alpha1-antitrypsin	Inhibiting neutrophil elastase (NE) in lower respiratory tract and thus preventing destruction of pulmonary tissue	Aventis Behring	Alpha1-proteinase inhibitor deficiency	2003
Zevalin (ibritumomab tiuxetan)	A radioimmuno-therapeutic agent consisting of a murine monoclonal anti-CD20 antibody linked by the chelator tiuxetan to the radioisotope yttrium-90 (Y 90)	Specific delivery of beta radiation to CD20-expressing tumor cells	Biogen Idec	Non-Hodgkin's lymphoma	2002

Chapter fourteen: Approved and commercialized nanomedicine

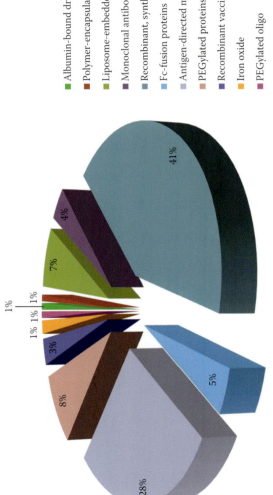

Figure 14.1 FDA-approved nanotherapeutics.

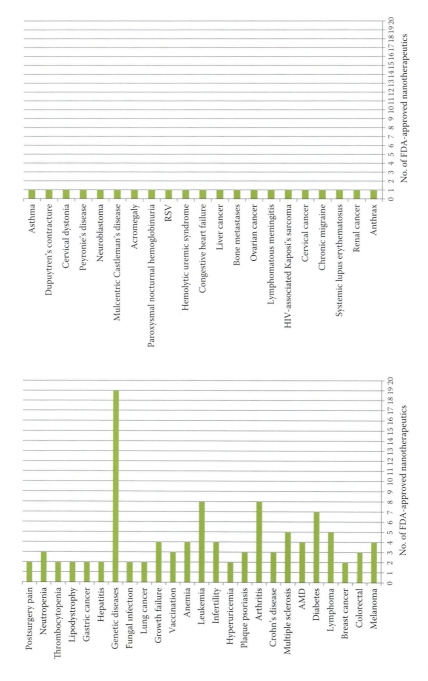

Figure 14.2 Disease targets for the FDA-approved nanotherapeutics.

chapter fifteen

Current safety issues
Biodegradability, reactivity, and clearance

The enormous surface area of the nanoparticles accompanied by high Gibbs free energy could lead to particle agglomeration or Ostwald ripening in order to minimize the energy either during formulation, thus generating larger particles, or after systemic administration, potentially causing vascular or lymphatic blockage. Coating of the particle surface with a hydrophilic polymer, such as polyethylene glycol (PEG), or protein, such as albumin, or with surfactants is highly advantageous in preventing the aggregation by elevating the activation energy of agglomeration and, thus, by adding electrostatic or steric barriers between the particles. In addition, nanoparticles can potentially interact with phospholipid bilayers, important membrane proteins, cytoskeletal networks, and even chromosomal DNA, causing various adverse effects. In the case of non-biodegradable or slowly biodegradable nanoparticles, their accumulation and persistent retention within cells or tissue may lead to a chronic inflammatory response.

15.1 Nanoparticle interaction with blood cells

Vascular-targeted nanocarriers demonstrated several adverse physiological effects including hemolysis, platelet activation and aggregation, leukocyte activation and adhesion, and complement activation (Huang et al. 2010).

15.1.1 Hemolysis of red blood cells

By interacting with red blood cells (RBCs), nanoparticles can induce hemolysis by disrupting the cell membrane, causing leakage of hemoglobin with an increased risk of anemia. Moreover, the debris released from damaged RBCs can initiate a phagocytic cascade, often leading to the rapid clearance of nanoparticles. Nanoparticles with net positive or negative surface charge usually induce hemolysis. Surface-active surfactants (e.g., polyoxyl-20-ether) used for nanoparticle stabilization can also cause hemolysis. Hemolytic activities can be reduced by grafting PEG chains onto the nanoparticle surface. Besides, smaller nanoparticles were shown to have more hemolytic potential than the larger ones.

15.1.2 Platelet aggregation and activation

Nanoparticles can influence aggregation and activation of platelets, potentially leading to uncontrollable blood coagulation with the ultimate result of stroke or heart attack in some cases. Gold nanoparticles and carbon nanotubes were shown to induce platelet activation. The surface charge and size of nanoparticles can facilitate aggregation of the platelets and consequentially their activation. For instance, similar levels of platelet activation were noticed by cationic and anionic polystyrene nanoparticles, whereas 20-nm carboxylated polystyrene nanoparticles were shown to trigger more platelet activation than 200-nm nanoparticles of similar surface characteristics (Mayer et al. 2009).

15.1.3 Macrophage uptake and immune responses

When nanoparticles, such as those formulated or surface coated with foreign proteins, and viral vectors enter the human body, they are ultimately taken up by macrophages, leading to humoral and cell-mediated immunity. Repeated administration of the same nanoparticles triggers a robust immune response with dangerous side effects. Depending on the size, shape, and surface characteristics of nanoparticles, complement pattern recognition molecules such as C1q, C-reactive protein, ficolins, mannose binding lectin, and properdin can interact with the nanoparticles, triggering complement activation via a classical, lectin, or alternative pathway and resulting in rapid recognition of the nanoparticle surface by phagocytic cells with subsequent clearance. Complement activation may also induce inflammatory responses through liberation of anaphylatoxins (e.g., C3a and C5a) and chemoattractant species (e.g., C5a) (Moghimi 2014).

15.2 Deformation of cellular membrane

15.2.1 Disturbance of phospholipid bilayer

The surface charge of nanoparticles can locally contribute to either solid–liquid or liquid–solid transition in the cell membrane. Negatively charged nanoparticles bound to a fluid area of the membrane were shown to induce gelation, whereas positively charged nanoparticles turned a gelled area into a fluid state for easier penetration. Furthermore, cationic polymeric nanoparticles were reported to induce "holes" in the region of the cell membrane, which possess reduced lipid or protein levels, leading to

Chapter fifteen: Current safety issues 289

the leakage of cytosolic enzymes. This rationalizes why cationic particles are more toxic than net neutral or anionic counterparts of the same size (Wu et al. 2013).

15.2.2   Interactions with membrane proteins and blocking of ion channels

Nanoparticles can potentially interact with cell membrane–associated proteins, such as receptors, cell–cell adhesion molecules, or transporters, interrupting the vital cellular functions. If the diameter of nanoparticles is similar to the size of a channel, they can even block the channel. For example, single-walled carbon nanotubes with the diameter distributions peaked at ~0.9 and 1.3 nm blocked K+ channel subunits depending on the shape and dimensions of the nanoparticles (Park et al. 2003). In addition, silver nanoparticles were demonstrated to induce conformational changes of the ion channel, thus altering the probability of channel opening and leading to nanotoxicity-associated neuronal dysfunction (Liu et al. 2009). Semiconductor nanomaterials, on the other hand, caused impairment of the ion channel structure and function indirectly via oxidative stress damage. Thus, the oxidative stress induced by CdSe quantum dots (Qds) was reported to activate N-type calcium channels, leading to the influx of extracellular calcium and concurrently a rapid increase of intracellular calcium concentration and thus triggering the pathway of apoptosis (Tang et al. 2008).

15.3   Lysosomal rupture and release of contents

Nanoparticles capable of lysosomal rupture after cellular uptake will result in the release of lysosomal contents into cytosol, particularly proteolytic enzymes that could cleave the vital cellular proteins. According to the proton sponge hypothesis, nanoparticles carrying polyamine groups (secondary or tertiary amines) with high proton binding affinity could lead to buffering and subsequent chloride influx, resulting in osmotic swelling and lysosomal rupture.

15.4   Disruption of cytoskeleton

After internalization, nanomaterials that escape endosomes can interact with the cytoskeleton, an interconnected network of filamentous proteins, such as microtubules, actin filaments and intermediate filaments, and regulatory proteins, thereby interrupting its functions in providing the mechanical stability and integrity of biological cells and transporting intracellular cargo. Depending on the shape, nanoparticles could lead to different effects on the cytoskeleton. For example, although mesoporous silica nanoparticles (100 nm) did not disturb the organization of actin filaments in the cytoskeleton, mesoporous silica nanorods with large aspect ratio of length to width (4:1) were found to disorganize the actin filaments with poorly formed filament bundles (Huang et al. 2010). In addition, gold nanoparticles, metal-oxide nanomaterials, such as ZnO or Fe_2O_3, and Qds were reported to cause alteration in the cellular cytoskeletal network.

15.5 Damage to nuclear DNA and proteins

Tiny nanoparticles, such as gold nanomaterials of ~1.4 nm, have the potential ability to penetrate the nuclear compartment and damage the chromosomal DNA through electrostatic binding. Other than directly binding to DNA, nanomaterials, such as those of ZnO and TiO_2, may indirectly cause DNA damage by promoting oxidative stress and inflammatory responses (Meng et al. 2009). DNA damage leads to the activation of p53, which, in turn, either activates DNA repair mechanism or induces transcriptional activation of pro-apoptotic factors, leading to apoptosis in case of irreparable DNA damage. The generation of reactive oxygen species (ROS) induced by ZnO was found to be closely associated with the oxidation and breakage of DNA inside the nucleus (Heng et al. 2010). TiO_2 nanoparticles were shown to cause DNA damage via the generation of ROS and activation of cell inflammatory response (Trouiller et al. 2009). Qds, on the other hand, after entering the nucleus through the nuclear pore, induce nuclear protein aggregation, inhibition of gene transcription, and, finally, cell proliferation (Nabiev et al. 2007).

Considering the potential toxic effects of nanoparticles as outlined above, it is necessary to study the mechanisms by which the body clears the nanomaterials. The clearance rate of the nanoparticles is an important parameter to evaluate their toxic effects. Moreover, the organ distribution of the nanoparticles should be carried out in preclinical and clinical trials to predict their potential hazards to the brain, liver, kidney, spleen, heart, lungs, and so on. Since the toxicity of many nanoscale materials will not be fully apparent until their exposure is felt by a diverse population, a long-term postmarket tracking or a surveillance system should be adopted.

References

Adibkia K, Omidi Y, Siahi M, Javadzadeh A, Barzegar-Jalali M, Barar J, Maleki N, Mohammadi G, Nokhodchi A. Inhibition of endotoxin-induced uveitis by methylprednisolone acetate nanosuspension in rabbits. *J Ocul Pharmacol Therap*. 2007;23:421–32.

Aditya NP, Patankar S, Madhusudhan B, Murthy RS, Souto EB. Arthemeter-loaded lipid nanoparticles produced by modified thin-film hydration: Pharmacokinetics, toxicological and in vivo anti-malarial activity. *Eur J Pharm Sci*. 2010;40:448–55.

Aditya NP, Vathsala PG, Vieira V, Murthy RS, Souto EB. Advances in nanomedicines for malaria treatment. *Adv Colloid Interface Sci*. 2013 Dec;201–202:1–17.

Aggarwal D, Garg A, Kaur IP. Development of a topical niosomal preparation of acetazolamide: Preparation and evaluation. *J Pharm Pharmacol*. 2004 Dec;56(12):1509–17.

Aggarwal P, Hall JB, McLeland CB, Dobrovolskaia MA, McNeil SE. Nanoparticle interaction with plasma proteins as it relates to particle biodistribution, biocompatibility and therapeutic efficacy. *Adv Drug Deliv Rev*. 2009;61:428–37.

Ahmad A, Othman I, Zaini A, Chowdhury EH. Oral nano-insulin therapy: Current progress on nanoparticle-based devices for intestinal epithelium-targeted insulin delivery. *J Nanomed Nanotechnol*. 2012;S4(007):1–11.

Ahmad A, Othman I, Zain AZM, Chowdhury EH. Recent advances in insulin therapy for diabetes. *Int J Diabetes Clin Res*. 2014;1(1):1–13.

Ahmed EM. Hydrogel: Preparation, characterization, and applications: A review. *J Adv Res*. 2015 Mar;6(2):105–121.

Akbarzadeh A, Rezaei-Sadabady R, Davaran S, Joo SW, Zarghami N, Hanifehpour Y, Samiei M, Kouhi M, Nejati-Koshki K. Liposome: Classification, preparation, and applications. *Nanoscale Res Lett*. 2013 Feb 22;8(1):102.

Akhtar F, Rizvi MM, Kar SK. Oral delivery of curcumin bound to chitosan nanoparticles cured *Plasmodium yoelii* infected mice. *Biotechnol Adv*. 2012;30:310–20.

Alderson RF, Toki BE, Roberge M, Geng W, Basler J, Chin R, Liu A et al. Characterization of a CC49-based single-chain fragment-beta-lactamase fusion protein for antibody-directed enzyme prodrug therapy (ADEPT). *Bioconjug Chem*. 2006 Mar–Apr;17(2):410–8.

Allémann E, Rousseau J, Brasseur N, Kudrevich SV, Lewis K, van Lier JE. Photodynamic therapy of tumours with hexadecafluoro zinc phthalocyanine formulated in PEG-coated poly(lactic acid) nanoparticles. *Int J Cancer*. 1996 Jun 11;66(6):821–4.

Ambruosi A, Gelperina S, Khalansky A, Tanski S, Theisen A, Kreuter J. Antitumor effect of doxorubicin loaded in poly(butyl cyanoacrylate) nanoparticles in rat glioma model: Influence of formulation parameters. *J Microencapsulation*. 2006;23:582–92.

Angst MS, Drover DR. Pharmacology of drugs formulated with DepoFoam: A sustained release drug delivery system for parenteral administration using multivesicular liposome technology. *Clin Pharmacokinet*. 2006;45(12):1153–76.

Aouadi M, Tesz GJ, Nicoloro SM, Wang M, Chouinard M, Soto E, Ostroff GR, Czech MP. Orally delivered siRNA targeting macrophage Map4k4 suppresses systemic inflammation. *Nature*. 2009 Apr 30;458(7242):1180–4.

Apte A, Koren E, Koshkaryev A, Torchilin VP. Doxorubicin in TAT peptide-modified multifunctional immunoliposomes demonstrates increased activity against both drug-sensitive and drug-resistant ovarian cancer models. *Cancer Biol Ther*. 2014 Jan;15(1):69–80.

Atkinson TJ, Fudin J, Jahn HL, Kubotera N, Rennick AL, Rhorer M. What's new in NSAID pharmacotherapy: Oral agents to injectables. *Pain Med*. 2013 Dec;14 Suppl 1:S11–7.

Bachmann MF, Jennings GT. Vaccine delivery: A matter of size, geometry, kinetics and molecular patterns. *Nat Rev Immunol*. 2010 Nov;10(11):787–96.

Badawi AA, El-Laithy HM, El Qidra RK, El Mofty H, El dally M. Chitosan based nanocarriers for indomethacin ocular delivery. *Arch Pharm Res*. 2008; 31:1040–9.

Bakhtiar A, Sayyad M, Rosli R, Maruyama A, Chowdhury EH. Intracellular delivery of potential therapeutic genes: Prospects in cancer gene therapy. *Curr Gene Ther*. 2014;14(4):247–57.

Balazs DA, Godbey W. Liposomes for use in gene delivery. *J Drug Deliv*. 2011; 2011:326497.

Balmayor ER, Azevedo HS, Reis RL. Controlled delivery systems: From pharmaceuticals to cells and genes. *Pharm Res*. 2011 Jun;28(6):1241–58.

Banai S, Chorny M, Gertz SD, Fishbein I, Gao J, Perez L, Lazarovichi G, Gazit A, Levitzki A, Golomb G. Locally delivered nanoencapsulated tyrphostin (AGL-2043) reduces neointima formation in balloon-injured rat carotid and stented porcine coronary arteries. *Biomaterials*. 2005 Feb;26(4):451–61.

Barhate G, Gautam M, Gairola S, Jadhav S, Pokharkar V. Enhanced mucosal immune responses against tetanus toxoid using novel delivery system comprised of chitosan-functionalized gold nanoparticles and botanical adjuvant: Characterization, immunogenicity, and stability assessment. *J Pharm Sci*. 2014 Nov;103(11):3448–56.

Barnes T, Moots R. Targeting nanomedicines in the treatment of rheumatoid arthritis: Focus on certolizumab pegol. *Int J Nanomed*. 2007;2(1):3–7.

Bauman JA, Li SD, Yang A, Huang L, Kole R. Anti-tumor activity of spliceswitching oligonucleotides. *Nucleic Acids Res*. 2010 Dec;38(22):8348–56.

Behera AK, Kumar M, Lockey RF, Mohapatra SS. Adenovirus-mediated interferon gamma gene therapy for allergic asthma: Involvement of interleukin 12 and STAT4 signaling. *Hum Gene Ther*. 2002 Sep 20;13(14):1697–709.

Benedetti MS, Whomsley R, Poggesi I, Cawello W, Mathy FX, Delporte ML, Papeleu P, Watelet JE. Drug metabolism and pharmacokinetics. *Drug Metab Rev*. 2009;41(3) 344–90.

Bhavsar MD, Amiji MM. Oral IL-10 gene delivery in a microsphere-based formulation for local transfection and therapeutic efficacy in inflammatory bowel disease. *Gene Ther*. 2008 Sep;15(17):1200–9.

References

Bi R, Shao W, Wang Q, Zhang N. Solid lipid nanoparticles as insulin inhalation carriers for enhanced pulmonary delivery. *J Biomed Nanotechnol*. 2009 Feb;5(1):84–92.

Bissett SL, Mattiuzzo G, Draper E, Godi A, Wilkinson DE, Minor P, Page M, Beddows S. Pre-clinical immunogenicity of human papillomavirus alpha-7 and alpha-9 major capsid proteins. *Vaccine*. 2014 Nov 12;32(48):6548–55.

Black M, Trent A, Tirrell M, Olive C. Advances in the design and delivery of peptide subunit vaccines with a focus on toll-like receptor agonists. *Expert Rev Vaccines*. 2010 Feb;9(2):157–73.

Blanco E, Ferrari M. Emerging nanotherapeutic strategies in breast cancer. *Breast*. 2014 Feb;23(1):10–8.

Bochot A, Fattal E, Boutet V, Deverre JR, Jeanny JC, Chacun H, Couvreur P. Intravitreal delivery of oligonucleotides by sterically stabilized liposomes. *Invest Ophthalmol Vis Sci*. 2002 Jan;43(1):253–9.

Bohrer MP, Baylis C, Humes HD, Glassock RJ, Robertson CR, Brenner BM. Permselectivity of the glomerular capillary wall. Facilitated filtration of circulating polycations. *J Clin Invest*. 1978;61:72–8.

Boudes PF. Gene therapy as a new treatment option for inherited monogenic diseases. *Eur J Intern Med*. 2014 Jan;25(1):31–6.

Bourges JL, Gautier SE, Delie F, Bejjani RA, Jeanny JC, Gurny R, BenEzra D, Behar-Cohen F. Ocular drug delivery targeting the retina and retinal pigment epithelium using polylactide nanoparticles. *Invest Ophthalmol Vis Sci*. 2003;44:3562–9.

Byrappa K, Ohara S, Adschiri T. Nanoparticles synthesis using supercritical fluid technology—Towards biomedical applications. *Adv Drug Deliv Rev*. 2008; 60:299–327.

Cabral H, Murakami M, Hojo H, Terada Y, Kano MR, Chung UI, Nishiyama N, Kataoka K. Targeted therapy of spontaneous murine pancreatic tumors by polymeric micelles prolongs survival and prevents peritoneal metastasis. *Proc Natl Acad Sci USA*. 2013 Jul 9;110(28):11397–402.

Canton I, Battaglia G. Endocytosis at the nanoscale. *Chem Soc Rev*. 2012 Apr 7;41(7):2718–39.

Caporale N, Kolstad KD, Lee T, Tochitsky I, Dalkara D, Trauner D, Kramer R, Dan Y, Isacoff EY, Flannery JG. LiGluR restores visual responses in rodent models of inherited blindness. *Mol Ther*. 2011 Jul;19(7):1212–9.

Carlmark A, Hawker C, Hult A, Malkoch M. New methodologies in the construction of dendritic materials. *Chem Soc Rev*. 2009 Feb;38(2):352–62.

Carmona S, Jorgensen MR, Kolli S, Crowther C, Salazar FH, Marion PL, Fujino M et al. Controlling HBV replication in vivo by intravenous administration of triggered PEGylated siRNA-nanoparticles. *Mol Pharm*. 2009;6:706–17.

Cavalli R, Gasco M, Chetoni P, Burgalassi S, Saettone M. Solid lipid nanoparticles (SLN) as ocular delivery system for tobramycin. *Int J Pharm*. 2002; 238:241–5.

Chacko A-M, Hood ED, Zern BJ, Muzykantov VR. Targeted Nanocarriers for Imaging and Therapy of Vascular Inflammation. *Curr Opin Colloid Interface Sci*. 2011;16:215–27.

Chandolu V, Dass CR. Treatment of lung cancer using nanoparticle drug delivery systems. *Curr Drug Discov Technol*. 2013 Jun;10(2):170–6.

Chen J, Shao R, Zhang XD, Chen C. Applications of nanotechnology for melanoma treatment, diagnosis, and theranostics. *Int J Nanomed*. 2013;8:2677–88.

Chen SC, Wu YC, Mi FL, Lin YH, Yu LC, Sung HW. A novel pH-sensitive hydrogel composed of *N,O*-carboxymethyl chitosan and alginate cross-linked by genipin for protein drug delivery. *J Control Release*. 2004 Apr 28;96(2):285–300.

Chen Y, Sen J, Bathula SR, Yang Q, Fittipaldi R, Huang L. Novel cationic lipid that delivers siRNA and enhances therapeutic effect in lung cancer cells. *Mol Pharm*. 2009 May-Jun;6(3):696–705.

Cheng FY, Su CH, Wu PC, Yeh CS. Multifunctional polymeric nanoparticles for combined chemotherapeutic and near-infrared photothermal cancer therapy in vitro and in vivo. *Chem Commun (Camb.)*. 2010;46:3167–9.

Chereddy KK, Coco R, Memvanga PB, Ucakar B, des Rieux A, Vandermeulen G, Préat V. Combined effect of PLGA and curcumin on wound healing activity. *J Control Release*. 2013 Oct 28;171(2):208–15.

Chimanuka B, Gabriëls M, Detaevernier MR, Plaizier-Vercammen JA. Preparation of beta-artemether liposomes, their HPLC-UV evaluation and relevance for clearing recrudescent parasitaemia in *Plasmodium chabaudi* malaria-infected mice. *J Pharm Biomed Anal*. 2002 Apr 1;28(1):13–22.

Chowdhury EH, Akaike T. A bio-recognition device developed onto nano-crystals of carbonate apatite for cell-targeted gene delivery. *Biotechnol Bioeng*. 2005 May 20;90(4):414–21.

Chowdhury EH, Kunou M, Nagaoka M, Kundu AK, Hoshiba T, Akaike T. High-efficiency gene delivery for expression in mammalian cells by nanoprecipitates of Ca-Mg phosphate. *Gene*. 2004 Oct 27;341:77–82.

Chowdhury EH, Maruyama A, Kano A, Nagaoka M, Kotaka M, Hirose S, Kunou M, Akaike T. pH-sensing nano-crystals of carbonate apatite: Effects on intracellular delivery and release of DNA for efficient expression into mammalian cells. *Gene*. 2006 Jul 5;376(1):87–94.

Chowdhury EH. Nuclear targeting of viral and non-viral DNA. *Expert Opin Drug Deliv*. 2009 Jul;6(7):697–703.

Chowdhury EH. pH-sensitive nano-crystals of carbonate apatite for smart and cell-specific transgene delivery. *Expert Opin Drug Deliv*. 2007 May;4(3):193–6.

Chowdhury EH. Strategies for tumor-directed delivery of siRNA. *Expert Opin Drug Deliv*. 2011 Mar;8(3):389–401.

Christie JG, Kompella UB. Ophthalmic light sensitive nanocarrier systems. *Drug Discov Today*. 2008 Feb;13(3–4):124–34.

Chuang EY, Lin KJ, Su FY, Chen HL, Maiti B, Ho YC, Yen TC, Panda N, Sung HW. Calcium depletion-mediated protease inhibition and apical-junctional-complex disassembly via an EGTA-conjugated carrier for oral insulin delivery. *J Control Release*. 2013 Aug 10;169(3):296–305.

Cohen K, Emmanuel R, Kisin-Finfer E, Shabat D, Peer D. Modulation of drug resistance in ovarian adenocarcinoma using chemotherapy entrapped in hyaluronan-grafted nanoparticle clusters. *ACS Nano*. 2014 Mar 25;8(3): 2183–95.

Cortivo R, Vindigni V, Iacobellis L, Abatangelo G, Pinton P, Zavan B. Nanoscale particle therapies for wounds and ulcers. *Nanomedicine (Lond)*. 2010 Jun;5(4):641–56.

Corvo ML, Marinho HS, Marcelino P, Lopes RM, Vale CA, Marques CR, Martins LC, Laverman P, Storm G, Martins MB. Superoxide dismutase enzymosomes: Carrier capacity optimization, in vivo behaviour and therapeutic activity. *Pharm Res*. 2015 Jan;32(1):91–102.

References 295

Cui F, Shi K, Zhang L, Tao A, Kawashima Y. Biodegradable nanoparticles loaded with insulin-phospholipid complex for oral delivery: Preparation, in vitro characterization and in vivo evaluation. *J Control Release*. 2006 Aug 28;114(2): 242–50.

Cui FD, Tao AJ, Cun DM, Zhang LQ, Shi K. Preparation of insulin loaded PLGA-Hp55 nanoparticles for oral delivery. *J Pharm Sci*. 2007 Feb;96(2):421–7.

Dalkara D, Kolstad KD, Guerin KI, Hoffmann NV, Visel M, Klimczak RR, Schaffer DV, Flannery JG. AAV mediated GDNF secretion from retinal glia slows down retinal degeneration in a rat model of retinitis pigmentosa. *Mol Ther*. 2011 Sep;19(9):1602–8.

Danhier F, Ansorena E, Silva JM, Coco R, Le Breton A, Préat V. PLGA-based nanoparticles: An overview of biomedical applications. *J Control Release*. 2012 Jul 20;161(2):505–22. doi: 10.1016/j.jconrel.2012.01.043.

Danhier F, Lecouturier N, Vroman B, Jérôme C, Marchand-Brynaert J, Feron O, Préat V. Paclitaxel-loaded PEGylated PLGA-based nanoparticles: In vitro and in vivo evaluation. *J Control Release*. 2009;133(1):11–7.

Del Amo EM, Urtti A. Current and future ophthalmic drug delivery systems. A shift to the posterior segment. *Drug Discov Today*. 2008 Feb;13(3–4):135–43.

Deleavey GF, Damha MJ. Designing chemically modified oligonucleotides for targeted gene silencing. *Chem Biol*. 2012 Aug 24;19(8):937–54.

D'Emanuele A, Attwood D. Dendrimer-drug interactions. *Adv Drug Deliv Rev*. 2005 Dec 14;57(15):2147–62.

Deng WG, Kawashima H, Wu G, Jayachandran G, Xu K, Minna JD, Roth JA, Ji L. Synergistic tumor suppression by coexpression of FUS1 and p53 is associated with down-regulation of murine double minute-2 and activation of the apoptotic protease-activating factor 1-dependent apoptotic pathway in human non-small cell lung cancer cells. *Cancer Res*. 2007 Jan 15;67(2):709–17.

Deng Y, Wang CC, Choy KW, Du Q, Chen J, Wang Q, Li L, Chung TK, Tang T. Therapeutic potentials of gene silencing by RNA interference: Principles, challenges, and new strategies. *Gene*. 2014 Apr 1;538(2):217–27.

Desai NP, Trieu V, Hwang LY, Wu R, Soon-Shiong P, Gradishar WJ. Improved effectiveness of nanoparticle albumin-bound (nab) paclitaxel versus polysorbate-based docetaxel in multiple xenografts as a function of HER2 and SPARC status. *Anticancer Drugs*. 2008;19:899–909.

Desai PR, Marepally S, Patel AR, Voshavar C, Chaudhuri A, Singh M. Topical delivery of anti-TNFα siRNA and caps aicin via novel lipid–polymer hybrid nanoparticles efficiently inhibits skin inflammation in vivo. *J Control Release*. 2013 Aug 28;170(1):51–63.

Desale SS, Cohen SM, Zhao Y, Kabanov AV, Bronich TK. Biodegradable hybrid polymer micelles for combination drug therapy in ovarian cancer. *J Control Release*. 2013 Nov 10;171(3):339–48.

D'Hallewin MA, Kochetkov D, Viry-Babel Y, Leroux A, Werkmeister E, Dumas D, Gräfe S, Zorin V, Guillemin F, Bezdetnaya L. Photodynamic therapy with intratumoral administration of lipid-based mTHPC in a model of breast cancer recurrence. *Lasers Surg Med*. 2008 Oct;40(8):543–9.

Diao L, Meibohm B. Pharmacokinetics and pharmacokinetic-pharmacodynamic correlations of therapeutic peptides. *Clin Pharmacokinet*. 2013 Oct;52(10):855–68.

Drummond DC, Noble CO, Guo Z, Hayes ME, Connolly-Ingram C, Gabriel BS, Hann B et al. Development of a highly stable and targetable nanoliposomal formulation of topotecan. *J Control Release*. 2010 Jan 4;141(1):13–21.

Du S-L, Pan H, Lu W-Y, Wang J, Wu J, Wang J-Y. Cyclic Arg-Gly-Asp peptide labeled liposomes for targeting drug therapy of hepatic fibrosis in rats. *J Pharmacol Exp Ther*. 2007;322:560–8.

Duan X, Li Y. Physicochemical characteristics of nanoparticles affect circulation, biodistribution, cellular internalization, and trafficking. *Small*. 2013 May 27; 9(9–10):1521–32.

Fadeel B, Garcia-Bennett AE. Better safe than sorry: Understanding the toxicological properties of inorganic nanoparticles manufactured for biomedical applications. *Adv Drug Deliv Rev*. 2010 Mar 8;62(3):362–74.

Faramarzi MA, Sadighi A. Insights into biogenic and chemical production of inorganic nanomaterials and nanostructures. *Adv Colloid Interface Sci*. 2013 Mar;189–190:1–20.

Farokhzad OC, Cheng J, Teply BA, Sherifi I, Jon S, Kantoff PW, Richie JP, Langer R. Targeted nanoparticle–aptamer bioconjugates for cancer chemotherapy *in vivo*. *Proc Natl Acad Sci USA*. 2006;103:6315–20.

Felnerova D, Viret JF, Glück R, Moser C. Liposomes and virosomes as delivery systems for antigens, nucleic acids and drugs. *Curr Opin Biotechnol*. 2004 Dec;15(6):518–29.

Feng Q, Yu MZ, Wang JC, Hou WJ, Gao LY, Ma XF, Pei XW et al. Synergistic inhibition of breast cancer by co-delivery of VEGF siRNA and paclitaxel via vapreotide-modified core-shell nanoparticles. *Biomaterials*. 2014 Jun;35(18): 5028–38.

Feng Z, Zhao G, Yu L, Gough D, Howell SB. Preclinical efficacy studies of a novel nanoparticle-based formulation of paclitaxel that out-performs Abraxane. *Cancer Chemother Pharmacol*. 2010 Apr;65(5):923–30.

Fleige E, Quadir MA, Haag R. Stimuli-responsive polymeric nanocarriers for the controlled transport of active compounds: Concepts and applications. *Adv Drug Deliv Rev*. 2012 Jun 15;64(9):866–84.

Frandsen JL, Ghandehari H. Recombinant protein-based polymers for advanced drug delivery. *Chem Soc Rev*. 2012 Apr 7;41(7):2696–706.

Fu Q, Sun J, Zhang W, Sui X, Yan Z, He Z. Nanoparticle albumin-bound (NAB) technology is a promising method for anti-cancer drug delivery. *Recent Pat Anticancer Drug Discov*. 2009;4:262–72.

Fukasawa M, Shimizu Y, Shikata K, Nakata M, Sakakibara R, Yamamoto N, Hatanaka M, Mizuochi T. Liposome oligomannose-coated with neoglycolipid, a new candidate for a safe adjuvant for induction of CD8+ cytotoxic T lymphocytes. *FEBS Lett*. 1998 Dec 28;441(3):353–6.

Gao H, Qian J, Yang Z, Pang Z, Xi Z, Cao S, Wang Y et al. Whole-cell SELEX aptamer-functionalised poly(ethyleneglycol)-poly(ε-caprolactone) nanoparticles for enhanced targeted glioblastoma therapy. *Biomaterials*. 2012 Sep; 33(26):6264–72.

Gao ZG, Tian L, Hu J, Park IS, Bae YH. Prevention of metastasis in a 4T1 murine breast cancer model by doxorubicin carried by folate conjugated pH sensitive polymeric micelles. *J Control Release*. 2011 May 30;152(1):84–9.

Garg A, Tisdale AW, Haidari E, Kokkoli E. Targeting colon cancer cells using PEGylated liposomes modified with a fibronectin-mimetic peptide. *Int J Pharm*. 2009;366:201–10.

References
297

Gentile F, Ferrari M, Decuzzi P. The transport of nanoparticles in blood vessels: The effect of vessel permeability and blood rheology. *Ann Biomed Eng*. 2008;36:254–61.

Ghosh A, Banerjee T, Bhandary S, Surolia A. Formulation of nanotized curcumin and demonstration of its antimalarial efficacy. *Int J Nanomed*. 2014;9:5373–87.

Gianella A, Jarzyna PA, Mani V, Ramachandran S, Calcagno C, Tang J, Kann B et al. Multifunctional nanoemulsion platform for imaging guided therapy evaluated in experimental cancer. *ACS Nano*. 2011;5:4422–33.

Glade-Bender J, Kandel JJ, Yamashiro DJ. VEGF blocking therapy in the treatment of cancer. *Expert Opin Biol Ther*. 2003;3:263–76.

Grange C, Geninatti-Crich S, Esposito G, Alberti D, Tei L, Bussolati B, Aime S, Camussi G. Combined delivery and magnetic resonance imaging of neural cell adhesion molecule-targeted doxorubicin-containing liposomes in experimentally induced Kaposi's sarcoma. *Cancer Res*. 2010;70:2180–90.

Gulyaev AE, Gelperina SE, Skidan IN, Antropov AS, Kivman GY, Kreuter J. Significant transport of doxorubicin into the brain with polysorbate 80-coated nanoparticles. *Pharm Res*. 1999 Oct;16(10):1564–9.

Günzburg WH, Salmons B. Virus vector design in gene therapy. *Mol Med Today*. 1995 Dec;1(9):410–7.

Guo S, Huang Y, Jiang Q, Sun Y, Deng L, Liang Z, Du Q et al. Enhanced gene delivery and siRNA silencing by gold nanoparticles coated with charge-reversal polyelectrolyte. *ACS Nano*. 2010 Sep 28;4(9):5505–11.

Gupta A, Avci P, Sadasivam M, Chandran R, Parizotto N, Vecchio D, de Melo WC, Dai T, Chiang LY, Hamblin MR. Shining light on nanotechnology to help repair and regeneration. *Biotechnol Adv*. 2013 Sep–Oct;31(5):607–31.

Gupta M, Agrawal U, Vyas SP. Nanocarrier-based topical drug delivery for the treatment of skin diseases. *Expert Opin Drug Deliv*. 2012 Jul;9(7):783–804.

Gupta M, Tiwari S, Vyas SP. Influence of various lipid core on characteristics of SLNs designed for topical delivery of fluconazole against cutaneous candidiasis. *Pharm Dev Technol*. 2013 May–Jun;18(3):550–9.

Gupta M, Vyas SP. Development, characterization and in vivo assessment of effective lipidic nanoparticles for dermal delivery of fluconazole against cutaneous candidiasis. *Chem Phys Lipids*. 2012 May;165(4):454–61.

Gupta N, Ibrahim HM, Ahsan F. Peptide-micelle hybrids containing fasudil for targeted delivery to the pulmonary arteries and arterioles to treat pulmonary arterial hypertension. *J Pharm Sci*. 2014 Nov;103(11):3743–53.

Gupta V, Gupta N, Shaik IH, Mehvar R, McMurtry IF, Oka M, Nozik-Grayck E, Komatsu M, Ahsan F. Liposomal fasudil, a rho-kinase inhibitor, for prolonged pulmonary preferential vasodilation in pulmonary arterial hypertension. *J Control Release*. 2013 Apr 28;167(2):189–99.

Hadinoto K, Sundaresan A, Cheow WS. Lipid–polymer hybrid nanoparticles as a new generation therapeutic delivery platform: A review. *Eur J Pharm Biopharm*. 2013 Nov;85(3 Pt A):427–43.

Han K, Miah MA, Shanmugam S, Yong CS, Choi HG, Kim JA, Yoo BK. Mixed micellar nanoparticle of amphotericin B and poly styrene-block-poly ethylene oxide reduces nephrotoxicity but retains antifungal activity. *Arch Pharm Res*. 2007 Oct;30(10):1344–9.

Hatakeyama H, Ito E, Akita H, Oishi M, Nagasaki Y, Futaki S, Harashima H. A pH-sensitive fusogenic peptide facilitates endosomal escape and greatly enhances the gene silencing of siRNA-containing nanoparticles in vitro and in vivo. *J Control Release*. 2009;139:127–32.

Haughney SL, Petersen LK, Schoofs AD, Ramer-Tait AE, King JD, Briles DE, Wannemuehler MJ, Narasimhan B. Retention of structure, antigenicity, and biological function of pneumococcal surface protein A (PspA) released from polyanhydride nanoparticles. *Acta Biomater*. 2013 Sep;9(9):8262–71.

Heng BC, Zhao XX, Xiong SJ, Ng KW, Boey FYC, Loo JSC. Toxicity of zinc oxide (ZnO) nanoparticles on human bronchial epithelial cells (BEAS-2B) is accentuated by oxidative stress. *Food Chem Toxicol*. 2010;48:1762–6.

Ho YP, Leong KW. Quantum dot-based theranostics. *Nanoscale*. 2010 Jan;2(1):60–8.

Hoare TR, Kohane DS. Hydrogels in drug delivery: Progress and challenges. *Polymer*. 2008 Apr 15;49(8):1993–2007.

Homem de Bittencourt PI Jr, Lagranha DJ, Maslinkiewicz A, Senna SM, Tavares AM, Baldissera LP, Janner DR et al. LipoCardium: Endothelium-directed cyclopentenone prostaglandin-based liposome formulation that completely reverses atherosclerotic lesions. *Atherosclerosis*. 2007 Aug;193(2):245–58.

Hu CM, Zhang L, Aryal S, Cheung C, Fang RH, Zhang L. Erythrocyte membrane-camouflaged polymeric nanoparticles as a biomimetic delivery platform. *Proc Natl Acad Sci USA*. 2011;108:10980–5.

Hu S, Zhang Y. Endostar-loaded PEG-PLGA nanoparticles: In vitro and in vivo evaluation. *Int J Nanomed*. 2010 Nov 24;5:1039–48.

Huang HY, Kuo WT, Chou MJ, Huang YY. Co-delivery of anti-vascular endothelial growth factor siRNA and doxorubicin by multifunctional polymeric micelle for tumor growth suppression. *J Biomed Mater Res A*. 2011;97:330–8.

Huang RB, Mocherla S, Heslinga MJ, Charoenphol P, Eniola-Adefeso O. Dynamic and cellular interactions of nanoparticles in vascular-targeted drug delivery (review). *Mol Membr Biol*. 2010 Aug;27(4–6):190–205.

Huang XL, Teng X, Chen D, Tang FQ, He JQ. The effect of the shape of mesoporous silica nanoparticles on cellular uptake and cell function. *Biomaterials*. 2010;31:438–48.

Huang YY, Wang CH. Pulmonary delivery of insulin by liposomal carriers. *J Control Release*. 2006 Jun 12;113(1):9–14.

Hussain Z, Katas H, Mohd Amin MC, Kumolosasi E, Buang F, Sahudin S. Self-assembled polymeric nanoparticles for percutaneous co-delivery of hydrocortisone/hydroxytyrosol: An ex vivo and in vivo study using an NC/Nga mouse model. *Int J Pharm*. 2013 Feb 28;444(1–2):109–19.

Hwang J, Rodgers K, Oliver JC, Schluep T. α-Methylprednisolone conjugated cyclodextrin polymer-based nanoparticles for rheumatoid arthritis therapy. *Int J Nanomed*. 2008;3:359–71.

Ideta R, Tasaka F, Jang WD, Nishiyama N, Zhang GD, Harada A, Yanagi Y, Tamaki Y, Aida T, Kataoka K. Nanotechnology-based photodynamic therapy for neovascular disease using a supramolecular nanocarrier loaded with a dendritic photosensitizer. *Nano Lett*. 2005 Dec;5(12):2426–31.

Igarashi A, Konno H, Tanaka T, Nakamura S, Sadzuka Y, Hirano T, Fujise Y. Liposomal photofrin enhances therapeutic efficacy of photodynamic therapy against the human gastric cancer. *Toxicol Lett*. 2003 Nov 30;145(2):133–41.

Inapagolla R, Guru BR, Kurtoglu YE, Gao X, Lieh-Lai M, Bassett DJ, Kannan RM. In vivo efficacy of dendrimer-methylprednisolone conjugate formulation for the treatment of lung inflammation. *Int J Pharm*. 2010 Oct 31;399 (1–2):140–7.

References

Isacchi B, Bergonzi MC, Grazioso M, Righeschi C, Pietretti A, Severini C, Bilia AR. Artemisinin and artemisinin plus curcumin liposomal formulations: Enhanced antimalarial efficacy against *Plasmodium berghei*-infected mice. *Eur J Pharm Biopharm*. 2012 Apr;80(3):528–34.

Ishida T, Kiwada H. Accelerated blood clearance (ABC) phenomenon upon repeated injection of PEGylated liposomes. *Int J Pharm* 2008;354:56–62.

Ishihara T, Kubota T, Choi T, Higaki M. Treatment of experimental arthritis with stealth-type polymeric nanoparticles encapsulating betamethasone phosphate. *J Pharmacol Exp Ther*. 2009;329:412–7.

Itaka K, Kataoka K. Progress and prospects of polyplex nanomicelles for plasmid DNA delivery. *Curr Gene Ther*. 2011 Dec;11(6):457–65.

Ito H, Koefoed M, Tiyapatanaputi P, Gromov K, Goater JJ, Carmouche J, Zhang X et al. Remodeling of cortical bone allografts mediated by adherent rAAV-RANKL and VEGF gene therapy. *Nat Med*. 2005 Mar;11(3):291–7.

Jain RK. Normalization of tumor vasculature: An emerging concept in antiangiogenic therapy. *Science*. 2005;307:58–62.

Jain RK, Stylianopoulos T. Delivering nanomedicine to solid tumors. *Nat Rev Clin Oncol*. 2010 Nov;7(11):653–64.

Jang B, Park J-Y, Tung C-H, Kim I-H, Choi Y. Gold nanorod–photosensitizer complex for near-infrared fluorescence imaging and photodynamic/photothermal therapy in vivo. *ACS Nano*. 2011;5:1086–94.

Jensen SA, Day ES, Ko CH, Hurley LA, Luciano JP, Kouri FM, Merkel TJ et al. Spherical nucleic acid nanoparticle conjugates as an RNAi-based therapy for glioblastoma. *Sci Transl Med*. 2013 Oct 30;5(209):209ra152.

Jiang M, Liu Z, Xiang Y, Ma H, Liu S, Liu Y, Zheng D. Synergistic antitumor effect of AAV-mediated TRAIL expression combined with cisplatin on head and neck squamous cell carcinoma. *BMC Cancer*. 2011;11:54.

Jiang YY, Liu C, Hong MH, Zhu SJ, Pei YY. Tumor cell targeting of transferrin–PEG–TNF-alpha conjugate via a receptor-mediated delivery system: Design, synthesis, and biological evaluation. *Bioconjug Chem*. 2007;18:41–9.

Li J, Wang Y, Zhu Y, Oupický D. Recent advances in delivery of drug–nucleic acid combinations for cancer treatment. *J Control Release*. 2013;172:589–600.

Kadota K, Huang CL, Liu D, Yokomise H, Haba R, Wada H. Combined therapy with a thymidylate synthase-inhibiting vector and S-1 has effective antitumor activity against 5-FU-resistant tumors. *Int J Oncol*. 2011 Feb;38(2):355–63.

Kaida S, Cabral H, Kumagai M, Kishimura A, Terada Y, Sekino M, Aoki I, Nishiyama N, Tani T, Kataoka K. Visible drug delivery by supramolecular nanocarriers directing to single-platformed diagnosis and therapy of pancreatic tumor model. *Cancer Res*. 2010;70:7031–41.

Kameyama N, Matsuda S, Itano O, Ito A, Konno T, Arai T, Ishihara K, Ueda M, Kitagawa Y. Photodynamic therapy using an anti-EGF receptor antibody complexed with verteporfin nanoparticles: A proof of concept study. *Cancer Biother Radiopharm*. 2011 Dec;26(6):697–704.

Kaminskas LM, Ascher DB, McLeod VM, Herold MJ, Le CP, Sloan EK, Porter CJ. PEGylation of interferon α2 improves lymphatic exposure after subcutaneous and intravenous administration and improves antitumour efficacy against lymphatic breast cancer metastases. *J Control Release*. 2013 Jun 10;168(2):200–8.

Kaminskas LM, McLeod VM, Porter CJ, Boyd BJ. Association of chemotherapeutic drugs with dendrimer nanocarriers: An assessment of the merits of covalent conjugation compared to noncovalent encapsulation. *Mol Pharm.* 2012 Mar 5;9(3):355–73.

Kaminskas LM, Porter CJ. Targeting the lymphatics using dendritic polymers (dendrimers). *Adv Drug Deliv Rev.* 2011 Sep 10;63(10–11):890–900.

Kanekiyo M, Wei CJ, Yassine HM, McTamney PM, Boyington JC, Whittle JR, Rao SS, Kong WP, Wang L, Nabel GJ. Self-assembling influenza nanoparticle vaccines elicit broadly neutralizing H1N1 antibodies. *Nature.* 2013 Jul 4;499(7456):102–6.

Karmali PP, Simberg D. Interactions of nanoparticles with plasma proteins: Implication on clearance and toxicity of drug delivery systems. *Expert Opin Drug Deliv.* 2011 Mar;8(3):343–57.

Kaur IP, Aggarwal D, Singh H, Kakkar S. Improved ocular absorption kinetics of timolol maleate loaded into a bioadhesive niosomal delivery system. *Graefes Arch Clin Exp Ophthalmol.* 2010 Oct;248(10):1467–72

Kawaguchi AT, Fukumoto D, Haida M, Ogata Y, Yamano M, Tsukada H. Liposome-encapsulated hemoglobin reduces the size of cerebral infarction in the rat: Evaluation with photochemically induced thrombosis of the middle cerebral artery. *Stroke.* 2007 May;38(5):1626–32.

Kawaguchi AT, Kurita D, Furuya H, Yamano M, Ogata Y, Haida M. Liposome-encapsulated hemoglobin alleviates brain edema after permanent occlusion of the middle cerebral artery in rats. *Artif Organs.* 2009 Feb;33(2):153–8.

Kay MA. State-of-the-art gene-based therapies: The road ahead. *Nat Rev Genet.* 2011 May;12(5):316–28.

Kim YC, Chiang B, Wu X, Prausnitz MR. Ocular delivery of macromolecules. *J Control Release.* 2014 Sep 28;190C:172–81.

Kirch J, Guenther M, Doshi N, Schaefer UF, Schneider M, Mitragotri S, Lehr CM. Mucociliary clearance of micro- and nanoparticles is independent of size, shape and charge—An ex vivo and in silico approach. *J Control Release.* 2012 Apr 10;159(1):128–34.

Komarova Y, Malik AB. Regulation of endothelial permeability via paracellular and transcellular transport pathways. *Annu Rev Physiol.* 2010;72:463–93.

Koo H, Lee H, Lee S, Min KH, Kim MS, Lee DS, Choi Y, Kwon IC, Kim K, Jeong SY. In vivo tumor diagnosis and photodynamic therapy via tumoral pH-responsive polymeric micelles. *Chem Commun (Camb.).* 2010;46:5668–70.

Koo OM, Rubinstein I, Önyüksel H. Actively targeted low-dose camptothecin as a safe, long-acting, disease-modifying nanomedicine for rheumatoid arthritis. *Pharm Res.* 2011;28:776–87.

Kotterman MA, Schaffer DV. Engineering adeno-associated viruses for clinical gene therapy. *Nat Rev Genet.* 2014 Jul;15(7):445–51.

Krebs MD, Salter E, Chen E, Sutter KA, Alsberg E. Calcium phosphate-DNA nanoparticle gene delivery from alginate hydrogels induces in vivo osteogenesis. *J Biomed Mater Res A.* 2010 Mar 1;92(3):1131–8.

Kretzer IF, Maria DA, Maranhão RC. Drug-targeting in combined cancer chemotherapy: Tumor growth inhibition in mice by association of paclitaxel and etoposide with a cholesterol-rich nanoemulsion. *Cell Oncol (Dordr).* 2012 Dec; 35(6):451–60.

Kubowicz P, Żelaszczyk D, Pękala E. RNAi in clinical studies. *Curr Med Chem.* 2013;20(14):1801–16.

Kumar M, Kong X, Behera AK, Hellermann GR, Lockey RF, Mohapatra SS. Chitosan IFN-gamma-pDNA nanoparticle (CIN) therapy for allergic asthma. *Genet Vaccines Ther*. 2003 Oct 27;1(1):3.

Kumar TR, Soppimath K, Nachaegari SK. Novel delivery technologies for protein and peptide therapeutics. *Curr Pharm Biotechnol*. 2006 Aug;7(4):261–76.

Kursa M, Walker GF, Roessler V, Ogris M, Roedl W, Kircheis R, Wagner E. Novel shielded transferrin–polyethylene glycol–polyethylenimine/DNA complexes for systemic tumor-targeted gene transfer. *Bioconjug Chem*. 2003;14:222–31.

Labiris NR, Dolovich MB. Pulmonary drug delivery. Part I: Physiological factors affecting therapeutic effectiveness of aerosolized medications. *Br J Clin Pharmacol*. 2003;56:588–99.

Landen CN Jr, Chavez-Reyes A, Bucana C, Schmandt R, Deavers MT, Lopez-Berestein G, Sood AK. Therapeutic EphA2 gene targeting in vivo using neutral liposomal small interfering RNA delivery. *Cancer Res*. 2005 Aug 1; 65(15):6910–8.

Lanford RE, Hildebrandt-Eriksen ES, Petri A, Persson R, Lindow M, Munk ME, Kauppinen S, Ørum H. Therapeutic silencing of microRNA-122 in primates with chronic hepatitis C virus infection. *Science*. 2010 Jan 8;327(5962):198–201.

Laroui H, Dalmasso G, Nguyen HT, Yan Y, Sitaraman SV, Merlin D. Drug-loaded nanoparticles targeted to the colon with polysaccharide hydrogel reduce colitis in a mouse model. *Gastroenterology*. 2010 Mar;138(3):843–53.e1–2.

Laroui H, Theiss AL, Yan Y, Dalmasso G, Nguyen HT, Sitaraman SV, Merlin D. Functional TNFα gene silencing mediated by polyethyleneimine/TNFα siRNA nanocomplexes in inflamed colon. *Biomaterials*. 2011 Feb;32(4): 1218–28.

Lee DW, Shirley SA, Lockey RF, Mohapatra SS. Thiolated chitosan nanoparticles enhance anti-inflammatory effects of intranasally delivered theophylline. *Respir Res*. 2006 Aug 24;7:112.

Lee GY, Qian WP, Wang L, Wang YA, Staley CA, Satpathy M, Nie S, Mao H, Yang L. Theranostic nanoparticles with controlled release of gemcitabine for targeted therapy and MRI of pancreatic cancer. *ACS Nano*. 2013 Mar 26;7(3):2078–89.

Lee IH, An S, Yu MK, Kwon HK, Im SH, Jon S. Targeted chemoimmunotherapy using drug-loaded aptamer–dendrimer bioconjugates. *J Control Release*. 2011 Nov 7;155(3):435–41.

Lee SJ, Koo H, Lee DE, Min S, Lee S, Chen X, Choi Y et al. Tumor-homing photosensitizer-conjugated glycol chitosan nanoparticles for synchronous photodynamic imaging and therapy based on cellular on/off system. *Biomaterials*. 2011 Jun;32(16):4021–9.

Leonavičienė L, Kirdaitė G, Bradūnaitė R, Vaitkienė D, Vasiliauskas A, Zabulytė D, Ramanavičienė A, Ramanavičius A, Ašmenavičius T, Mackiewicz Z. Effect of gold nanoparticles in the treatment of established collagen arthritis in rats. *Medicina (Kaunas)*. 2012;48(2):91–101.

Li SD, Chen YC, Hackett MJ, Huang L. Tumor-targeted delivery of siRNA by self-assembled nanoparticles. *Mol Ther*. 2008 Jan;16(1):163–9.

Li Y, Wang J, Wientjes MG, Au JL. Delivery of nanomedicines to extracellular and intracellular compartments of a solid tumor. *Adv Drug Deliv Rev*. 2012 Jan;64(1):29–39.

Li Z, Pan LL, Zhang FL, Zhu XL, Liu Y, Zhang ZZ. 5-Aminolevulinic acid-loaded fullerene nanoparticles for in vitro and in vivo photodynamic therapy. *Photochem Photobiol*. 2014 Sep–Oct;90(5):1144–9.

Liberman A, Mendez N, Trogler WC, Kummel AC. Synthesis and surface functionalization of silica nanoparticles for nanomedicine. *Surf Sci Rep*. 2014 Sep;69(2–3):132–58.

Liu H, Moynihan KD, Zheng Y, Szeto GL, Li AV, Huang B, Van Egeren DS, Park C, Irvine DJ. Structure-based programming of lymph-node targeting in molecular vaccines. *Nature*. 2014 Mar 27;507(7493):519–22.

Liu J, Gong T, Fu H, Wang C, Wang X, Chen Q, Zhang Q, He Q, Zhang Z. Solid lipid nanoparticles for pulmonary delivery of insulin. *Int J Pharm*. 2008 May 22;356(1–2):333–44.

Liu J, Ohta S, Sonoda A, Yamada M, Yamamoto M, Nitta N, Murata K, Tabata Y. Preparation of PEG-conjugated fullerene containing Gd3+ ions for photodynamic therapy. *J Control Release*. 2007 Jan 22;117(1):104–10.

Liu ZW, Ren GG, Zhang T, Yang Z. Action potential changes associated with the inhibitory effects on voltage-gated sodium current of hippocampal CA1 neurons by silver nanoparticles. *Toxicology*. 2009;264:179–84.

Ljunggren HG, Malmberg KJ. Prospects for the use of NK cells in immunotherapy of human cancer. *Nat Rev Immunol*. 2007 May;7(5):329–39.

Lo Prete AC, Maria DA, Rodrigues DG, Valduga CJ, Ibañez OC, Maranhão RC. Evaluation in melanoma-bearing mice of an etoposide derivative associated to a cholesterol-rich nano-emulsion. *J Pharm Pharmacol*. 2006 Jun;58(6):801–8.

Longo JP, Leal SC, Simioni AR, de Fátima Menezes Almeida-Santos M, Tedesco AC, Azevedo RB. Photodynamic therapy disinfection of carious tissue mediated by aluminum-chloride-phthalocyanine entrapped in cationic liposomes: An in vitro and clinical study. *Lasers Med Sci*. 2012 May;27(3):575–84.

Lovell JF, Jin CS, Huynh E, Jin H, Kim C, Rubinstein JL, Chan WC, Cao W, Wang LV, Zheng G. Porphysome nanovesicles generated by porphyrin bilayers for use as multimodal biophotonic contrast agents. *Nat Mater*. 2011 Apr;10(4):324–32.

Lu W, Sun Q, Wan J, She Z, Jiang XG. Cationic albumin-conjugated pegylated nanoparticles allow gene delivery into brain tumors via intravenous administration. *Cancer Res*. 2006 Dec 15;66(24):11878–87.

Lu Y, Kawakami S, Yamashita F, Hashida M. Development of an antigen-presenting cell-targeted DNA vaccine against melanoma by mannosylated liposomes. *Biomaterials*. 2007 Jul;28(21):3255–62.

Madani SY, Naderi N, Dissanayake O, Tan A, Seifalian AM. A new era of cancer treatment: Carbon nanotubes as drug delivery tools. *Int J Nanomed*. 2011;6:2963–79.

Madeira C, Loura LM, Prieto M, Fedorov A, Aires-Barros MR. Effect of ionic strength and presence of serum on lipoplexes structure monitorized by FRET. *BMC Biotechnol*. 2008 Feb 26;8:20.

Maeda H, Nakamura H, Fang J. The EPR effect for macromolecular drug delivery to solid tumors: Improvement of tumor uptake, lowering of systemic toxicity, and distinct tumor imaging in vivo. *Adv Drug Deliv Rev*. 2013 Jan;65(1):71–9.

Maeda H. The enhanced permeability and retention (EPR) effect in tumor vasculature: The key role of tumor-selective macromolecular drug targeting. *Adv Enzym Regul*. 2001;41:189–207.

Maeda H. Tumor-selective delivery of macromolecular drugs via the EPR effect: Background and future prospects. *Bioconjug Chem*. 2010 May 19;21(5):797–802.

Maeda H, Wu J, Sawa T, Matsumura Y, Hori K. Tumor vascular permeability and the EPR effect in macromolecular therapeutics: A review. *J Control Release*. 2000;65:271–84.

Malik MA, Wani MY, Hashim MA. Microemulsion method: A novel route to synthesize organic and inorganic nanomaterials: 1st nano update. *Arabian J Chem*. 2012 Oct;5(4);397–417.

Manosroi A, Kongkaneramit L, Manosroi J. Stability and transdermal absorption of topical amphotericin-B liposome formulations. *Int J Pharm*. 2004;270:279–86.

Marano RJ, Toth I, Wimmer N, Brankov M, Rakoczy PE. Dendrimer delivery of an anti-VEGF oligonucleotide into the eye: A long-term study into inhibition of laser-induced CNV, distribution, uptake and toxicity. *Gene Ther*. 2005 Nov;12(21):1544–50.

Marin E, Briceño MI, Caballero-George C. Critical evaluation of biodegradable polymers used in nanodrugs. *Int J Nanomed*. 2013;8:3071–90.

Lobatto ME, Fuster V, Fayad ZA, Mulder WJM. Perspectives and opportunities for nanomedicine in the management of atherosclerosis. *Nat Rev Drug Discov*. 2011;10(11):835–52.

Markovsky E, Baabur-Cohen H, Eldar-Boock A, Omer L, Tiram G, Ferber S, Ofek P, Polyak D, Scomparin A, Satchi-Fainaro R. Administration, distribution, metabolism and elimination of polymer–drug conjugates. *J Control Release*. 2011;161:446–60.

Martins S, Sarmento B, Ferreira DC, Souto EB. Lipid-based colloidal carriers for peptide and protein delivery—Liposomes versus lipid nanoparticles. *Int J Nanomed*. 2007;2(4):595–607.

Mayer A, Vadon M, Rinner B, Novak A, Wintersteiger R, Frohlich E. 2009. The role of nanoparticle size in hemocompatibility. *Toxicology*. 258:139–147.

McAllaster JD, Cohen MS. Role of the lymphatics in cancer metastasis and chemotherapy applications. *Adv Drug Deliv Rev*. 2011;63:867–75.

McCarroll J, Teo J, Boyer C, Goldstein D, Kavallaris M, Phillips PA. Potential applications of nanotechnology for the diagnosis and treatment of pancreatic cancer. *Front Physiol*. 2014 Jan 24;5:2.

McCarthy JR, Korngold E, Weissleder R, Jaffer FA. A light-activated theranostic nanoagent for targeted macrophage ablation in inflammatory atherosclerosis. *Small*. 2010;6:2041–9.

McGee Sanftner LH, Abel H, Hauswirth WW, Flannery JG. Glial cell line derived neurotrophic factor delays photoreceptor degeneration in a transgenic rat model of retinitis pigmentosa. *Mol Ther*. 2001 Dec;4(6):622–9.

Meng H, Xia T, George S, Nel AE. A predictive toxicological paradigm for the safety assessment of nanomaterials. *ACS Nano*. 2009;3:1620–7.

Merisko-Liversidge E, Liversidge GG. Nanosizing for oral and parenteral drug delivery: A perspective on formulating poorly-water soluble compounds using wet media milling technology. *Adv Drug Deliv Rev*. 2011 May 30;63(6):427–40.

Merlot AM, Kalinowski DS, Richardson DR. Unraveling the mysteries of serum albumin-more than just a serum protein. *Front Physiol*. 2014 Aug 12;5:299.

Micallef MJ, Tanimoto T, Kohno K, Ikeda M, Kurimoto M. Interleukin 18 induces the sequential activation of natural killer cells and cytotoxic T lymphocytes to protect syngeneic mice from transplantation with Meth A sarcoma. *Cancer Res*. 1997;57:4557–63.

Minshall RD, Tiruppathi C, Vogel SM, Niles WD, Gilchrist A, Hamm HE, Malik AB. Endothelial cell-surface gp60 activates vesicle formation and trafficking via G(i)-coupled Src kinase signaling pathway. *J Cell Biol*. 2000 Sep 4;150 (5):1057–70.

Mishra V, Mahor S, Rawat A, Gupta PN, Dubey P, Khatri K, Vyas SP. Targeted brain delivery of AZT via transferrin anchored pegylated albumin nanoparticles. *J Drug Target*. 2006 Jan;14(1):45–53.

Mizuarai S, Ono K, You J, Kamihira M, Iijima S. Protamine-modified DDAB lipid vesicles promote gene transfer in the presence of serum. *J Biochem (Tokyo)*. 2001;129(1):125–32.

Mo R, Jiang T, Di J, Wanyi T Gu Z. Emerging micro- and nanotechnology based synthetic approaches for insulin delivery. *Chem Soc Rev*. 2014;43:3595.

Mody VV, Cox A, Shah S, Singh A, Bevins W, Parihar H. Magnetic nanoparticle drug delivery systems for targeting tumor. *Appl Nanosci*. 2014;4:385–392.

Moghimi SM. Cancer nanomedicine and the complement system activation paradigm: Anaphylaxis and tumour growth. *J Control Release*. 2014 Sep 28;190: 556–62.

Monto AS, Ansaldi F, Aspinall R, McElhaney JE, Montaño LF, Nichol KL, Puig-Barberà J, Schmitt J, Stephenson I. Influenza control in the 21st century: Optimizing protection of older adults. *Vaccine*. 2009 Aug 13;27(37):5043–53.

Moros M, Mitchell SG, Grazú V, de la Fuente JM. The fate of nanocarriers as nanomedicines in vivo: Important considerations and biological barriers to overcome. *Curr Med Chem*. 2013;20(22):2759–78.

Morrissey DV, Lockridge JA, Shaw L, Blanchard K, Jensen K, Breen W, Hartsough K et al. Potent and persistent in vivo anti-HBV activity of chemically modified siRNAs. *Nat Biotechnol*. 2005 Aug;23(8):1002–7.

Möschwitzer JP. Drug nanocrystals in the commercial pharmaceutical development process. *Int J Pharm*. 2013 Aug 30;453(1):142–56.

Motavaf M, Safari S, Alavian SM. Therapeutic potential of RNA interference: A new molecular approach to antiviral treatment for hepatitis C. *J Viral Hepat*. 2012 Nov;19(11):757–65.

Moulari B, Pertuit D, Pellequer Y, Lamprecht A. The targeting of surface modified silica nanoparticles to inflamed tissue in experimental colitis. *Biomaterials*. 2008 Dec;29(34):4554–60.

Moyle PM, Toth I. Modern subunit vaccines: Development, components, and research opportunities. *ChemMedChem*. 2013 Mar;8(3):360–76.

Moya-Horno I, Cortés J. The expanding role of pertuzumab in the treatment of HER2-positive breast cancer. *Breast Cancer (Dove Med Press)*. 2015 May 21;7: 125–32.

Mühlfeld C, Rothen-Rutishauser B, Blank F, Vanhecke D, Ochs M, Gehr P. Interactions of nanoparticles with pulmonary structures and cellular responses. *Am J Physiol Lung Cell Mol Physiol*. 2008 May;294(5):L817–29.

Mura S, Couvreur P. Nanotheranostics for personalized medicine. *Adv Drug Deliv Rev*. 2012 Oct;64(13):1394–416.

Muramatsu K, Maitani Y, Takayama K, Nagai T. The relationship between the rigidity of the liposomal membrane and the absorption of insulin after nasal administration of liposomes modified with an enhancer containing insulin in rabbits. *Drug Dev Ind Pharm*. 1999 Oct;25(10):1099–105.

Nabiev I, Mitchell S, Davies A, Williams Y, Kelleher D, Moore R, Gun'ko YK et al. Nonfunctionalized nanocrystals can exploit a cell's active transport machinery delivering them to specific nuclear and cytoplasmic compartments. *Nano Lett*. 2007;7:3452–61.

References

Nakamura K, Abu Lila AS, Matsunaga M, Doi Y, Ishida T, Kiwada H. A double-modulation strategy in cancer treatment with a chemotherapeutic agent and siRNA. *Mol Ther*. 2011 Nov;19(11):2040–7.

Nandakumar DN, Nagaraj VA, Vathsala PG, Rangarajan P, Padmanaban G. Curcumin-artemisinin combination therapy for malaria. *Antimicrob Agents Chemother*. 2006 May;50(5):1859–60.

Nguyen HN, Wey SP, Juang JH, Sonaje K, Ho YC, Chuang EY, Hsu CW, Yen TC, Lin KJ, Sung HW. The glucose-lowering potential of exendin-4 orally delivered via a pH-sensitive nanoparticle vehicle and effects on subsequent insulin secretion in vivo. *Biomaterials*. 2011 Apr;32(10):2673–82.

Nguyen PD, O'Rear EA, Johnson AE, Patterson E, Whitsett TL, Bhakta R. Accelerated thrombolysis and reperfusion in a canine model of myocardial infarction by liposomal encapsulation of streptokinase. *Circ Res*. 1990 Mar;66(3):875–8.

Nicolazzi C, Mignet N, de la Figuera N, Cadet M, Ibad RT, Seguin J, Scherman D, Bessodes M. Anionic polyethyleneglycol lipids added to cationic lipoplexes increase their plasmatic circulation time. *J Control Release*. 2003;88(3):429–43.

Nishiyama N, Nakagishi Y, Morimoto Y, Lai PS, Miyazaki K, Urano K, Horie S et al. Enhanced photodynamic cancer treatment by supramolecular nanocarriers charged with dendrimer phthalocyanine. *J Control Release*. 2009 Feb 10;133(3):245–51.

Nitsch MJ, Banakar UV. Implantable drug delivery. *J Biomater Appl*. 1994;8:247–84.

Noble CO, Krauze MT, Drummond DC, Forsayeth J, Hayes ME, Beyer J, Hadaczek P et al. Pharmacokinetics, tumor accumulation and antitumor activity of nanoliposomal irinotecan following systemic treatment of intracranial tumors. *Nanomedicine (Lond)*. 2014 Jul;9(14):2099–108.

Nunes A, Al-Jamal KT, Kostarelos K. Therapeutics, imaging and toxicity of nanomaterials in the central nervous system. *J Control Release*. 2012;161:290–306.

Nurhasni H, Cao J, Choi M, Kim I, Lee BL, Jung Y, Yoo JW. Nitric oxide-releasing poly(lactic-co-glycolic acid)-polyethylenimine nanoparticles for prolonged nitric oxide release, antibacterial efficacy, and in vivo wound healing activity. *Int J Nanomed*. 2015 Apr 22;10:3065–80.

Ohlin M, Borrebaeck CA. Characteristics of human antibody repertoires following active immune responses in vivo. *Mol Immunol*. 1996 May–Jun;33(7–8):583–92.

Ohtani S, Iwamaru A, Deng W, Ueda K, Wu G, Jayachandran G, Kondo S et al. Tumor suppressor 101F6 and ascorbate synergistically and selectively inhibit non-small cell lung cancer growth by caspase-independent apoptosis and autophagy. *Cancer Res*. 2007 Jul 1;67(13):6293–303.

Ojea-Jiménez I, Comenge J, García-Fernández L, Megson ZA, Casals E, Puntes VF. Engineered inorganic nanoparticles for drug delivery applications. *Curr Drug Metab*. 2013 Jun;14(5):518–30.

Opanasopit P, Nishikawa M, Hashida M. Factors affecting drug and gene delivery: Effects of interaction with blood components. *Crit Rev Ther Drug Carrier Syst*. 2002;19:191–233.

Opanasopit P, Sakai M, Nishikawa M, Kawakami S, Yamashita F, Hashida M. Inhibition of liver metastasis by targeting of immunomodulators using mannosylated liposome carriers. *J Control Release*. 2002 Apr 23;80(1–3): 283–94.

Orr MT, Fox CB, Baldwin SL, Sivananthan SJ, Lucas E, Lin S, Phan T et al. Adjuvant formulation structure and composition are critical for the development of an effective vaccine against tuberculosis. *J Control Release*. 2013 Nov 28;172(1):190–200.

Owais M, Varshney GC, Choudhury A, Chandra S, Gupta CM. Chloroquine encapsulated in malaria-infected erythrocyte-specific antibody-bearing liposomes effectively controls chloroquine-resistant *Plasmodium berghei* infections in mice. *Antimicrob Agents Chemother*. 1995;39:180–4.

Özbaş-Turan S, Akbuğa J. Plasmid DNA-loaded chitosan/TPP nanoparticles for topical gene delivery. *Drug Deliv*. 2011 Apr;18(3):215–22.

Özcan I, Azizoğlu E, Senyiğit T, ÖzyazıcıM, Özer Ö. Enhanced dermal delivery of diflucortolone valerate using lecithin/chitosan nanoparticles: In-vitro and in-vivo evaluations. *Int J Nanomed*. 2013;8:461–75.

Ozpolat B, Sood AK, Lopez-Berestein G. Liposomal siRNA nanocarriers for cancer therapy. *Adv Drug Deliv Rev*. 2014 Feb;66:110–6.

Pan XQ, Zheng X, Shi G, Wang H, Ratnam M, Lee RJ. Strategy for the treatment of acute myelogenous leukemia based on folate receptor beta-targeted liposomal doxorubicin combined with receptor induction using all-trans retinoic acid. *Blood*. 2002 Jul 15;100(2):594–602.

Pandit J, Garg M, Jain NK. Miconazole nitrate bearing ultraflexible liposomes for the treatment of fungal infection. *J Liposome Res*. 2014 Jun;24(2):163–9.

Parhi R, Suresh P. Preparation and characterization of solid lipid nanoparticles—A review. *Curr Drug Discov Technol*. 2012 Mar;9(1):2–16.

Park JW, Hong K, Kirpotin DB, Colbern G, Shalaby R, Baselga J, Shao Y et al. Anti-HER2 immunoliposomes: Enhanced efficacy attributable to targeted delivery. *Clin Cancer Res*. 2002 Apr;8(4):1172–81.

Park KH, Chhowalla M, Iqbal Z, Sesti F. Single-walled carbon nanotubes are a new class of ion channel blockers. *J Biol Chem*. 2003 Dec 12;278(50):50212–6.

Pasut G, Veronese FM, State of the art in PEGylation: The great versatility achieved after forty years of research. *J Control Release*. 2012;161:461–72.

Pathak P, Nagarsenker M. Formulation and evaluation of lidocaine lipid nanosystems for dermal delivery. *AAPS PharmSciTech*. 2009;10(3):985–92.

Peer D, Park EJ, Morishita Y, Carman CV, Shimaoka M. Systemic leukocyte-directed siRNA delivery revealing cyclin D1 as an anti-inflammatory target. *Science*. 2008 Feb 1;319(5863):627–30.

Peeters PA, Brunink BG, Eling WM, Crommelin DJ. Therapeutic effect of chloroquine(CQ)-containing immunoliposomes in rats infected with *Plasmodium berghei* parasitized mouse red blood cells: Comparison with combinations of antibodies and CQ or liposomal CQ. *Biochim Biophys Acta*. 1989;981:269–76.

Perche F, Benvegnu T, Berchel M, Lebegue L, Pichon C, Jaffrès PA, Midoux P. Enhancement of dendritic cells transfection in vivo and of vaccination against B16F10 melanoma with mannosylated histidylated lipopolyplexes loaded with tumor antigen messenger RNA. *Nanomedicine*. 2011 Aug;7(4): 445–53.

Pertuit D, Moulari B, Betz T, Nadaradjane A, Neumann D, Ismaïli L, Refouvelet B, Pellequer Y, Lamprecht A. 5-amino salicylic acid bound nanoparticles for the therapy of inflammatory bowel disease. *J Control Release*. 2007 Nov 20;123(3):211–8.

Photos PJ, Bacakova L, Discher B, Bates FS, Discher DE. Polymer vesicles in vivo: Correlations with PEG molecular weight. *J Control Release*. 2003;90:323–34.

References

Pichai MV, Ferguson LR. Potential prospects of nanomedicine for targeted therapeutics in inflammatory bowel diseases. *World J Gastroenterol.* 2012 Jun 21;18(23):2895–901.

Pirollo KF, Rait A, Zhou Q, Hwang SH, Dagata JA, Zon G, Hogrefe RI, Palchik G, Chang EH. Materializing the potential of small interfering RNA via a tumor-targeting nanodelivery system. *Cancer Res.* 2007 Apr 1;67(7):2938–43.

Pissuwan D, Niidome T, Cortie MB. The forthcoming applications of gold nanoparticles in drug and gene delivery systems. *J Control Release.* 2011 Jan 5; 149(1):65–71.

Popovska O, Simonovska J, Kavrakovski Z, Rafajlovska V. An overview: Methods for preparation and characterization of liposomes as drug delivery systems. *Int J Pharm Phytopharmacol Res.* 2013;3(2): 2250–1029.

Postma NS, Crommelin DJ, Eling WM, Zuidema J. Treatment with liposome-bound recombinant human tumor necrosis factor-alpha suppresses parasitemia and protects against *Plasmodium berghei* k173-induced experimental cerebral malaria in mice. *J Pharmacol Exp Ther.* 1999;288:114–20.

Prados J, Melguizo C, Ortiz R, Perazzoli G, Cabeza L, Alvarez PJ, Rodriguez-Serrano F, Aranega A. Colon cancer therapy: Recent developments in nanomedicine to improve the efficacy of conventional chemotherapeutic drugs. *Anticancer Agents Med Chem.* 2013 Oct;13(8):1204–16.

Prausnitz MR, Mitragotri S, Langer R. Current status and future potential of transdermal drug delivery. *Nat Rev Drug Discov.* 2004 Feb;3(2):115–24.

Provenzano PP, Cuevas C, Chang AE, Goel VK, Von Hoff DD, Hingorani SR. Enzymatic targeting of the stroma ablates physical barriers to treatment of pancreatic ductal adenocarcinoma. *Cancer Cell.* 2012 Mar 20;21(3):418–29.

Raizada MK, Der Sarkissian S. Potential of gene therapy strategy for the treatment of hypertension. *Hypertension.* 2006 Jan;47(1):6–9.

Ramteke S, Ganesh N, Bhattacharya S, Jain NK. Amoxicillin, clarithromycin, and omeprazole based targeted nanoparticles for the treatment of H. pylori. *J Drug Target.* 2009 Apr;17(3):225–34.

Rawat M, Singh D, Saraf S, Saraf S. Lipid carriers: A versatile delivery vehicle for proteins and peptides. *Yakugaku Zasshi.* 2008 Feb;128(2):269–80.

Reddy GR, Bhojani MS, McConville P, Moody J, Moffat BA, Hall DE, Kim G et al. Vascular targeted nanoparticles for imaging and treatment of brain tumors. *Clin Cancer Res.* 2006;12:6677–86.

Reddy LH, Couvreur P. Nanotechnology for therapy and imaging of liver diseases. *J Hepatol.* 2011 Dec;55(6):1461–6.

Roby A, Erdogan S, Torchilin VP. Enhanced in vivo antitumor efficacy of poorly soluble PDT agent, meso-tetraphenylporphine, in PEG-PE-based tumor-targeted immunomicelles. *Cancer Biol Ther.* 2007 Jul;6(7):1136–42.

Rodriguez PL, Harada T, Christian DA, Pantano DA, Tsai RK, Discher DE. Minimal "self" peptides that inhibit phagocytic clearance and enhance delivery of nanoparticles. *Science.* 2013 Feb 22;339(6122):971–5.

Rogers LM, Veeramani S, Weiner GJ. Complement in monoclonal antibody therapy of cancer. *Immunol Res.* 2014 Aug;59(1–3):203–10.

Rösler A, Vandermeulen GW, Klok HA. Advanced drug delivery devices via self-assembly of amphiphilic block copolymers. *Adv Drug Deliv Rev.* 2001 Dec 3;53(1):95–108.

Rubinstein I, Weinberg GL. Nanomedicines for chronic non-infectious arthritis: The clinician's perspective. *Nanomedicine.* 2012 September;8S1:S77–82.

Sailor MJ, Park JH. Hybrid nanoparticles for detection and treatment of cancer. *Adv Mater*. 2012 Jul 24;24(28):3779–802.

Sakai T, Ishihara T, Higaki M, Akiyama G, Tsuneoka H. Therapeutic effect of stealth-type polymeric nanoparticles with encapsulated betamethasone phosphate on experimental autoimmune uveoretinitis. *Invest Ophthalmol Vis Sci*. 2011 Mar 18;52(3):1516–21.

Sakai T, Kohno H, Ishihara T, Higaki M, Saito S, Matsushima M, Mizushima Y, Kitahara K. Treatment of experimental autoimmune uveoretinitis with poly (lactic acid) nanoparticles encapsulating betamethasone phosphate. *Exp Eye Res*. 2006 Apr;82(4):657–63.

Sakuma T, Barry MA, Ikeda Y. Lentiviral vectors: Basic to translational. *Biochem J*. 2012 May 1;443(3):603–18.

Salama R, Traini D, Chan HK, Young PM. Recent advances in controlled release pulmonary therapy. *Curr Drug Deliv*. 2009 Aug;6(4):404–14.

Samad A, Alam MI, Saxena K. Dendrimers: A class of polymers in the nanotechnology for the delivery of active pharmaceuticals. *Curr Pharm Des*. 2009;15 (25):2958–69.

Sato Y, Murase K, Kato J, Kobune M, Sato T, Kawano Y, Takimoto R et al. Resolution of liver cirrhosis using vitamin A-coupled liposomes to deliver siRNA against a collagen-specific chaperone. *Nat Biotechnol*. 2008;26:431–42.

Scanlon KJ. Anti-genes: siRNA, ribozymes and antisense. *Curr Pharm Biotechnol*. 2004 Oct;5(5):415–20.

Schneider T, Becker A, Ringe K, Reinhold A, Firsching R, Sabel BA. Brain tumor therapy by combined vaccination and antisense oligonucleotide delivery with nanoparticles. *J Neuroimmunol*. 2008;195:21–7.

Scholz C, Wagner E. Therapeutic plasmid DNA versus siRNA delivery: Common and different tasks for synthetic carriers. *J Control Release*. 2012 Jul 20;161 (2):554–65.

Schütz CA, Juillerat-Jeanneret L, Mueller H, Lynch I, Riediker M, NanoImpactNet Consortium. Therapeutic nanoparticles in clinics and under clinical evaluation. *Nanomedicine (Lond)*. 2013 Mar;8(3):449–67.

Sethi V, Rubinstein I, Onyuksel H. Vasoactive intestinal peptide (VIP) loaded sterically stabilized micelles (SSM) for improved therapy of collagen induced arthritis (CIA) in mice. *Pharm Sci*. 2002;4(Suppl.):T2036.

Shah V, Taratula O, Garbuzenko OB, Taratula OR, Rodriguez-Rodriguez L, Minko T. Targeted nanomedicine for suppression of CD44 and simultaneous cell death induction in ovarian cancer: An optimal delivery of siRNA and anticancer drug. *Clin Cancer Res*. 2013 Nov 15;19(22):6193–204.

Shantha Kumar TR, Soppimath K, Nachaegari SK. Novel delivery technologies for protein and peptide therapeutics. *Curr Pharm Biotechnol*. 2006;7:261–276.

Sharma A, Mayhew E, Straubinger RM. Antitumor effect of taxol-containing liposomes in a taxol-resistant murine tumor model. *Cancer Res*. 1993 Dec 15;53(24):5877–81.

Sharma A, Sharma S, Khuller GK. Lectin-functionalized poly (lactide-co-glycolide) nanoparticles as oral/aerosolized antitubercular drug carriers for treatment of tuberculosis. *J Antimicrob Chemother*. 2004 Oct;54(4):761–6.

Shen H, Rodriguez-Aguayo C, Xu R, Gonzalez-Villasana V, Mai J, Huang Y, Zhang G et al. Enhancing chemotherapy response with sustained EphA2 silencing using multistage vector delivery. *Clin Cancer Res*. 2013 Apr 1;19(7):1806–15.

References

Sheng WY, Huang L. Cancer immunotherapy and nanomedicine. *Pharm Res.* 2011 Feb;28(2):200–14.

Shi P, Aluri S, Lin YA, Shah M, Edman M, Dhandhukia J, Cui H, MacKay JA. Elastin-based protein polymer nanoparticles carrying drug at both corona and core suppress tumor growth in vivo. *J Control Release.* 2013 Nov 10;171(3):330–8.

Shi P, Gustafson JA, MacKay JA. Genetically engineered nanocarriers for drug delivery. *Int J Nanomed.* 2014 Mar 26;9:1617–26.

Shim G, Lee S, Kim YB, Kim CW, Oh YK. Enhanced tumor localization and retention of chlorin e6 in cationic nanolipoplexes potentiate the tumor ablation effects of photodynamic therapy. *Nanotechnology.* 2011 Sep 7;22(36):365101.

Singh KK, Vingkar SK. Formulation, antimalarial activity and biodistribution of oral lipid nanoemulsion of primaquine. *Int J Pharm.* 2008;347:136–43.

Singh SR, Grossniklaus HE, Kang SJ, Edelhauser HF, Ambati BK, Kompella UB. Intravenous transferrin, RGD peptide and dual-targeted nanoparticles enhance anti-VEGF intraceptor gene delivery to laser-induced CNV. *Gene Ther.* 2009 May;16(5):645–59.

Smith JD, Morton LD, Ulery BD. Nanoparticles as synthetic vaccines. *Curr Opin Biotechnol.* 2015 Apr 6;34:217–224.

Smyth MJ, Hayakawa Y, Cretney E, Zerafa N, Sivakumar P, Yagita H, Takeda K. IL-21 enhances tumor-specific CTL induction by anti-DR5 antibody therapy. *J Immunol.* 2006;176:6347–55.

Soman NR, Baldwin SL, Hu G, Marsh JN, Lanza GM, Heuser JE, Arbeit JM, Wickline SA, Schlesinger PH. Molecularly targeted nanocarriers deliver the cytolytic peptide melittin specifically to tumor cells in mice, reducing tumor growth. *J Clin Invest.* 2009;119:2830–42.

Sonaje K, Chen YJ, Chen HL, Wey SP, Juang JH, Nguyen HN, Hsu CW, Lin KJ, Sung HW. Enteric-coated capsules filled with freeze-dried chitosan/poly(gamma-glutamic acid) nanoparticles for oral insulin delivery. *Biomaterials.* 2010 Apr;31(12):3384–94.

Sonaje K, Lin YH, Juang JH, Wey SP, Chen CT, Sung HW. In vivo evaluation of safety and efficacy of self-assembled nanoparticles for oral insulin delivery. *Biomaterials.* 2009 Apr;30(12):2329–39.

Sonavane G, Tomoda K, Makino K. Biodistribution of colloidal gold nanoparticles after intravenous administration: Effect of particle size. *Colloids Surf B Biointerfaces.* 2008 Oct 15;66(2):274–80.

Steiniger SC, Kreuter J, Khalansky AS, Skidan IN, Bobruskin AI, Smirnova ZS, Severin SE et al. Chemotherapy of glioblastoma in rats using doxorubicin-loaded nanoparticles. *Int J Cancer.* 2004 May 1;109(5):759–67.

Sterman DH, Recio A, Carroll RG, Gillespie CT, Haas A, Vachani A, Kapoor V et al. A phase I clinical trial of single-dose intrapleural IFN-beta gene transfer for malignant pleural mesothelioma and metastatic pleural effusions: High rate of antitumor immune responses. *Clin Cancer Res.* 2007;13:4456–66.

Stylianopoulos T, Poh MZ, Insin N, Bawendi MG, Fukumura D, Munn LL, Jain RK. Diffusion of particles in the extracellular matrix: The effect of repulsive electrostatic interactions. *Biophys J.* 2010 Sep 8;99(5):1342–9.

Su B, Cengizeroglu A, Farkasova K, Viola JR, Anton M, Ellwart JW, Haase R, Wagner E, Ogris M. Systemic TNFalpha gene therapy synergizes with liposomal doxorubicine in the treatment of metastatic cancer. *Mol Ther.* 2012; 21:300–8.

Su FY, Lin KJ, Sonaje K, Wey SP, Yen TC, Ho YC, Panda N, Chuang EY, Maiti B, Sung HW. Protease inhibition and absorption enhancement by functional nanoparticles for effective oral insulin delivery. *Biomaterials.* 2012 Mar;33(9):2801–11.

Sun B, Ji Z, Liao YP, Wang M, Wang X, Dong J, Chang CH et al. Engineering an effective immune adjuvant by designed control of shape and crystallinity of aluminum oxyhydroxide nanoparticles. *ACS Nano.* 2013 Dec 23;7(12):10834–49.

Sun S, Liang N, Kawashima Y, Xia D, Cui F. Hydrophobic ion pairing of an insulin-sodium deoxycholate complex for oral delivery of insulin. *Int J Nanomed.* 2011;6:3049–56.

Swai H, Semete B, Kalombo L, Chelule P, Kisich K, Sievers B. Nanomedicine for respiratory diseases. *Wiley Interdiscip Rev Nanomed Nanobiotechnol.* 2009 May–Jun;1(3):255–63.

Tabata Y, Murakami Y, Ikada Y. Photodynamic effect of polyethylene glycol-modified fullerene on tumor. *Jpn J Cancer Res.* 1997 Nov;88(11):1108–16.

Tagami T, Foltz WD, Ernsting MJ, Lee CM, Tannock IF, May JP, Li S-D. MRI monitoring of intratumoral drug delivery and prediction of the therapeutic effect with a multifunctional thermosensitive liposome. *Biomaterials.* 2011; 32:6570–8.

Tahara K, Samura S, Tsuji K, Yamamoto H, Tsukada Y, Bando Y, Tsujimoto H, Morishita R, Kawashima Y. Oral nuclear factor-κB decoy oligonucleotides delivery system with chitosan modified poly(D,L-lactide-co-glycolide) nanospheres for inflammatory bowel disease. *Biomaterials.* 2011 Jan;32(3):870–8.

Tang ML, Wang M, Xing TR, Zeng J, Wang HL, Ruan DY. Mechanisms of unmodified CdSe quantum dot-induced elevation of cytoplasmic calcium levels in primary cultures of rat hippocampal neurons. *Biomaterials.* 2008;29:4383–91.

Tang N, Du G, Wang N, Liu C, Hang H, Liang W. Improving penetration in tumors with nanoassemblies of phospholipids and doxorubicin. *J Natl Cancer Inst.* 2007 Jul 4;99(13):1004–15.

Taratula O, Garbuzenko OB, Kirkpatrick P, Pandya I, Savla R, Pozharov VP, He H, Minko T. Surface-engineered targeted PPI dendrimer for efficient intracellular and intratumoral siRNA delivery. *J Control Release.* 2009;140:284–93.

Tautzenberger A, Kovtun A, Ignatius A. Nanoparticles and their potential for application in bone. *Int J Nanomedicine.* 2012;7:4545–57.

Teijeiro-Osorio D, Remuñán-López C, Alonso MJ. New generation of hybrid poly/oligosaccharide nanoparticles as carriers for the nasal delivery of macromolecules. *Biomacromolecules.* 2009 Feb 9;10(2):243–9.

Theiss AL, Laroui H, Obertone TS, Chowdhury I, Thompson WE, Merlin D, Sitaraman SV. Nanoparticle-based therapeutic delivery of prohibitin to the colonic epithelial cells ameliorates acute murine colitis. *Inflamm Bowel Dis.* 2011 May;17(5):1163–76.

Tian J, Wong KK, Ho CM, Lok CN, Yu WY, Che CM, Chiu JF, Tam PK. Topical delivery of silver nanoparticles promotes wound healing. *ChemMedChem.* 2007 Jan;2(1):129–36.

Tietze LF, Schmuck K. Prodrugs for targeted tumor therapies: Recent developments in ADEPT, GDEPT and PMT. *Curr Pharm Des.* 2011;17(32):3527–47.

Tran MA, Gowda R, Sharma A, Park EJ, Adair J, Kester M, Smith NB, Robertson GP. Targeting V600EB-Raf and Akt3 using nanoliposomal-small interfering RNA inhibits cutaneous melanocytic lesion development. *Cancer Res.* 2008 Sep 15;68(18):7638–49.

References 311

Tran MA, Smith CD, Kester M, Robertson GP. Combining nanoliposomal ceramide with sorafenib synergistically inhibits melanoma and breast cancer cell survival to decrease tumor development. *Clin Cancer Res.* 2008 Jun 1;14(11):3571–81.

Tros de Ilarduya C, Sun Y, Düzgüneş N. Gene delivery by lipoplexes and polyplexes. *Eur J Pharm Sci.* 2010 Jun 14;40(3):159–70.

Trouiller B, Reliene R, Westbrook A, Solaimani P, Schiestl RH. Titanium dioxide nanoparticles induce DNA damage and genetic instability in vivo in mice. *Cancer Res.* 2009;69:8784–9.

Tsai CY, Shiau AL, Chen SY, Chen YH, Cheng PC, Chang MY, Chen DH, Chou CH, Wang CR, Wu CL. Amelioration of collagen-induced arthritis in rats by nanogold. *Arthritis Rheum.* 2007 Feb;56(2):544–54.

Ueda K, Kawashima H, Ohtani S, Deng WG, Ravoori M, Bankson J, Gao B et al. The 3p21.3 tumor suppressor NPRL2 plays an important role in cisplatin-induced resistance in human non-small-cell lung cancer cells. *Cancer Res.* 2006 Oct 1;66(19):9682–90.

Uhumwangho MU, Okor RS. Current trends in the production and biomedical applications of liposomes: A review. *J Med Biomed Res.* June 2005;4(1):9–21.

van den Hoven JM, Van Tomme SR, Metselaar JM, Nuijen B, Beijnen JH, Storm G. Liposomal drug formulations in the treatment of rheumatoid arthritis. *Mol Pharm.* 2011 Aug 1;8(4):1002–15.

van Riet E, Ainai A, Suzuki T, Kersten G, Hasegawa H. Combatting infectious diseases; nanotechnology as a platform for rational vaccine design. *Adv Drug Deliv Rev.* 2014 Jul 30;74C:28–34.

Vandamme T, Brobeck L. Poly(amidoamine) dendrimers as ophthalmic vehicles for ocular delivery of pilocarpine nitrate and tropicamide. *J Control Rel.* 2005;102:23–38.

Vandenbroucke K, de Haard H, Beirnaert E, Dreier T, Lauwereys M, Huyck L, Van Huysse J et al. Orally administered L. lactis secreting an anti-TNF nanobody demonstrate efficacy in chronic colitis. *Mucosal Immunol.* 2010 Jan;3(1):49–56.

Varshosaz J, Minaiyan M, Forghanian M. Prolonged hypocalcemic effect by pulmonary delivery of calcitonin loaded poly(methyl vinyl ether maleic acid) bioadhesive nanoparticles. *Med Res Int.* 2014;2014:932615.

Vauthier C, Bouchemal K. Methods for the preparation and manufacture of polymeric nanoparticles. *Pharm Res.* 2009 May;26(5):1025–58.

Veiseh O, Gunn JW, Zhang M. Design and fabrication of magnetic nanoparticles for targeted drug delivery and imaging. *Adv Drug Deliv Rev.* 2010 Mar 8;62(3):284–304.

Vercauteren D, Rejman J, Martens TF, Demeester J, De Smedt SC, Braeckmans K. On the cellular processing of non-viral nanomedicines for nucleic acid delivery: Mechanisms and methods. *J Control Release.* 2012 Jul 20;161(2):566–81.

Wacker M. Nanocarriers for intravenous injection—The long hard road to the market. *Int J Pharm.* 2013 Nov 30;457(1):50–62.

Wang C, Mallela J, Mohapatra S. Pharmacokinetics of polymeric micelles for cancer treatment. *Curr Drug Metab.* 2013 Oct;14(8):900–9.

Wang J, Sui M, Fan W. Nanoparticles for tumor targeted therapies and their pharmacokinetics. *Curr Drug Metab.* 2010;11:129–41.

Wang K, Liu L, Zhang T, Zhu YL, Qiu F, Wu XG, Wang XL, Hu FQ, Huang J. Oxaliplatin-incorporated micelles eliminate both cancer stem-like and bulk cell populations in colorectal cancer. *Int J Nanomed.* 2011;6:3207–18.

Wang N, Wang T, Zhang M, Chen R, Niu R, Deng Y. Mannose derivative and lipid A dually decorated cationic liposomes as an effective cold chain free oral mucosal vaccine adjuvant-delivery system. *Eur J Pharm Biopharm*. 2014 Sep;88(1):194–206.

Wang Y, Xu Z, Guo S, Zhang L, Sharma A, Robertson GP, Huang L. Intravenous delivery of siRNA targeting CD47 effectively inhibits melanoma tumor growth and lung metastasis. *Mol Ther*. 2013 Oct;21(10):1919–29.

Wang YX, Xuan S, Port M, Idee JM. Recent advances in superparamagnetic iron oxide nanoparticles for cellular imaging and targeted therapy research. *Curr Pharm Des*. 2013;19(37):6575–93.

Williams A, Goodfellow R, Topley N, Amos N, Williams B. The suppression of rat collagen-induced arthritis and inhibition of macrophage derived mediator release by liposomal methotrexate formulations. *Inflamm Res*. 2000 Apr;49(4): 155–61.

Williams AS, Topley N, Dojcinov S, Richards PJ, Williams BD. Amelioration of rat anti-gen-induced arthritis by liposomally conjugated methotrexate is accompanied by down-regulation of cytokine mRNA expression. *Rheumatology (Oxford)*. 2001 Apr;40(4):375–83.

Wilson DS, Dalmasso G, Wang L, Sitaraman SV, Merlin D, Murthy N. Orally delivered thioketal nanoparticles loaded with TNF-α-siRNA target inflammation and inhibit gene expression in the intestines. *Nat Mater*. 2010 Nov;9(11):923–8.

Winter PM, Neubauer AM, Caruthers SD, Harris TD, Robertson JD, Williams TA, Schmieder AH et al. Endothelial αvβ3 integrin-targeted fumagillin nanoparticles inhibit angiogenesis in atherosclerosis. *Arterioscler Thromb Vasc Biol*. 2006;26:2103–09.

Wold WS, Toth K. Adenovirus vectors for gene therapy, vaccination and cancer gene therapy. *Curr Gene Ther*. 2013 Dec;13(6):421–33.

Wong JP, Yang H, Blasetti KL, Schnell G, Conley J, Schofield LN. Liposome delivery of ciprofloxacin against intracellular Francisella tularensis infection. *J Control Release*. 2003 Oct 30;92(3):265–73.

Wu J, Nantz MH, Zern MA. Targeting hepatocytes for drug and gene delivery: Emerging novel approaches and applications. *Front Biosci*. 2002 Mar 1;7:d717–25.

Wu N, Ataai MM. Production of viral vectors for gene therapy applications. *Curr Opin Biotechnol*. 2000 Apr;11(2):205–8.

Wu YL, Putcha N, Ng KW, Leong DT, Lim CT, Loo SC, Chen X. Biophysical responses upon the interaction of nanomaterials with cellular interfaces. *Acc Chem Res*. 2013 Mar 19;46(3):782–91.

Xin H, Chen L, Gu J, Ren X, Wei Z, Luo J, Chen Y, Jiang X, Sha X, Fang X. Enhanced anti-glioblastoma efficacy by PTX-loaded PEGylated poly(ε-caprolactone) nanoparticles: In vitro and in vivo evaluation. *Int J Pharm*. 2010 Dec 15;402 (1–2):238–47.

Xin H, Sha X, Jiang X, Zhang W, Chen L, Fang X. Anti-glioblastoma efficacy and safety of paclitaxel-loading Angiopep-conjugated dual targeting PEG-PCL nanoparticles. *Biomaterials*. 2012 Nov;33(32):8167–76.

Xiong XB, Falamarzian A, Garg SM, Lavasanifar A. Engineering of amphiphilic block copolymers for polymeric micellar drug and gene delivery. *J Control Release*. 2011 Oct 30;155(2):248–61.

Yamamoto H, Kuno Y, Sugimoto S, Takeuchi H, Kawashima Y. Surface-modified PLGA nanosphere with chitosan improved pulmonary delivery of calcitonin by mucoadhesion and opening of the intercellular tight junctions. *J Control Release*. 2005 Feb 2;102(2):373–81.

Yang J, Lee C-H, Ko H-J, Suh J-S, Yoon H-G, Lee K, Huh Y-M, Haam S. Multifunctional magneto-polymeric nanohybrids for targeted detection and synergistic therapeutic effects on breast cancer. *Angew Chem Int Ed*. 2007;46:8836–9.

Yao W, Sun K, Mu H, Liang N, Liu Y, Yao C, Liang R, Wang A. Preparation and characterization of puerarin–dendrimer complexes as an ocular drug delivery system. *Drug Dev Ind Pharm*. 2010 Sep;36(9):1027–35.

Ye Q, Asherman J, Stevenson M, Brownson E, Katre NV. DepoFoam technology: A vehicle for controlled delivery of protein and peptide drugs. *J Control Release*. 2000 Feb 14;64(1–3):155–66.

Yi X, Shi X, Gao H. Cellular uptake of elastic nanoparticles. *Phys Rev Lett* 2011; 107:98101.

Yin H, Kanasty RL, Eltoukhy AA, Vegas AJ, Dorkin JR, Anderson DG. Non-viral vectors for gene-based therapy. *Nat Rev Genet*. 2014 Aug;15(8):541–55.

Yin L, Ding J, He C, Cui L, Tang C, Yin C. Drug permeability and mucoadhesion properties of thiolated trimethyl chitosan nanoparticles in oral insulin delivery. *Biomaterials*. 2009 Oct;30(29):5691–700.

Yin Q, Shen J, Chen L, Zhang Z, Gu W, Li Y. Overcoming multidrug resistance by co-delivery of Mdr-1 and survivin-targeting RNA with reduction-responsive cationic poly(β-amino esters). *Biomaterials*. 2012 Sep;33(27):6495–506.

Zarbin MA, Montemagno C, Leary JF, Ritch R. Regenerative nanomedicine and the treatment of degenerative retinal diseases. *Wiley Interdiscip Rev Nanomed Nanobiotechnol*. 2012 Jan–Feb;4(1):113–37.

Zhang C, Zhao L, Dong Y, Zhang X, Lin J, Chen Z. Folate-mediated poly(3-hydroxybutyrate-co-3-hydroxyoctanoate) nanoparticles for targeting drug delivery. *Eur J Pharm Biopharm*. 2010 Sep;76(1):10–6.

Zhang G, Guo B, Wu H, Tang T, Zhang BT, Zheng L, He Y et al. A delivery system targeting bone formation surfaces to facilitate RNAi-based anabolic therapy. *Nat Med*. 2012 Jan 29;18(2):307–14.

Zhang H, Hou L, Jiao X, Ji Y, Zhu X, Zhang Z. Transferrin-mediated fullerenes nanoparticles as Fe(2+)-dependent drug vehicles for synergistic anti-tumor efficacy. *Biomaterials*. 2015 Jan;37:353–66.

Zhang J, Wu L, Chan HK, Watanabe W. Formation, characterization, and fate of inhaled drug nanoparticles. *Adv Drug Deliv Rev*. 2011 May 30;63(6):441–55.

Zhang L, Chan JM, Gu FX, Rhee J-W, Wang AZ, Radovic-Moreno AF, Alexis F, Langer R, Farokhzad OC. Self-assembled lipid–polymer hybrid nanoparticles: A robust drug delivery platform. *ACS Nano*. 2008;2:1696–702.

Zhang N, Ping QN, Huang GH, Xu WF. Investigation of lectin-modified insulin liposomes as carriers for oral administration. *Int J Pharm*. 2005 Apr 27;294(1–2):247–59.

Zhang W, Yang H, Kong X, Mohapatra S, San Juan-Vergara H, Hellermann G, Behera S, Singam R, Lockey RF, Mohapatra SS. Inhibition of respiratory syncytial virus infection with intranasal siRNA nanoparticles targeting the viral NS1 gene. *Nat Med*. 2005 Jan;11(1):56–62.

Zhang Y, Pardridge WM. Blood-brain barrier targeting of BDNF improves motor function in rats with middle cerebral artery occlusion. *Brain Res*. 2006 Sep 21;1111(1):227–9.

Zhang Z, Tongchusak S, Mizukami Y, Kang YJ, Ioji T, Touma M, Reinhold B, Keskin DB, Reinherz EL, Sasada T. Induction of anti-tumor cytotoxic T cell responses through PLGA-nanoparticle mediated antigen delivery. *Biomaterials*. 2011 May;32(14):3666–78.

Zhou X, Wong LL, Karakoti AS, Seal S, McGinnis JF. Nanoceria inhibit the development and promote the regression of pathologic retinal neovascularization in the Vldlr knockout mouse. *PLoS One*. 2011 Feb 22;6(2):e16733.

Ziv-Polat O, Topaz M, Brosh T, Margel S. Enhancement of incisional wound healing by thrombin conjugated iron oxide nanoparticles. *Biomaterials*. 2010 Feb;31(4):741–7.

Index

Page numbers followed by f and t indicate figures and tables, respectively.

A

Abraxane, 17, 79, 164, 180
Acetone, 52, 53
Ad, 143
    PHB gene–carrying, 206
    thymidylate synthase shRNA-
        expressing, 177
Adagen, 219
Adeno-associated virus (AAV), 30–31,
        127–128, 128f, 143, 202–203; *see also*
        Viral vectors
Adenoviral vectors, 30, 127; *see also* Viral
        vectors
ADEPT, *see* Antibody-directed enzyme
        prodrug therapy
Adjuvants, 151, 153
    nanoparticles as, 154–156
Adrenoleukodystrophy (ALD) protein, 148
Adrenomedullin, 149
Aerosol administration, 124
Aerosol solvent extraction solvent (ASES), 49
Aggregation, 71
Airway hyperresponsiveness (AHR), 208
Albumin, 6, 17, 69, 138
    fusion, 133–134
    preventing opsonization of
        nanoparticles, 79
    transcellular trafficking of, 85
AlClPc (Aluminum chloride
        phthalocyanine), 161
Alipogene tiparvovec, 128
Allergic asthma
    nanomedicine for, 208–209; *see also*
        Obstructive respiratory diseases
    theophylline, 209
All-*trans* retinoic acid (ATRA), 104

Alpha-1 antitrypsin, 123
Aluminum-based nanoparticles (AlNPs),
    154
Aluminum chloride phthalocyanine
    (AlClPc), 161
Aluminum salts (alum), 151
Alveolar macrophages, 124
Alzheimer's disease (AD), 149–150
5-Aminolevulinic acid (5-ALA), 163–164,
    205
Amoxicillin, 216
Amphiphilic block copolymer (ABC)-drug
        interactions; *see also* Interactions/
        orientation of therapeutic drugs
    about, 62
    drug loading into polymeric micelles,
        63
    electrostatic complexation with DNA/
        siRNA, 63–64
    polymeric micellar drug conjugate, 63,
        64f
Amphiphilic phospholipids, 31
Angiogenesis, deregulated, 86
Angiopep-2, 172
Angiotensinogen (AGT), 149
Anionic liposomes, 69; *see also* Liposomes
Anisamide, 179
Anisamide–PEG–LPD nanoparticles, 179
Anthracyclines, 173
Antiangiogenic therapy and MRI, 170
Antiapoptotic proteins, 177
Antibody-dependent cell-mediated
        cytotoxicity (ADCC), 157–158
Antibody-directed enzyme prodrug
        therapy (ADEPT), 18, 18f
Antibody–enzyme conjugates (AEC), 18

315

_316_   _Index_

Antibody Fc portion, conjugation to, 133
Anticancer drugs, water-insoluble, 138
Anticancer vaccine, 155–156
Anticoagulant therapy, 193; _see also_
       Cardiovascular diseases
Antigene molecules, chemical
       modification, 145–147
Antigen-loaded DCs, 159
Anti-inflammatory therapy, 192–193, 203;
       see also Cardiovascular diseases
Antiopsonization effect of PEG, 81, 82f
Antioxidant therapy, 203
Antirestenotic therapy, 193; _see also_
       Cardiovascular diseases
Antisense ODNs, 144
Apatite-based nanoparticles, 35–36
Apolipoprotein, 69
Apolipoprotein A-I (APOA1), 192
Apolipoprotein E (ApoE), 81
Aptamer-mediated targeting, 100, 101f
Arginine–glycine–aspartic acid (RGD), 99
ARM-loaded lipid nanoparticles, 214; _see_
       also Malaria
Artemisinin- and curcumin-loaded
       liposomes, 213; _see also_ Malaria
Arthritis, 189–191
     betamethasone-encapsulated
       PEGylated PLGA nanoparticles,
       191
     CPT–SSM–VIP, 190–191
     gold nanoparticles, 191
     liposomally conjugated MTX (G-MLV),
       189
     PEGylated cyclodextrin–
       methylprednisolone conjugate, 191
     SOD enzymosomes, 190
     VIP–SSM, 190
Artificial augmentation of EPR effect, 95
AS03, 153
Ascentra, 136
Asialofetuin, 99
Atrial natriuretic peptide, 149
Atrigel, 136
Azathioprine, 209

B

Band gap energy, 37
β–artemether-loaded liposomes, 213; _see_
       also Malaria
Basic fibroblast growth factor (bFGF), 202
B-cell epitopes, 151
Betamethasone, 191

Betamethasone-encapsulated PEGylated
       PLGA nanoparticles, 191
Bevacizumab (Avastin), 203–204
Biliary excretion, 8
Bioadhesive niosomal formulation, 118
Bioavailability of dermal drugs, _see_
       Sustained release of dermal
       drugs
Bioavailability of ocular drugs
     overview, 114, 115f
     routes for controlled release of
       ophthalmic drugs
       corneal route, 114–115
       noncorneal route, 115
Bioavailability of ophthalmic drugs,
       enhanced
     about, 115–116
     dendrimers, 118
     liposome, 116–117
     micelles, 116
     nanosuspensions, 117
     niosomes, 118
     polymeric nanoparticles, 117
     solid lipid nanoparticles (SLN), 118
Bioavailability of oral nanoformulations,
       see Controlled release; Drug
       release from nanoparticles,
       strategies for
Bioavailability of pulmonary drugs, _see_
       Controlled release
1,2-bis(oleoyloxy)-3-(trimethylammonio)
       propane (DOTAP), 32
Bispecific antibodies (bAbs), 158
Bivalirudin (Hirulog), 193
Block copolymer, 39
Blood–brain barrier (BBB), 171
     to drug delivery, 101f
Blood capillaries, category of, 84, 85f
Blood circulation time, dysopsonin-
       enhanced, 79–81, 80f
Blood flow, 5
B-lymphocytes, 57
Bone regeneration, 217–218
Boranophosphate internucleotide linkages,
       145
Bottom-up approaches for nonviral
       vectors production, _see_ Nonviral
       vectors production, bottom-up
       approaches
Bradykinin, 88
Brain cancer, nanoparticles for therapeutic
       delivery in animal models of,
       171–173

Index

Bcl2L12-targeting spherical nucleic acids, 173
doxorubicin-loaded surfactant-coated nanoparticles, 172
hTRAIL gene-loaded cationic albumin–conjugated PEGylated PLA nanoparticles, 172
nanoliposomal irinotecan, 173
taxane-loaded poly(ε-caprolactone) nanoparticles, 171–172
Brain capillaries, 102
Brain-derived neurotrophic factor (BDNF), 102
Branched poly(ethyleneimine) (BPEI), 91, 92f
Breast cancer, 173–176
anti-HER2 liposome formulations of DOX and topotecan, 174
DOX-loaded folate-targeted pH-sensitive polymer micelle, 176
ERBB2 on, 158
mdr-1 and surviving siRNA-loaded poly(α-amino esters), 175
rapamycin-loaded nanoparticles of elastin-like polypeptide diblock copolymers, 174–175
sorafenib-loaded nanoliposomal ceramide, 175
Breast cancer resistance protein (BCRP), 6
Buccal routes, barriers to, 2–3

C

C60–5-ALA (5-Aminolevulinic acid (5-ALA)), 163–164
Calcitonin, 112
Calcium phosphate precipitation, 35, 56
Camptothecin (CPT)–sterically stabilized phospholipid micelles (SSMs)–Vasoactive intestinal peptide (VIP), 190–191
Cancer; see also specific entries
gene therapy for, 148–149
nanotheranostics in, see Nanotheranostics in cancer
Cancer immunotherapy, see Immunotherapy
Cancer stem-like cells (CSLCs), 178
Capsaicin, 122
Carbohydrate-mediated targeting, 99
Carbonate apatite nanoparticle, 66
Carbon dioxide, supercritical fluid, 49

Carbon nanotubes (CNT), 38
Carboxyl (C)-amidation, 134
Carcinogens, 156
Cardiotoxicity, 173
Cardiovascular diseases, 192–193
gene therapy for, 149
nanotheranostics in, 170
nanotherapeutics in atherosclerosis, 192–193
anticoagulant therapy, 193
anti-inflammatory therapy, 192–193
antirestenotic therapy, 193
modulating HDL levels, 192
Cationic charge neutralization, 69
Cationic lipids, 32–33
Cationic liposomes, 130; see also Liposomes
Cationic polymers, 91
CD19, 158
CD30, 158
CD44, 183
CD47, 83
CD47–SIRP interaction, 83
Cell components and nanoparticles, interactions between, 76
Cell surface receptors/facilitated uptake, targeting of
aptamer-mediated targeting, 100, 101f
carbohydrate-mediated targeting, 99
folate-mediated targeting, 102, 104
monoclonal antibody-mediated targeting, 97–99, 98f
peptide-mediated targeting, 99–100
transferrin receptor (TfR)-mediated targeting, 100, 102, 103f
Cellular uptake
by active transport, 6
of nanoparticles, physicochemical properties on
surface charge/hydrophilicity/ligand attachment, 23, 23f
by passive diffusion, 6
Ce6 nanolipoplexes, 161
Cerium oxide nanoparticles, 203
Cervarix, 155
C60–HA (Hyaluronic acid), 163
Charge and hydrophilicity, 81–82
Chemical enhancers, 4
Chemical modification
with functional groups, 7
of proteins/peptides, see Synthetic peptides
Chemical reduction, 55

318   *Index*

Chemical synthesis/engineering
   drug nanoparticle production, top-
      down approaches
      high-pressure homogenization
        (HPH), 43–44
      wet milling, 43–44
   nonviral vectors production, bottom-up
      approaches
      assembly of polymeric micelles, 54
      dendrimers, 50
      inorganic nanoparticles, synthesis
        of, 55–56
      lipid–polymer hybrid nanoparticles,
        54–55
      liposome, 44–46
      nanocapsules/nanospheres, 50–54
      nanostructured lipid carriers (NLC),
        50
      solid lipid nanoparticles, 46–50
Chemo/photothermal therapy and MRI,
      166–167
Chemotherapy
   with genes or siRNAs promoting
      apoptosis, 140–141
   and optical imaging, 167–168
   with siRNAs, overcoming multidrug
      resistance, 140
   and ultrasonograpy, 169–170
Chemotherapy and MRI
   liposomes, 166
   multifunctional magneto-polymeric
      nanohybrids (MMPNs), 166
   polymeric micelles, 165–166
Chemotherapy drugs, 98, 106
   resistance to, 140
Chitosan-modified AuNPs, 154
Chloroquine-encapsulated
      immunoliposomes, 213–214; *see
      also* Malaria
Choroidal neovascularization (CNV), 116
Chronic inflammatory skin diseases, 119
Cisplatin (CDDP), 180, 181
   hybrid micelles carrying, 182–183
Clarithromycin, 216
Clearance via alveolar macrophages or
      dendritic cells, 124
Cold high-pressure homogenization, 46
Collagen, 87
Collagen-induced arthritis (CIA), 190; *see
      also* Arthritis
Colorectal cancer (CRC), nanoparticles for
      therapeutic delivery in animal
      models of, 176–178

   endostar-loaded PEG–PLGA
      nanoparticles, 177
   oxaliplatin-encapsulated chitosan
      micelle, 178
   PEG-coated Bcl-2 siRNA lipoplex, 177
   taxol-containing liposomes, 176–177
   thymidylate synthase shRNA-
      expressing Ad, 177
Complement-dependent cytotoxicity
      (CDC), 158–159
Constitutive mammalian promoters, 129
Controlled delivery of proteins from
      injectable formulations, *see*
      Synthetic peptides
Controlled release
   of drug, pharmaceutical techniques
      for; *see also* Nanotechnology
      approaches
      about, 109, 111
      drug retention/uptake at intestinal
        epithelium, promotion of, 112
      drug stability against digestive
        enzymes/bile salts, enhancement
        of, 111
   of ophthalmic drugs, *see* Bioavailability
      of ophthalmic drugs, enhanced
   of pulmonary drugs
      liposomes and lipid–core micelles,
        124–125
      polymeric nanoparticles, 125–126
      solid lipid nanoparticles (SLN), 125
   of pulmonary drugs from
      nanoformulations
      clearance via alveolar macrophages
        or dendritic cells, 124
      mucociliary escalator, 123–124
      transport to blood or lymphatic
        system, 124
Convection method, 45
Conventional small drugs, *see* Intracellular
      transport vehicles
Convergent method, dendrimer synthesis,
      50, 51f
Corneal route, 114–115
Corona formation, 67, 76
Corticosteroids, 205
C60–PEG, 163, 167
Cremophor EL (CRM) emulsions, 162
Critical micelle concentration (CMC), 45
Crohn's disease, 205
Crosslinked dextran-coated iron oxide
      (CLIO) nanoparticles, 170
CTL epitopes, 151

Index

C-type lectin receptors, 153–154
Curcumin-entrapped chitosan
    nanoparticles, 213; see also
    Malaria
Curcumin nanoparticles, 212; see also
    Malaria
Curvature of nanoparticles, 68f, 69–70
Cyclin D1 (CyD1), 206–207
Cyclodextrin polymer (CDP)–siRNA
    formulation, 131
Cyclopentenone prostaglandins, 192–193
Cyclosporine, 209
Cystic fibrosis transmembrane conductance
    regulator (CFTR), 127
Cystic fibrosis treatment, 123
Cytarabine (DepoCyt), 137
Cytokine interleukin-2 (IL-2), 97
Cytokines, 156–157
Cytomegalovirus (CMV) retinitis, 114, 117
Cytotoxic agents, 13
Cytotoxic T cells, 155
Cytotoxic T lymphocyte (CTL), 99

D

DACHPt, 184
DC-SIGN, 154
Degenerative retinal diseases, 202–204
    anti-inflammatory therapy, 203
    antioxidant therapy, 203
    inhibiting choroidal
        neovascularization, 203–204
    neurotrophic factor therapy, 202–203
    retinal gene therapy, 204
Dendrimer–drug interactions, 61–62, 62f;
        see also Interactions/orientation
        of therapeutic drugs
Dendrimer phthalocyanine (DPc), 162
Dendrimers, 34, 50, 118
Dendritic cells (DCs), 124, 153
Dendritic photosensitizer, 116
Dendrons, 50
Dephosphorylation of Akt, 175
DepoFoam technology, 136
Dermal drugs, sustained release of, see
        Sustained release of dermal
        drugs
Dermal patch, 119f
Desmopressin, 135
Detergent removal, 45
Diabetes mellitus (DM), 193–199
    nasal insulin delivery using
        nanotechnologies, 197–198

oral insulin delivery using
    nanotechnologies, 194–196
pulmonary insulin delivery, 198
subcutaneous insulin delivery, 199
transdermal insulin delivery, 199
Diflucortolone valerate (DFV), 122
Dimethylnitrosamine (DMN), 211
Dioctadecylamidoglycylspermine (DOGS),
    33
Diphtheria vaccines, 154
Discontinuous capillary, 84
Disulfide-rich peptides, 135
Divergent method, dendrimer synthesis,
    50, 51f
Diversity of bioactive nanoparticles
    genetically engineered drug
        nanocarriers, 39–40
    hybrid particles
        inorganic hybrid nanoparticles, 39
        lipid–polymer hybrid nanoparticles
            (LPN), 38–39
        organic–inorganic hybrid
            nanoparticles, 39
    nonviral vectors
        inorganic carriers, 35–38, 36f
        lipid-based, 31–33, 32f
        polymer-based, 33–35
    surface functionalization and ligand
        attachment, 40–41, 40f
    viral vectors
        DNA virus vectors, 30–31
        retrovirus vector, 29–30
DLD-1 xenograft mouse model, 177
DNA vaccines, 151–152
DNA virus vectors, see Viral vectors
Docetaxel (DTX), 172
DOTAP [1,2-bis(oleoyloxy)-3-
        (trimethylammonio)propane],
        130, 179
DOTAP/cholesterol–tumor suppressor
        plasmid complexes, 179–180
Double emulsion method, 48, 49f
DOX, see Doxorubicin (DOX)
DOX-carrying lipid-based
        nanoformulations, 182
DOX-encapsulated PEG–PE micelle, 181
Doxil (doxorubicin-loaded nanoparticles),
        137, 164
DOX-loaded folate-targeted pH-sensitive
        polymer micelle, 176
Doxorubicin (DOX), 137, 166, 173
Drug delivery via lymphatic system,
        95–97

Drug entrapment into polymeric micelles, 63
Drug metabolites, 8
Drug nanocarriers, genetically engineered, 39–40
Drug nanosuspension, 11f
Drug release from nanoparticles, strategies for
    pH- and enzyme-dependent strategy, 113–114
    time-dependent strategy, 112–113
Drug retention/uptake at intestinal epithelium, promotion of, 112
Drugs and nucleic acids, combinations of, *see* Intracellular transport vehicles
Drugs from body, elimination of; *see also* Small-molecule drugs in blood, fates of
    biliary excretion, 8
    chemical modification with functional groups, 7
    renal excretion, 7–8
Drug solubility/stability, enhancement of, 9–10, 11f
Drug stability against digestive enzymes/bile salts, enhancement of, 111
Drug transport; *see also* Small-molecule drugs in blood, fates of
    from blood capillaries to extracellular fluid
        blood flow, influence of, 5
        plasma level/molecular size of drug, influences of, 5
        protein binding, influence of, 6
    from extracellular fluid to cells
        cellular uptake by active transport, 6
        cellular uptake by passive diffusion, 6
        uptake by nonresponsive cells, 7
        uptake by tumor cells, 6–7
Dry eye symptoms treatment, 117
Dry milling, 43
DSGLA, 179
Dysopsonin-enhanced blood circulation time, 79–81, 80f
Dysopsonins, 79

E

EDC [1-ethyl-3-(3-dimethylaminopropyl) carbodiimide], 41
Efflux transporters, 6

Elasticity of nanoparticles, 79
Electroporation method, 120
Electrostatic complexation with DNA/siRNA, 63–64
Emergence of nanotherapeutics
    DNA/RNA-based macromolecular drugs, 17
    macromolecules for image-guided drug delivery, 19–20, 20f
    macromolecules for prodrug therapy
        about, 17, 18f
        ADEPT, 18
        GDEPT, 18–19
    macromolecules for vaccine delivery, 19
    nanoparticles for photodynamic therapy (PDT), 19
    protein-based macromolecular drugs
        chemically modified/carrier-bound proteins, 15
        proteins as drug carriers, 17
        proteins as independent therapeutic drugs, 15
    small-molecule drugs, administration of
        intramuscular/subcutaneous routes, barriers to, 4
        intranasal route, barriers to, 3
        oral route, barriers to, 1–2, 2f
        pulmonary route, barriers to, 3–4
        rectal route, barriers to, 3
        sublingual/buccal routes, barriers to, 2–3
        transdermal route, barriers to, 4
    small-molecule drugs formulations, problems with, 8
    small-molecule drugs in blood, fates of
        drug transport from blood capillaries to extracellular fluid, 5–6
        drug transport from extracellular fluid to cells, 6–7
        elimination of drugs from body, 7–8
        overview, 9f
        plasma drug concentration on therapeutic action intensity, 4–5
    small-molecule drugs with macromolecules, alteration of pharmacokinetics of
        active targeting through receptor-mediated delivery, 13, 14f

Index

enhancement of drug solubility and stability, 9–10, 11f
fast drug release, facilitating, 12
magnetic targeting, 13, 15, 16f
objectives, 10f
passive targeting by exploiting leaky vasculature architecture, 13, 14f
prolongation of retention time in blood, 10, 12
sustained release, enabling, 12–13
Emulsion droplets, gelation of, 52–53
Endocytosis, receptor-mediated, 13
Endogenous gene expression, silencing, 144–147
chemical modifications of antigene molecules, 145–147
Endogenous membrane coating, 82
Endogenous miRNA functions, inhibition of, 147
Endosomal buffering, 105
Endosomal escape; see also Nanoparticles in drug transport
about, 104–105
endosomal buffering, 105
fusogenic lipids or peptides, 105
of nucleic acid, 103f, 104f
Endostar, 177
Endothelial escape, passive targeting to facilitate, 95, 96f
Endothelial nitric oxide synthase, 149
Engerix b (GlaxoSmithKline biologicals), 152
Enhanced permeability and retention (EPR) effect, 13, 86, 87f, 95
Enzyme-dependent strategy, for drug release from nanoparticles, 113–114
EphA2, liposomal
siRNA-encapsulated neutral, 181
siRNA-loaded silica particles, 181–182
Epidermal growth factor receptor (EGFR), 178–179
EPR effect, see Enhanced permeability and retention (EPR) effect
Epstein–Barr virus, 156
ERBB2, 158
Excretory organs, 7
Exopeptidase-mediated proteolysis, 134
Extended recombinant polypeptide (XTEN), 132
Extracellular matrix (ECM), 88
Extracellular matrix (ECM)-nanoparticle interactions, 75, 76f

Extracellular transport vehicles, see Intracellular transport vehicles
Extravasation from blood through vascular endothelium; see also Pharmacokinetics/ biodistribution of nanoparticles
deregulated vascular endothelium, 86, 87f
permeability of vascular endothelia, 84, 85f
routes of traffic across continuous endothelium, 85, 86f
vascular endothelium as target for drug delivery, 87
Extrusion, 45

F

Fabrication strategies
chemical synthesis/engineering
drug nanoparticle production, top-down approaches, 43–44
nonviral vectors production, bottom-up approaches, 44–56
recombinant DNA/hybridoma/phage display techniques
monoclonal antibodies, generation of, 56–58
protein-based nanoparticles, synthesis of, 56–57
viral vectors, production of, 58–59
Fast drug release, facilitating, 12
Fenestrated capillary, 84
Ferritin, 153
Ferritin–hemagglutinin (HA) fusion proteins, 153
Fibrinogen, 69
First-pass effect, 1
Flagella, 153
Flip-flop mechanism, 105
Flourochromes, 117
5-Fluorouracil (5-FU), 176
Folate-mediated targeting, 102, 104
Folate receptors (FR), 102
Folic acid, 176
Fomivirsen, 145
Food and Drug Administration–approved drugs, 43
Foslip (liposomal mTHPC), 161
Fullerene-based nanomaterials, 163–164; see also Nanoscale drug delivery systems in PDT

C60–5-ALA, 163–164
C60–HA, 163
C60–PEG, 163
Fumagillin, 170
Fusogenic lipids or peptides, 105

G

Gancyclovir, 148–149
Gardasil, 155
Gas antisolvent/supercritical antisolvent (GAS/SAS), 49
Gastrointestinal (GI) tract, 1
Gd-DOTA monoamide, 166
GDEPT, *see* Gene-directed enzyme prodrug therapy (GDEPT)
GDNF gene-carrying AAV, 202–203
Gelation of emulsion droplets, 52–53
Gemcitabine (Gem), 183–184
Gemtuzumab, 98
Gene-directed enzyme prodrug therapy (GDEPT), 18–19, 18f
Gene silencing tools, 140
Gene therapy, 143–151
    applications, 147–151
    inhibition of endogenous miRNA functions, 147
    knockdown of endogenous gene expression, 144–147
    transgene expression, 143
Gene therapy applications, 147–151
    gene therapy for cancer, 148–149
    gene therapy for cardiovascular diseases, 149
    gene therapy for HIV, 150
    gene therapy for neurodegenerative diseases, 149–150
    gene therapy for viral hepatitis, 150–151
    gene therapy of hereditary diseases, 147
Genetic abnormalities, 156
Glioblastoma multiforme (GBM), 171
Glomerular filtration reduction, 12; *see also* Retention time in blood, prolongation of
Glomerulus, 7
Glucocorticoids, 192
Glybera, 128
Glycosaminoglycan (GAG), 75, 76
GNR–photosensitizer complex, 169
Gold nanoparticles (AuNPs), 36–37, 36f, 154, 191; *see also* Nanoparticles
Gold nanorod (GNR), 167

Gold–sulfur interaction, 36
Granzyme, 158
Growth hormone–releasing hormone (GHRH), 135
Gut-associated lymphatic vessels, 96f

H

Hagedorn insulin, 194
Helicobacter pylori, 156, 216
Hemagglutinin (HA), 105, 152
Hematopoietic stem cells (HSCs)
    transferring the *ABCD1* gene in, 148
Hemophilia B, 148
Hepatic fibrosis and infections, 209–212
    hepatic fibrosis, 210–211
    hepatic infections, 211–212
Hepatitis B vaccines, 155
Hepatitis B virus (HBV), 152, 155, 156
Hepatitis C virus (HCV), 150–151
Herpes simplex virus 1, 143
Herpes simplex virus (HSV), 31; *see also* Viral vectors
Hexadecafluoro zinc phthalocyanine (ZnPcF16), 162
High-density lipoprotein (HDL), 192
High-pressure homogenization (HPH), 43–44, 46; *see also* Top-down approaches, for drug nanoparticle production
Hodgkin's lymphoma
    CD30 on, 158
Hot high-pressure homogenization, 46
HT-29 colon cancer model, 177
Human epidermal growth factor receptor 2 (HER2), 173–174
Human immunodeficiency virus (HIV), 96, 143
    gene therapy for, 150
Human papillomavirus (HPV), 156
    vaccines, 155
Huntington's disease (HD), 149–150
Hyaluronic acid (HA), 163
Hybrid micelles carrying cisplatin and PTX, 182–183
Hybridoma technology for monoclonal antibodies production, 57, 57f; *see also* Recombinant DNA technology
Hydrocortisone (HC), 122
Hydrogels, 34–35
Hydrophilic coating of nanoparticle, 76
Hydrophilicity, charge and, 81–82

Index

Hydroxyapatite, 105
Hydroxytyrosol (HT), 122
Hypercholesterolemia, 149
    treatment, 130
Hypersensitivity, nanocarrier removal by,
    70–71

I

IFN-α/β, 156
IFN-γ, 157, 158
IL-18, 157
IL-2 therapy, 156–157
Image-guided drug delivery,
    macromolecules for, 19–20, 20f
Image-guided therapy, 164–170
    nanotheranostics in cancer, 165–170
        combined MR/optical imaging and
            therapy, 169
        MRI and chemo/photothermal
            therapy, 166–167
        MRI and chemotherapy, 165–166
        MRI and PDT, 167
        optical imaging and chemotherapy,
            167–168
        optical imaging and PDT, 168
        optical imaging and PTT, 168
        optical imaging and PTT/PDT, 169
        ultrasonograpy and chemotherapy,
            169–170
    nanotheranostics in cardiovascular
        diseases, 170
        MRI and antiangiogenic therapy, 170
        optical imaging and PDT, 170
    personalized medicine, 164
Immune activation
    non-specific, 156–157
    tumor-specific, 157–159
Immune surveillance, 156
Immunosuppressive agents, 205
Immunotherapy, cancer, 156–159
    nonspecific immune activation,
        156–157
        cytokines, 156–157
        interferons, 157
        nanoparticles in nonspecific
            immune activation, 157
        TLR agonist, 157
    tumor-specific immune activation,
        157–159
        antibody-dependent cell-mediated
            cytotoxicity, 157–158
        antigen-loaded DCs, 159

complement-dependent cytotoxicity,
    158–159
monoclonal Ab or siRNA, 159
Inflammatory bowel diseases (IBDs),
    205–207
    β₇ integrin-targeted, CyD1 siRNA-
        loaded liposomes, 206–207
    IL-10 gene-encapsulated NiMOS, 207
    Map4k4 siRNA-encapsulated glucan
        shells, 206
    mesalamine (5-ASA)-loaded
        nanoparticles, 207
    NF-κB decoy ODN-loaded chitosan–
        PLGA nanoparticles, 206
    PHB gene–carrying Ad and PHB-
        entrapped PLA nanoparticles,
        206
    TNF-α–neutralizing nanobodies,
        205–206
    TNF-α siRNA-encapsulated polymeric
        nanoparticles, 205
Inflexal (Crucell), 152
Infliximab, 205
Influenza-derived hemagglutinin (HA),
    155
Influenza virus, 152
Inhibiting choroidal neovascularization,
    203–204
Inhibition of endogenous miRNA
    functions, 147
Injectable in situ gel forming solution, 136
Inorganic carriers, see Nonviral vectors
Inorganic nanoparticle-drug interactions,
    65–66, 66f
Inorganic nanoparticles, 89, 154; see also
    Nanoparticles
Inorganic nanoparticles, synthesis of
    chemical reduction, 55
    microemulsions, 56
    nanoprecipitation, 56
    sol-gel process, 55
    spray-drying, 55–56
In situ polymerization, 53
Insulin; see also Diabetes mellitus (DM)
    detemir, 194
    glargine, 194
    Hagedorn, 194
    Lente, 194
    nasal, see Nasal insulin delivery using
        nanotechnologies
    oral, see Oral insulin delivery using
        nanotechnologies
    pulmonary insulin delivery, 198

subcutaneous insulin delivery, 199
transdermal insulin delivery, 199
Insulin therapy, 194
αvβ3-integrin, 170
Interactions/orientation of therapeutic
drugs
amphiphilic block copolymer (ABC)-
drug interactions
about, 62
drug loading into polymeric
micelles, 63
electrostatic complexation with
DNA/siRNA, 63–64
polymeric micellar drug conjugate,
63, 64f
dendrimer–drug interactions, 61–62, 62f
inorganic nanoparticle–drug
interactions, 65–66, 66f
liposome–drug interactions, 64–65
Interfacial polycondensation, 53
Interferons (IFN), 75, 157
nanotheranostics in cancer, 165–170
nanotheranostics in cardiovascular
diseases, 170
personalized medicine, 164
Interstitial fluid, interactions of
nanoparticles with, 74–75, 74f
Interstitial fluid pressure (IFP), 88, 88f, 184
Intracellular transport vehicles
conventional small drugs
liposomes, 137–138
nanoparticle albumin-bound (nab)
technology, 138
polymeric nanoparticles, 138
drugs and nucleic acids, combinations
of
chemotherapy with genes or siRNAs
promoting apoptosis, 140–141
chemotherapy with siRNAs,
overcoming multidrug
resistance, 140
drugs with nucleic acids with
antiangiogenic functions,
combining, 141
overview, 138, 139f
gene/siRNA/ODN
adeno-associated virus (AAV),
127–128, 128f
adenoviral vectors, 127
nonviral vectors, 128–131
retroviral vectors, 126–127
synthetic peptides/recombinant
proteins

chemical modifications of proteins/
peptides, 132–135
controlled delivery of proteins from
injectable formulations, 135–137
protein delivery, current obstacles
to, 132
Intradermal injection, 186
Intramuscular/subcutaneous routes,
barriers to, 4
Intranasal route, barriers to, 3
Intrathecal injection of DepoCyt, 137
Intravitreal injection, 116
Ionic gelation, 54
Irinotecan, 137, 173
Irritable bowel syndrome, European
Medicines Agency for, 135

J

Jet injectors, 152

K

Kallikrein, 149
KB tumors, 168
Killer cell immunoglobulin-like receptor
(KIR), 159
Kupffer cells, 81

L

Lapatinib, 173–174
Leakage of drugs, 71, 73f
Leaky vasculature architecture, passive
targeting by, 13, 14f
Lectins, mannose-binding, 70
Lente insulin, 194
Lentiviral vector, 150
production, 59f
Lentiviruses, 126
Lewis lung carcinoma (LLC), 162
mouse model, 181
Lewy bodies, 150
Ligand coating, 82–83
Linear poly(ethyleneimine) (PEI), 91, 92f
Lipid-based DNA/siRNA vectors, 130; see
also Nonviral vectors
Lipid-based nanocarriers, 98, 194–196
Lipid-based nanoparticles in PDT,
160–161
Ce6 nanolipoplexes, 161
liposomal AlClPc, 161
liposomal mTHPC (foslip), 161

Index

liposomal photofrin, 161
liposome-encapsulated verteporfin, 160–161
Lipid-based nonviral vectors, *see* Nonviral vectors
Lipid-based protein delivery, 136–137
Lipid carriers, nanostructured, 33
Lipid–core micelles, 124–125
Lipidization, 135
Lipid nanoemulsions (LNE), 33
  PTX/ETP-loaded, 186–187
Lipid–polymer hybrid nanoparticles, 122
  one-step method, 55
  two-step method, 54
Lipodox, 182
Lipophilic drugs, 52
Lipoplex stability, determinants of, 92, 94
Lipopolyplexes, 94
Lipoprotein lipase (LPL) deficiency, 127
Liposomally conjugated MTX (G-MLV), 189
Liposomes, 31–33, 32f, 152, 155, 166
  anionic, 69
  cationic, 179
  conventional small drugs, 137–138
  detergent removal, 45
  and enhanced bioavailability of ophthalmic drugs, 116–117
  EphA2 siRNA-encapsulated neutral, 181
  EphA2 siRNA-loaded silica particles, 181–182
  extrusion, 45
  and lipid–core micelles, 124–125
  neural cell adhesion molecule (NCAM)-targeted, 165–166
  reverse-phase evaporation, 46
  sonication, 44–45
  and sustained release of dermal drugs, 120–121
  taxol-containing, 176–177
  thin-film hydration, 44
  via endocytosis, 89
Liposome-coupled TNF-α, 214; *see also* Malaria
Liposome–drug interactions, 64–65, 65f
Liposome–polycation–DNA (LPD), 178–179
Liver hepatocytes, 6
Locked nucleic acid (LNA), 146
Lung cancer, nanoparticles for therapeutic delivery in animal models of, 178–181
  Bcl-x SSO-loaded anisamide–PEG–LPD nanoparticles, 179

DOTAP/cholesterol–tumor suppressor plasmid complexes, 179–180
DOX-encapsulated PEG–PE micelle, 181
EGFR siRNA-containing anisamide–PEG–LPD nanoparticles, 178–179
poly-(γ-l-glutamylglutamine)–PTX nanoparticles, 180
Lupron Depot (leuprolide acetate), 136
Luteinizing hormone–releasing hormone (LHRH), 100
Lymph, interactions of nanoparticles with, 74–75, 74f

M

Macromolecular drugs
  categories of, 20f
  DNA/RNA-based, 17
  protein-based
    chemically modified/carrier-bound proteins, 15
    proteins as drug carriers, 17
    proteins as independent therapeutic drugs, 15
Macromolecules
  for image-guided drug delivery, 19–20, 20f
  for prodrug therapy
    about, 17, 18f
    ADEPT, 18
    GDEPT, 18–19
  for vaccine delivery, 19
Macrophages, nanocarrier removal by, 70–71
Magnetic drug targeting, principle of, 16f
Magnetic nanoparticles, 38
Magnetic targeting, 13, 15, 16f
Malaria, 212–215
  ARM-loaded lipid nanoparticles, 214
  artemisinin- and curcumin-loaded liposomes, 213
  β–artemether-loaded liposomes, 213
  chloroquine-encapsulated immunoliposomes, 213–214
  curcumin-entrapped chitosan nanoparticles, 213
  curcumin nanoparticles, 212
  liposome-coupled TNF-α, 214
  primaquine-loaded nanoemulsions, 215
Mannose receptors, 99
MART-1 melanoma antigen, 187
Melittin-loaded nanoemulsion, 169–170

326   Index

Melting dispersion, 47–48
Mesoporous silica nanoparticles (MSNP), 37, 76
Meth-A fibrosarcoma cells, 163
Methylprednisolone, 125
Methylprednisolone acetate (MPA), 117
Metronidazole, 216
MF59, 153
Micelles, 33, 116, 155–156
    assembly of polymeric, 54
    DOX-encapsulated PEG-PE, 181
    hybrid, carrying cisplatin and PTX, 182–183
    lipid–core, 124–125
    polymeric, 165–166: see also Polymeric micelles
    polymeric, DACHPt-loaded, 184
Microemulsions, 46, 56
Microneedles, 152
Mineralizers, 55
Molecular size of drug, 5
Monoclonal Ab or siRNA, 159
Monoclonal antibodies, generation of
    hybridoma technique, 57, 57f
    phage display technology, 58, 58f
Monoclonal antibody-mediated targeting, 97–99, 98f
Monodisperse gold nanoparticles, 36
Mononuclear phagocytic system (MPS), 34, 70, 71, 72f, 77
Morphine (DepoDur), 137
Morpholino phosphoramidates, 146
MPLA (monophosphoryl lipid A)-plus-alum, 153
MRI
    and antiangiogenic therapy, 170
    and chemo/photothermal therapy, 166–167
    and PDT, 167; see also Photodynamic therapy (PDT)
MRI and chemotherapy, 165–166
    liposomes, 166
    multifunctional magneto-polymeric nanohybrids (MMPNs), 166
    polymeric micelles, 165–166
MR/optical imaging and therapy, combined, 169
MTHPC–Foscan1, 161
MTX, 209
Mucoadhesive PLGA, 125
Mucoadhesive polymer, 112
Mucociliary escalator, 123–124, 123f
Multidrug resistance (MDR), 2, 175

Multidrug resistance-associated protein (MRP), 2, 6
Multifunctional magneto-polymeric nanohybrids (MMPNs), 166
Multifunctional polymeric micelles, 34
Multilamellar vesicle (MLV), 44
Muramyl dipeptide (MDP), 99
Murine mammary carcinoma model, 176
Myocet, 137

N

Nab-paclitaxel, 17
Nanocapsules/nanospheres; see also Fabrication strategies
    about, 50–51
    one-step procedure
        ionic gelation, 54
        nanoprecipitation or solvent displacement, 53
        polyelectrolyte complexes, formation of, 53–54
    two-step procedure
        gelation of emulsion droplets, 52–53
        polymer precipitation by solvent removal, 52
        in situ polymerization, 53
Nanoemulsion method, 39
Nanoliposomal irinotecan, 173
Nanomedicine
    for allergic asthma, 208–209; see also Obstructive respiratory diseases
    for respiratory syncytial virus, 209; see also Obstructive respiratory diseases
Nanomedicine, approved and commercialized, 253–286
    FDA-approved nanotherapeutics, 285f
    disease targets for, 286f
    nanotherapeutics in clinical trials, 254–284
Nanomedicine in clinical trials
    different phases of clinical trials, 219–220
    monoclonal antibodies as therapeutics in selected clinical trials, 220, 244t–252t
    nanoparticulate drug delivery systems in clinical trials, 221t–243t
Nanoparticles
    as adjuvants and vaccine carriers, 154–156
    albumin-bound (nab) technology, 138

Index

in nonspecific immune activation, 157
for photodynamic therapy (PDT), 19
physicochemical characteristics of,
153–154; *see also* Vaccine delivery
Nanoparticles in drug transport
cell surface receptors/facilitated uptake,
targeting
aptamer-mediated targeting, 100, 101f
carbohydrate-mediated targeting, 99
folate-mediated targeting, 102, 104
monoclonal antibody–mediated
targeting, 97–99, 98f
peptide-mediated targeting, 99–100
transferrin receptor (TfR)–mediated
targeting, 100, 102, 103f
drug delivery via lymphatic system,
95–97
endosomal escape
about, 104–105
endosomal buffering, 105
fusogenic lipids or peptides, 105
endothelial escape, passive targeting to
facilitate, 95, 96f
nuclear targeting, 106–107, 107f
nucleic-/protein-based drugs,
protection against degradation
lipoplex stability, determinants of,
92, 94
polyplex stability, determinants of,
91–92, 92f, 93f
Nanoparticles-in-microsphere oral system
(NiMOS), 207
Nanoprecipitation, 53, 56
Nanoscale drug delivery systems in PDT,
160–164
fullerene-based nanomaterials, 163–164
lipid-based carriers, 160–161
polymeric carriers, 162–163
Nanospheres, 51
Nanostructured lipid carriers (NLC), 50
Nanostructured lipid nanoparticles (NLC),
121
Nanosuspensions, 117
Nanotechnology approaches
controlled release/bioavailability of
oral nanoformulations
about, 109, 110f
drug release control, pharmaceutical
techniques for, 109, 111–112
drug release from nanoparticles,
strategies for, 112–114, 113f
intracellular/extracellular transport
vehicles

conventional small drugs, 137–138
drugs and nucleic acids,
combinations of, 138–141, 139f
gene/siRNA/ODN, 126–131
synthetic peptides/recombinant
proteins, 132–137
sustained release/bioavailability of
dermal drugs
about, 119–120, 119f
roles of nanoparticles in, 120–122
sustained release/bioavailability of
ocular drugs
controlled release/enhanced
bioavailability of ophthalmic
drugs, 115–118
routes for controlled release of
ophthalmic drugs, 114–115, 115f
sustained release/bioavailability of
pulmonary drugs
controlled release/bioavailability of
pulmonary drugs, 124–126
controlled release from
nanoformulations, challenges,
123–124, 123f
overview, 122–123
Nanotechnology in wound healing,
215–217
Nanotheranostics, 19
Nanotheranostics in cancer, 165–170; *see
also* Image-guided therapy
combined MR/optical imaging and
therapy, 169
MRI and chemo/photothermal therapy,
166–167
MRI and chemotherapy, 165–166
MRI and PDT, 167
optical imaging and chemotherapy,
167–168
optical imaging and PDT, 168
optical imaging and PTT, 168
optical imaging and PTT/PDT, 169
ultrasonograpy and chemotherapy,
169–170
Nanotheranostics in cardiovascular
diseases, 170; *see also* Image-
guided therapy
MRI and antiangiogenic therapy, 170
optical imaging and PDT, 170
Nanotherapeutics in atherosclerosis, 192;
see also Cardiovascular diseases
Nasal insulin delivery using
nanotechnologies, 197–198; *see
also* Diabetes mellitus (DM)

chitosan-based formulation, 198
liposome-based formulation, 197–198
Natural ligands, 99
N-[1-(2,3-dioleyloxy)propyl]-N,N,N-trimethylammonium chloride (DOTMA), 32
Negatively charged carriers, 74
Neoangiogenesis, 170
Nephron, 7
Nephrotoxicity, 138
Neural cell adhesion molecule (NCAM)–targeted liposomes, 165–166
Neuraminidase, 155
Neurodegenerative diseases, 199–202
  gene therapy for, 149–150
  nanoparticles as neuroprotective and therapeutic drugs for, 200–202
Neuroprotective and therapeutic drugs for neurodegenerative diseases, 200–202
  dendrimers, 201
  lactoferrin-conjugated PEG–PLGA nanoparticles, 201
  liposome-encapsulated hemoglobin (LEH), 202
  neurotensin polyplexes, 201–202
  poly(butyl cyanoacrylate) nanoparticles, 200
  transferrin receptor–targeted PEG–chitosan nanospheres, 201
  VP025 nanoparticles, 200
Neurotrophic factor therapy, 202–203
Nicotine patch, 119
Niosomes, 118
Nitric oxide (NO), 94
Nitrogen permease regulator-like 2 (NPRL2), 179–180
NK cells, tumor recognition by, 159
NKG2D, 156
Noncorneal route, 115
Nonimmunogenic polymers, 51
Nonlipid-based nanocarriers, 98
Nonphagocytic cells, uptake by, 81
Nonresponsive cells, uptake of drug by, 7
Non–small cell lung cancer (NSCLC), 178; see also Lung cancer
Nonsteroidal anti-inflammatory drugs (NSAID), 12
Nonviral vectors; see also Diversity of bioactive nanoparticles; Intracellular transport vehicles
  about, 128–129
  inorganic carriers

apatite-based nanoparticles, 35–36
carbon nanotubes (CNT), 38
gold (Au) nanoparticles, 36–37, 36f
magnetic nanoparticles, 38
quantum dots (QD), 37–38
silica nanoparticles, 37
lipid-based
  lipid nanoemulsions (LNE), 33
  liposome, 31–33, 32f
  nanostructured lipid carriers, 33
  solid lipid nanoparticles (SLN), 33
lipid-based DNA/siRNA vectors, 130
long-term/tissue-specific expression of, strategies for, 129–130
polymer-based
  dendrimers, 34
  from dispersion of preformed polymers, 35
  hydrogels, 34–35
  polymeric micelles, 33–34
polymeric DNA/siRNA vectors, 130–131, 131f
Nonviral vectors production, bottom-up approaches
assembly of polymeric micelles, 54
dendrimers, 50
inorganic nanoparticles, synthesis of
  chemical reduction, 55
  microemulsions, 56
  nanoprecipitation, 56
  sol–gel process, 55
  spray-drying, 55–56
lipid-polymer hybrid nanoparticles
  one-step method, 55
  two-step method, 54
liposome
  detergent removal, 45
  extrusion, 45
  reverse-phase evaporation, 46
  sonication, 44–45
  thin-film hydration, 44
nanocapsules/nanospheres
  about, 50–51
  one-step procedure, 53–54
  two-step procedure, 51–53
nanostructured lipid carriers (NLC), 50
solid lipid nanoparticles
  double emulsion, 48
  high-pressure homogenization (HPH), 46
  melting dispersion, 47–48
  microemulsion, 46
  solvent emulsification–diffusion, 47

Index

solvent emulsification–evaporation, 46–47
solvent injection, 48
supercritical fluid technology, 49
Notch signaling pathway, 140
N3′ phosphoramidate linkages, 145
NS5A, 150–151
Nuclear localization sequences (NLS), 106
Nuclear targeting, 106–107, 107f
Nucleic acids with antiangiogenic functions, 141
Nucleic-based drugs, protection against degradation
lipoplex stability, determinants of, 92, 94
polyplex stability, determinants of, 91–92, 92f, 93f
Nucleobase modification, 146–147

O

Obstructive respiratory diseases, 208–209
nanomedicine for allergic asthma, 208–209
nanomedicine for respiratory syncytial virus, 209
Octreotide, 134, 135
Ocular routes of delivery and barriers, 115f
Ofatumumab, 158–159
Oil-in-water emulsions, 153
Oligodeoxyribonucleotides (ODN)
intracellular delivery of, see Intracellular transport vehicles
serum stability of, enhanced, 17
Omeprazole, 216
One-microemulsion method, 56
One-step method
lipid-polymer hybrid nanoparticles, 55
nanocapsules and nanospheres, see Nanocapsules/nanospheres
Ophthalmic drugs
enhanced bioavailability of, see Bioavailability of ophthalmic drugs, enhanced
routes for controlled release of
corneal route, 114–115
noncorneal route, 115
water-soluble, 117
Opsonin-facilitated phagocytosis, 79, 80f
Opsonins, 79
Optical imaging
and chemotherapy, 167–168

and PDT, 168, 170; see also Photodynamic therapy (PDT)
and PTT, 168
and PTT/PDT, 169
Oral insulin delivery using nanotechnologies, 194–196
lipid-based nanocarriers, 194–196
polymeric nanoparticles, 196–197
Oral route, barriers to, 1–2, 2f
Oral vaccine formulations, 152
Organic anion transporter (OAT), 6
Organic cation transporter (OCT), 6
Organic ligands, 38
Orthotopic human pancreatic (BxPC3) xenograft model, 165–166
Ovalbumin (OVA)-induced airway inflammation, 208–209
Ovarian cancer, 181–183
EphA2 siRNA-encapsulated neutral liposomes, 181
hybrid micelles carrying cisplatin and PTX, 182–183
LHRH peptide- and PTX-conjugated dendrimer/CD44 siRNA, 183
liposomal EphA2 siRNA-loaded silica particles, 181–182
multifunctional DOX-carrying lipid-based nanoformulations, 182
Oxaliplatin (OXA), 178

P

Paclitaxel (PTX), 79, 167, 168, 171–172
hybrid micelles carrying, 182–183
hybrid micelles carrying cisplatin and, 182–183
PTX A, 173
PTX/ETP-loaded lipid nanoemulsions (LNE), 186–187
PTX/ETP-loaded lipid nanoemulsions in skin cancer, 185–186
Palmitic acid (PA), 111
PAMAM dendrimer, 125
Pancreatic cancer, 183–185
DACHPt-loaded polymeric micelle, 184
HER-2 siRNA-loaded immunoliposome, 185
PEGylated human recombinant hyaluronidase PH20, 184
uPAR-targeted, gemcitabine-loaded iron oxide nanoparticles, 183–184
Parkinson's disease (PD), 149–150

330   *Index*

Particles from gas saturated solution (PGSS), 49
Particle size in biodistribution of nanotherapeutics, 77–79, 78f
Passive diffusion, cellular uptake by, 6
Passive targeting by leaky vasculature architecture, 13, 14f
PAX, 181, 182
PDT, *see* Photodynamic therapy (PDT)
PEG, *see* Polyethylene glycol (PEG)
PEGylated cyclodextrin–methylprednisolone conjugate, 191
PEGylated human recombinant hyaluronidase PH20 (PEGPH20), 184
Pegylated IFN (IFN-α), 150
PEGylation, 132
Peptide-mediated targeting, 99–100
Peptide nucleic acid (PNA), 107, 146
Peptide terminus, modification of, 134–135
Perfluorooctyl bromide (PFOB), 169–170
Perforin, 156, 158
Permeability of vascular endothelia, 84, 85f
Peroxisomal membrane protein, 148
Personalized medicine, 164; *see also* Image-guided therapy
Pertuzumab, 173–174
P-glycoprotein (P-gp), 2, 175
Phage display technology for monoclonal antibodies production, 58, 58f; *see also* Recombinant DNA technology
Phagocytosis
of drug carriers, 12
opsonin-facilitated, 79, 80f
Pharmacokinetics/biodistribution of nanoparticles
cellular uptake/metabolism/excretion, 89
charge and hydrophilicity, influence of, 81–82
coating of CD47 as self marker, 83–84, 83f
endogenous membrane coating, influence of, 82
extravasation from blood through vascular endothelium
deregulated vascular endothelium, 86, 87f
permeability of vascular endothelia, 84, 85f
routes of traffic across continuous endothelium, 85, 86f
vascular endothelium as target for drug delivery, 87

ligand coating, 82–83
particle size, influence of, 77–79, 78f
plasticity of nanoparticles, influence of, 79
protein corona formation around nanoparticles
dysopsonin-enhanced blood circulation time, 79–81, 80f
opsonin-facilitated phagocytosis, 79, 80f
uptake by nonphagocytic cells, 81
transport across interstitium, 87, 88f
pH-dependent strategy, for drug release from nanoparticles, 113–114
Phosphodiester oligonucleotide, 117
Phosphonoacetate linkages, 145
Phosphorothioate linkage, 145
Photodynamic therapy (PDT), 159–164
MRI and, 167
nanoparticles for, 19
nanoscale drug delivery systems in, 160–164
fullerene-based nanomaterials, 163–164
lipid-based carriers, 160–161
polymeric carriers, 162–163
and optical imaging, 168
Photofrin, 161, 162, 163, 167
Photosensitizers (PS), 19, *see also* Nanoscale drug delivery systems in PDT
defined, 159
lipid-based carriers, 160–161
polymeric carriers, 162–163
Photothermal therapy (PTT), 166–167
and optical imaging, 168
PTT/PDT and optical imaging, 169
pH-responsive micelles, 176
Physicochemical characteristics of nanoparticles, 153–154; *see also* Vaccine delivery
PLA and PLA–PEG nanosteroids, 203
Plasma drug concentration on therapeutic action intensity, 4–5
Plasma level of drug, 5
Plasmid DNA, 143
Plasmids, 129
Plasmodium berghei, 212, 215
Plasmodium falciparum, 214
Plasmodium yoelii, 213
Plasticity of nanoparticles, 79
Platelet-derived growth factor (PDGF), 193
Platelet-derived growth factor antagonist (anti-PDGF), 88

Index

PLGA, *see* Poly(lactic-co-glycolic acid)
Pluronic F68, 172
Pneumococcal surface protein A (PspA), 155
Poly(β-amino esters) (PAEs), 175
Polyelectrolyte complexes, formation of, 53–54
Polyethylene glycol (PEG), 61, 79, 81, 82f, 92f, 94, 111
Poly(ethyleneimine) (PEI), 91, 131, 876
Poly(ethylene oxide)–poly(L-lysine) [PEO–P(Lys)], 63
Poly-(γ-l-glutamylglutamine)–PTX nanoparticles, 180
Poly(lactic-co-glycolic acid) (PLGA), 100, 111, 125
   nanoparticles, 186
Polylactide (PLA), 111
Poly(l-lysine) (PLL), 130
Poly-L-lysine nanoparticles, 71
Polymer-based nanoparticles, 155, 162–163; *see also* Nanoparticles
   DPc–polymeric micelle complex, 162
   PpIX-conjugated glycol chitosan, 162
   TPP-loaded PEG–PE, 162
   verteporfin–PMBN–antibody, 163
   ZnPcF16-incorporated PEG–PLA nanoparticles, 162
Polymer-based nonviral vectors, *see* Nonviral vectors
Polymeric DNA/siRNA vectors, 130–131, 131f; *see also* Nonviral vectors
Polymeric micellar drug conjugate, 63, 64f; *see also* Interactions/orientation of therapeutic drugs
Polymeric micelles, 10, 11f, 33–34, 165–166; *see also* Micelles
   assembly of, 54
   DACHPt-loaded, 184
   DOX-loaded folate-targeted pH-sensitive, 176
   drug loading into, 63
Polymeric microspheres, 135–136
Polymeric nanohybrids, 39
Polymeric nanoparticles, 117, 121–122, 125–126, 138, 196–197
Polymer precipitation by solvent removal, 52
Polymers, conjugation with, 132–133, 133f
Polyplex stability, determinants of, 91–92, 92f, 93f
Polysaccharides, 113
Polystyrene-blockpolyethylene oxide (PS-block-PEO), 138

Polystyrene nanoparticles, 69
Porphyrin, 168
Porphysomes, 168
Positively charged particles, 74
Posterior segment ocular diseases, 114
Prednisolone acetate valerate (PAV), 169
Prednisolone phosphate, 192
Primaquine (PQ), 215
Primaquine-loaded nanoemulsions, 215; *see also* Malaria
Primary binders, 70
Prodrug therapy, macromolecules for
   about, 17, 18f
   ADEPT, 18
   GDEPT, 18–19
Prohibitin 1 (PHB) protein, 206
Protamine, 179
Protein(s)
   binding, 6
   chemically modified/carrier-bound, 15
   corona, 67
   corona formation around nanoparticles; *see also* Pharmacokinetics/ biodistribution of nanoparticles
      dysopsonin-enhanced blood circulation time, 79–81, 80f
      opsonin-facilitated phagocytosis, 79, 80f
      uptake by nonphagocytic cells, 81
   delivery, current obstacles to, 132
   as drug carriers, 17
   folding, 151
   as independent therapeutic drugs, 15
   polymers, 39
Protein- and DNA-based prophylactic vaccines, 151–156
   DNA vaccines, 151–152
   subunit vaccines, 151
   vaccine delivery, 152–156
Protein-based drugs, protection against degradation, *see* Nucleic-based drugs, protection against degradation
Protein-based nanoparticles, synthesis of, 56–57
Proton sponge, 104f, 105
Protoporphyrin IX (PpIX), 162
PTT, *see* Photothermal therapy (PTT)
PTX, *see* Paclitaxel (PTX)
Pulmonary arterial hypertension (PAH), 124
Pulmonary drugs, controlled release/ bioavailability of, *see* Controlled release

Index

Pulmonary insulin delivery, 198
  liposome, 198
  PLGA nanoparticles, 198
  solid lipid nanoparticles, 198
Pulmonary route, barriers to, 3–4

Q

Quantum dots (QD), 37–38

R

Radioimmunotherapy, 98
Rapamycin (Rapa), 174–175
Rapid expansion of supercritical solution
  (RESS), 49
Receptor-mediated delivery, active
  targeting through, 13, 14f
Recombinant DNA technology
  monoclonal antibodies, generation of
    hybridoma technique, 57, 57f
    phage display technology, 58, 58f
  protein-based nanoparticles, synthesis
    of, 56–57
  viral vectors, production of, 58–59
Recombinant proteins, *see* Synthetic
  peptides
Rectal route, barriers to, 3
Red blood cell (RBC), 82, 83
Regeneration of tissues, 215–218
  bone regeneration, 217–218
  nanotechnology in wound healing,
    215–217
Renal excretion, 7–8
Renin–angiotensin system (RAS), 149
Respiratory syncytial virus, nanomedicine
  for, 209; *see also* Obstructive
  respiratory diseases
Retention time in blood, prolongation of;
  see also Small-molecule drugs
  about, 10
  glomerular filtration reduction, 12
  reticuloendothelial uptake reduction, 12
Reticuloendothelial uptake reduction, 12;
  see also Retention time in blood,
  prolongation of
Retinal gene therapy, 204
Retinal pigment epithelium (RPE), 115
Retinol binding protein, 210–211
Retrovirus vector, 29–30, 126–127; *see also*
  Viral vectors
Reverse-phase evaporation, 46
Reverse salting out technique, 52

Rheumatoid arthritis (RA), 137
Rho-kinase inhibitor (fasudil), 124
Ribavirin, 150
Ribozyme, 144
Rituximab, 159
RNA-induced silencing complex (RISC),
  144
Routes for controlled release of ophthalmic
  drugs
  corneal route, 114–115
  noncorneal route, 115

S

Safety issues, 287–290
  damage to nuclear DNA and proteins,
    290
  deformation of cellular membrane,
    288–289
    disturbance of phospholipid bilayer,
      288–289
    interactions with membrane
      proteins and blocking of ion
      channels, 289
  disruption of cytoskeleton, 289–290
  lysosomal rupture and release of
    contents, 289
  nanoparticle interaction with blood
    cells, 287–288
    hemolysis of red blood cells,
      287–288
    macrophage uptake and immune
      responses, 288
    platelet aggregation and activation,
      288
Salmon calcitonin, 135
Second-generation lipid nanoparticles, 33
Secreted protein acidic and rich in cysteine
  (SPARC), 17
Self-emulsifying drug delivery systems
  (SEDDS), 33
Serum protein-coated nanoparticles, fates
  of
  aggregation, 71
  dissociation of complex/leakage of
    drugs, 71, 73f
  removal by macrophage/thrombosis/
    hypersensitivity, 70–71
Serum proteins, 17
Serum proteins with affinity to
  nanoparticles
  about, 67, 68f
  size/curvature of nanoparticles, 69–70

Index

for specific chemical groups of nanoparticles, 70
surface charge, 69
surface hydrophobicity, 69
shRNA (short hairpin RNA), 106, 126, 128, 128f
cellular actions of, 144f
Signal regulatory protein alpha, 83
Silica nanoparticles, 37
Silver sulfadiazine, 217
Single-chain fragment (scFv), 97, 98
Single-walled nanotube (SWNT), 38
siRNAs (small interfering RNAs), 159
cellular actions of, 144f
cyclodextrin polymer (CDP)–siRNA formulation, 131
electrostatic complexation with, 63–64
intracellular delivery of, see Intracellular transport vehicles
mdr-1 and surviving siRNA-loaded poly(α-amino esters), 175
PEG-coated Bcl-2 siRNA lipoplex, 177
serum stability of, enhanced, 17
for silencing gene expression, 144
siRNA-encapsulated neutral liposomal EphA2, 181
siRNA-loaded silica particles liposomal EphA2, 181–182
Size of nanoparticles, 68f, 69–70
Skin cancer, 185–187
anti-CD47 siRNA-encapsulated liposome–protamine–HA nanoparticles, 187
antigenic peptide–encapsulated PLGA nanoparticles, 186
MART-1 mRNA-loaded mannosylated nanoparticles, 187
nanoliposomal siRNA targeting B-Raf and Akt3, 186
PTX/ETP-loaded lipid nanoemulsions, 185–186
Skin diseases, 119
Skin disruption methods, 152
Skin tumors, 119
Small drugs
conventional, see Intracellular transport vehicles
versus nanotherapeutics, 76f
Small-molecule drugs; see also Emergence of nanotherapeutics
administration of
intramuscular/subcutaneous routes, barriers to, 4

intranasal route, barriers to, 3
oral route, barriers to, 1–2, 2f
pulmonary route, barriers to, 3–4
rectal route, barriers to, 3
sublingual/buccal routes, barriers to, 2–3
transdermal route, barriers to, 4
alteration of pharmacokinetics of, with macromolecules
active targeting through receptor-mediated delivery, 13, 14f
enhancement of drug solubility and stability, 9–10, 11f
fast drug release, facilitating, 12
magnetic targeting, 13, 15, 16f
objectives, 10f
passive targeting by exploiting leaky vasculature architecture, 13, 14f
prolongation of retention time in blood, 10, 12
sustained release, enabling, 12–13
formulations, problems with, 8
Small-molecule drugs in blood, fates of; see also Emergence of nanotherapeutics
drug transport from blood capillaries to extracellular fluid
blood flow, influence of, 5
plasma level/molecular size of drug, influences of, 5
protein binding, influence of, 6
drug transport from extracellular fluid to cells
cellular uptake by active transport, 6
cellular uptake by passive diffusion, 6
uptake by nonresponsive cells, 7
uptake by tumor cells, 6–7
elimination of drugs from body
biliary excretion, 8
chemical modification with functional groups, 7
renal excretion, 7–8
overview, 9f
plasma drug concentration on therapeutic action intensity, 4–5
Small unilamellar vesicle (SUV), 44
Sol–gel process, 55
Solid lipid nanoparticles (SLN), 33
controlled release/bioavailability of pulmonary drugs, 125
double emulsion, 48

334   *Index*

high-pressure homogenization (HPH), 46
melting dispersion, 47–48
microemulsion, 46
and nanostructured lipid nanoparticles, 121
solvent emulsification-diffusion, 47
solvent emulsification-evaporation, 46–47
solvent injection, 48
supercritical fluid technology, 49
for topical ocular delivery, 118
Solution-enhanced dispersion by supercritical fluid (SEDS), 49
Solvent displacement, 53
Solvent emulsification-diffusion, 47
Solvent emulsification-evaporation, 46–47, 48f
Solvent injection, 48
Solvent removal, polymer precipitation by, 52
Sonication, 44–45
Sorafenib, 175
Spherical nucleic acids (SNAs), 173
Splice-switching oligonucleotides (SSOs), 179
Spontaneous emulsification method, 52
Spray-drying, 55–56
Stable nucleic acid–lipid particle (SNALP), 211–212
Staphylococcus aureus, 215–216
Stratum corneum, 4
Streptococcus pneumoniae strains, 155
*Streptococcus pyogenes*e, 156
Subcutaneous insulin delivery, 199
Sublingual routes, barriers to, 2–3
Subunit vaccines, 151; *see also* Protein- and DNA-based prophylactic vaccines
Suicide gene therapy, 13
Supercritical fluid extraction of emulsions (SFEE), 49
Supercritical fluid technology, 49
Superoxide dismutase (SOD) enzymosomes, 190
Superparamagnetic iron oxide, 13
Superparamagnetic iron oxide nanoparticles (SPION), 38
Surface charge, 69
Surface functionalization/ligand attachment, 40–41, 40f; *see also* Diversity of bioactive nanoparticles

Surface hydrophobicity, 69
Sustained drug release, 12–13; *see also* Small-molecule drugs
Sustained release of dermal drugs
about, 119–120, 119f
roles of nanoparticles in
lipid–polymer hybrid nanoparticles, 122
liposomes, 120–121
polymeric nanocarriers, 121–122
SLN and nanostructured lipid nanoparticles, 121
Sustained release of ocular drugs
controlled release/enhanced bioavailability of ophthalmic drugs
about, 115–116
dendrimers, 118
liposome, 116–117
micelles, 116
nanosuspensions, 117
niosomes, 118
polymeric nanoparticles, 117
solid lipid nanoparticles (SLN), 118
overview, 114, 115f
routes for controlled release of ophthalmic drugs
corneal route, 114–115
noncorneal route, 115
Sustained release of pulmonary drugs, *see* Controlled release
Synthetic peptides; *see also* Intracellular transport vehicles
chemical modifications of proteins/ peptides
albumin fusion, 133–134
amino acid substitution, 134
conjugation to antibody Fc portion, 133
conjugation with polymers, 132–133, 133f
disulfide-rich peptides, 135
lipidization, 135
peptide terminus, modification of, 134–135
controlled delivery of proteins from injectable formulations
injectable in situ gel forming solution, 136
lipid-based protein delivery, 136–137
polymeric microspheres, 135–136
protein delivery, current obstacles to, 132
α-synuclein, 150

Index

T

Taxanes, 173
Taxol-containing liposomes, 176–177
Taxol–Cremophor formulation, 177
Tegafur (TF), 177
Tesamorelin, 135
Tetanus vaccines, 154, 155
Tetracycline, 216
Meso-tetraphenylporphine (TPP), 162
α-thalassemia, 148
β-thalassemia, 148
Th epitopes, 151
Therapeutic genes, intracellular delivery of, *see* Intracellular transport vehicles
Thin-film hydration, 44
Thrombopoietin (TPO), 133
Thrombosis, nanocarrier removal by, 70–71
Thymidylate synthase shRNA, 177
Time-dependent strategy for drug release from nanoparticles, 112–113
TLR agonist, 157
TNF-α, 159
Tobramycin, 118
Toll-like receptors (TLRs), 153–154
Top-down approaches, for drug nanoparticle production
high-pressure homogenization (HPH), 43–44
wet milling, 43–44
Topotecan, 174
Traffic across continuous endothelium, routes of, 85, 86f
TRAIL (TNF-related apoptosis-inducing ligand), 140, 156
TRAIL-sensitive, 156–157
Transcellular transport, 85
Transcytosis, 85, 86f, 138
Transdermal drug delivery, 4
Transdermal insulin delivery, 199
Transdermal route, barriers to, 4
Transferrin receptor (TfR)-mediated targeting, 100, 102, 103f
Transport across interstitium, 87, 88f
Transport to blood or lymphatic system, 124
Trastuzumab, 173–174
Trastuzumab emtansine, 173–174
Trinitrobenzenesulfonic acid (TNBS), 207
Tryptophan (Trp), 134
Tumor-associated antigen, 158
Tumor cells, uptake of drug by, 6–7

Tumor necrosis factor (TNF)–related apoptosis-inducing ligand (TRAIL), 148–149
Tumor recognition by NK cells, 159
Tumor-specific antigens, 156
Tumor-specific immune activation, 157–159
antibodydependent cell-mediated cytotoxicity (ADCC), 157–158
antigen-loaded DCs, 159
complementdependent cytotoxicity (CDC), 158–159
monoclonal Ab or siRNA, 159
Two-microemulsion method, 56
Two-step method
lipid-polymer hybrid nanoparticles, 54
nanocapsules and nanospheres, *see* Nanocapsules/nanospheres

U

Ulcerative colitis (UC), 205
Ultraflexible liposomes, 121
Ultralente, 194
Ultrasonograpy and chemotherapy, 169–170
U87MG brain tumor model, 173
Unilamellar vesicle (ULV), 45
Uptake by nonphagocytic cells, 81
Uptake of drug
by nonresponsive cells, 7
by tumor cells, 6–7

V

Vaccine carriers
nanoparticles as, 154–156
Vaccine delivery, 152–156; *see also* Protein- and DNA-based prophylactic vaccines
administration site, 152
macromolecules for, 19
nanoparticles as adjuvants and vaccine carriers, 154–156
physicochemical characteristics of nanoparticles, 153–154
Variable interactions of nanoparticles
extracellular matrix (ECM)-nanoparticle interactions, 75, 76f
with interstitial fluid/lymph, 74–75, 74f
nanoparticles and cell components, interactions between, 76
serum protein-coated nanoparticles, fates of
aggregation, 71

dissociation of complex/leakage of drugs, 71, 73f
removal by macrophage/thrombosis/hypersensitivity, 70–71
serum proteins with affinity to nanoparticles
about, 67, 68f
size/curvature of nanoparticles, 69–70
for specific chemical groups of nanoparticles, 70
surface charge, 69
surface hydrophobicity, 69
Vascular endothelial growth factors (VEGF), 86, 96, 117
Vascular endothelium
deregulated, 86, 87f
permeability of, 84, 85f
Vasoactive intestinal peptide (VIP)–sterically stabilized phospholipid micelles (SSMs), 190
Vasodilators, overexpression of, 149
Verteporfin, 160–161
Verteporfin–PMBN–antibody, 163
Viral proteins, 155
Viral vectors; *see also* Diversity of bioactive nanoparticles
DNA virus vectors
adeno-associated viral vectors (AAV), 30–31

adenoviral vectors, 30
herpes simplex virus (HSV), 31
production of, 58–59
retrovirus vector, 29–30
Virus-like particles (VLPs), 152, 155
Vision restoration, 204

W

Weak acids, 6
Wet AMD (age-related macular degeneration), 161
Wet milling, 43–44; *see also* Top-down approaches, for drug nanoparticle production
Wound healing, nanotechnology in, 215–217

X

X-linked adrenoleukodystrophy (X-ALD), 148
X-linked severe combined immune deficiency (X-SCID), 147
XTEN (extended recombinant polypeptide), 132
Xylocaine gel, 121

Z

Zevalin, 98